IgY-Technology: Production and Application of Egg Yolk Antibodies

Xiao-Ying Zhang • Ricardo S. Vieira-Pires •
Patricia M. Morgan • Rüdiger Schade
Editors

IgY-Technology: Production and Application of Egg Yolk Antibodies

Basic Knowledge for a Successful Practice

Springer

Editors
Xiao-Ying Zhang
College of Biological Science and Engineering
Shaanxi University of Technology
Hanzhong, Shaanxi, China

Ricardo S. Vieira-Pires
Center for Neuroscience and Cell Biology
University of Coimbra
Coimbra, Portugal

Patricia M. Morgan
School of Natural Sciences
National University of Ireland Galway
Galway, Ireland

Rüdiger Schade
Institute of Pharmacology
Charité – Universitätsmedizin Berlin
Berlin, Germany

ISBN 978-3-030-72686-7 ISBN 978-3-030-72688-1 (eBook)
https://doi.org/10.1007/978-3-030-72688-1

This Springer imprint is published by the registered company Springer Nature Switzerland AG.
The registered company address is: Gewerbestrasse 11, 6330 Cham, Switzerland

Preface

Adopting new scientific methodologies can be a slow process. Researchers gain expertise with a particular model or set of techniques, and it often requires groundbreaking advances to garner support for new ideas. IgY technology has developed quite slowly given its potential, but we consider that now is the right time to provide scientists and practitioners with an overview of IgY and IgY technology research and to underpin the urgent requirement for standardization and shared perspectives of this technology. We have also attempted to provide the theoretical background combined with the practical approaches to support new researchers in the field as well as review the advances to date and set out the challenges ahead for more established researchers.

The book in divided into three parts. Part I describes the biological basis of IgY and IgY technology. IgY is a fascinating molecule important in the evolution of immunoglobulins and the immune system. Birds are phylogenetically distinct from mammals and have several unique features, and given the predominant use of mammalian immunoglobulins in research, we have emphasized the differences between mammalian and avian systems.

Part II of the book covers the core methods employed in IgY technology: not only the established methods such as hen keeping, immunization and antibody monitoring, but also some novel methodological needs, such as IgY delivery and dosage form design. The demand for proteins used as therapeutics is soaring. The methodologies for the production of monoclonal IgY and transgenic hens are discussed.

Part III of the book summarizes and discusses the applications of IgY technology. The real impact of any technology is in its beneficial contributions to society. The potential of IgY in maintaining animal health has in some part been realized, but opportunities still need to be exploited for treatment of human diseases. The ability to generate polyclonal IgY, which does not need expensive equipment, may address critical needs where costs are a factor. It is also instructive to see how basic research combined with innovation and entrepreneurial spirit is driving the development of new products and processes.

Through this monograph, it is our anticipation and hope to create a shared outlook among the new researchers and established practitioners of IgY technology. The global spread of IgY technology is evidenced by our contributors to this book. It would be advantageous to formally establish an "IgY Society" to link the globally spread researchers, to link academia and industry, and to draw others with different research backgrounds and skills into the pursuit of the many benefits of IgY technology. We hope that we have communicated our enthusiasm for this area of research, and we hope that this book will play some part in expanding the membership of the community of IgY researchers.

Hanzhong, Shaanxi, China — Xiao-Ying Zhang
Coimbra, Portugal — Ricardo S. Vieira-Pires
Galway, Ireland — Patricia M. Morgan
Berlin, Germany — Rüdiger Schade

Acknowledgements

Bringing this book to completion would not have been possible without the generous participation of the contributors to this book. The editors are grateful to them for their interest and commitment to the project and their willingness to prioritize work on this book. The editors would like to thank Professor Schade for his influence as lead author on the first manual on IgY technology entitled *Chicken Egg Yolk Antibodies, Production and Application IgY-Technology* which inspired the growing interests in this area and us as editors. One of us (Xiaoying Zhang) has greatly benefited from his personal mentorship and for generating some of the initial ideas for this work. We would like to acknowledge the support provided by Springer during the development of this book.

On a personal level, one of us (PMM) would like to acknowledge the support and encouragement provided by Gerry Morgan, throughout the planning and execution of this book. It would not have been possible without him. RVP would like to thank his family for their long-term support particularly during the journey of writing this book.

Dr. Xiaoying Zhang
Dr. Patricia M. Morgan
Dr. Ricardo S. Vieira-Pires

Funding Acknowledgements

The work of A. Wigdorovitz, C.G. Vega, Marina Bok, V. Parreño, was co-funded by Argentine Sector Fund (FONARSEC) Ministry of Science and Technology of the Nation and Technological agreement VETANCO—INTA N 20794.

The Germ-Free avian work of Ovagen Group Ltd. received funding from the European Union's Horizon 2020 Research and Innovation Programme, OvaVAx Grant agreement ID: 858390.

The work of P. Chacana and C.L. Leiva was funded by Argentinian national funds via the National Institute of Agricultural Technology (INTA) and the National Scientific and Technical Research Council (CONICET).

The work of B. Chelliappan was supported by the Department of Science & Technology (DST)—INSPIRE Fellowship Program (2013–2018), Government of India.

The work of M. Freire was developed within the scope of the project CICECO—Aveiro Institute of Materials, UIDB/50011/2020 & UIDP/50011/2020, financed by national funds through the Portuguese Foundation for Science and Technology/MCTES.

The work of C.A. Harley was supported by the National Institutes of Health grant R01NS081320.

The work of A. Ferreira Júnior was supported by the National Council for Scientific and Technological Development (CNPq), Coordination of Superior Level Staff Improvement (CAPES), and Fundação de Amparo à Pesquisa do Estado de Goiás. The author wishes to acknowledge Jandra Pacheco dos Santos, Gracielle Teles Pádua, and Denise Teixeira for collaborating in the organisation of Chap. 10.

C. H. Leow and team would like to thank the funding sponsored by Prototype Research Grant Scheme (PRGS) from the Ministry of Higher Education (MOHE) Grant no.: 203/CIPPM/6740051, and University Sains Malaysia RUI Grant no.: RUI 1001/CIPPM/8011050 to complete Chap. 13. Warren Lee was a Ph.D. candidate supported by the USM Fellowship Scheme RU(1001/CIPS/AUPE001) at the time this work was conducted.

The work of A. Michael was funded by University Grants Commission Minor Project [File No F.MRP-079/02], Department of Biotechnology under Small Business Innovation Research Initiative (SBIRI) [BT/SBIRI/743/49- B14/2010], University Grants Commission Major Project (41-1175/2012(SR), Indian Council of Medical Research (ICMR)—(5/8-1(22)/2011-12-ECD-II, Biotechnology Industry

Research Assistance Council (BIRAC)-Promoting Academic Research Conversion to Enterprise (PACE) (BT/TEMP3381/PACE- 12/17).

P.M. Morgan would like to acknowledge the contribution of the James Hardiman Library, NUI Galway, who provided online resources and library support crucial to the completion of this project.

T. Ochôa-Pires is the recipient of a DOCTORATES 4 COVID-19 FCT Ph.D. fellowship (2020.10150.BD), supported by Portuguese national funds via the Foundation for Science and Technology (FCT) and by the European Social Fund (ESF) through the Human Potential Operational Programme (POPH).

G. Papadakos would like to thank Lissa Herron, Joe Mee, Jacqui Matthews, and his colleagues in Roslin Technologies Ltd. for helpful discussions and for proofreading Chap. 14.

M.E. Rosa acknowledges FCT—Fundação para a Ciência e a Tecnologia, I.P. for the doctoral grant SFRH/BD/136995/2018.

J. Salimian dedicates his contributions in honour of the memory of the late Professor Emeritus Ahmad Massoud, Department of Immunology, Tehran University of Medical Sciences.

The work of A. P.M. Tavares was developed within the scope of the project CICECO—Aveiro Institute of Materials, UIDB/50011/2020 & UIDP/50011/2020, financed by national funds through the Portuguese Foundation for Science and Technology/MCTES. The author acknowledges the FCT Investigator Programme and Exploratory Project (IF/01634/2015) with financing from the European Social Fund and the Human Potential Operational Programme.

The work of R.S. Vieira-Pires was co-funded by Portuguese national funds via the Foundation for Science and Technology (FCT) and by the European Social Fund (ESF) through Human Potential Operational Programme (POPH) under grant IF/00123/2013 and the European Regional Development Fund (ERDF) through Operational Programme for Competitiveness and Internationalisation (COMPETE 2020) under the projects POCI-01-0145-FEDER-030550 (PTDC/BTM-SAL/30550/2017; BYDRUG), UIDB/04539/2020, and UIDP/04539/2020 (CIBB). Private funds from the company HBT—Saúde e Biotecnologia Lda (Portugal) also supported the work.

X.Y. Zhang is grateful for the Distinguished Professorship awarded by the Shaanxi University of Technology that provides an excellent environment in which to conduct research. X.Y. Zhang also thanks the University of Guelph for providing an adjunct faculty position. The work of X.Y. Zhang (and team) was supported by the China Natural Science Foundation (31873006, 31572556) and the National Key Research and Development Program of China (2017YFD0501405).

Contents

Contributors

Hyeong Chul Ahn ADBioTech Co., Chuncheon-si, South Korea

Marina Bok INCUINTA, CICVyA-INTA, Comodoro Rivadavia, Argentina

Catherine D. Caulfield Ovagen Group Ltd., Carrentrila, Ballina, Ireland

Pablo Chacana UEDD INTA-CONICET IPVet, Instituto Nacional de Tecnología Agropecuaria, Comodoro Rivadavia, Argentina

Brindha Chelliappan Department of Microbiology, PSG College of Arts & Science, Bharathiar University, Coimbatore, Tamil Nadu, India

Chinnappan Gandhimathi Sri Ramakrishna Dental College and Hospital, Coimbatore, Tamil Nadu, India

Ankit Choraria PSG College of Arts and Science, Coimbatore, Tamil Nadu, India

Fazle Elahi ADBioTech Co., Chuncheon-si, South Korea

Álvaro Ferreira Júnior School of Veterinary and Zootechnics, Universidade Federal de Goiás, Goiânia, Brazil

Mara G. Freire CICECO-Aveiro Institute of Materials, Department of Chemistry, University of Aveiro, Aveiro, Portugal

Shikun Ge College of Biological Science and Engineering, Shaanxi University of Technology, Hanzhong, China

Carol A. Harley i3S - Instituto de Investigação e Inovação em Saúde and IBMC – Instituto do Biologia Molecular e Celular, University of Porto, Porto, Portugal

Anders O. Larsson Department of Medical Sciences, Uppsala University, Uppsala, Sweden

Carlos Leónidas Leiva UEDD INTA-CONICET IPVet, Instituto Nacional de Tecnología Agropecuaria, Argentina

Chiuan Herng Leow Institute for Research in Molecular Medicine (INFORMM), Universiti Sains Malaysia, Pulau Pinang, Malaysia

Ben Li College of Biological Science and Engineering, Shaanxi University of Technology, Hanzhong, China

Antonysamy Michael Department of Microbiology, PSG College of Arts & Science, Bharathiar University, Coimbatore, Tamil Nadu, India

Leonard M. Moran Ovagen Group Ltd., Carrentrila, Ballina, Ireland

Patricia M. Morgan Emerita Lecturer, School of Natural Sciences, National University of Ireland Galway, Galway, Ireland

Martin G. Murphy Ovagen Group Ltd., Carrentrila, Ballina, Ireland

Tiago Ôchoa-Pires CNC - Center for Neuroscience and Cell Biology, University of Coimbra, Coimbra, Portugal

Greg Papadakos Roslin Technologies, Edinburgh, UK

Viviana Parreño INCUINTA, CICVyA-INTA, CONICET, Comodoro Rivadavia, Argentina

Sentila Rajan Bharathiar University, Coimbatore, Tamil Nadu, India

Marguerita Rosa CICECO – Aveiro Institute of Materials, Department of Chemistry, University of Aveiro, Aveiro, Portugal

Jafar Salimian Applied Microbiology Research Center, Chemical Injuries Research Center, Systems Biology and Poisoning Institute, Baqiyatallah University of Medical Sciences, Tehran, Iran

Rüdiger Schade Institute of Pharmacology, Charité – Universitätsmedizin Berlin, Berlin, Germany

Meenatchisundaram Subramani Nehru Arts and Science College, Coimbatore, Tamil Nadu, India

Ana Paula M. Tavares CICECO-Aveiro Institute of Materials, Department of Chemistry, University of Aveiro, Aveiro, Portugal

Celina G. Vega Centro de Investigación en Ciencias Veterinarias y Agronómicas, INTA, Buenos Aires, Argentina
Bioinnovo; INCUINTA, CICVyA-INTA, CONICET, Comodoro Rivadavia, Argentina

Ricardo S. Vieira-Pires CNC – Center for Neuroscience and Cell Biology, University of Coimbra, Coimbra, Portugal

Andrés Wigdorovitz Bioinnovo; INCUINTA, CICVyA-INTA, CONICET, Comodoro Rivadavia, Argentina

Rao Wu College of Biological Science and Engineering, Shaanxi University of Technology, Hanzhong, China

Long Xu CBMA, University of Minho, Portugal; College of Biological Science and Engineering, Shaanxi University of Technology, Hanzhong, China

Xiaoying Zhang College of Biological Science and Engineering, Shaanxi University of Technology, Hanzhong, China

Fagang Zhong State Key Laboratory for Sheep Genetic Improvement and Healthy Production, Xinjiang Academy of Agricultural and Reclamation Science, Shihezi, China

Part I

Biological Basis of IgY Technology

1 Development of IgY Technology: A Historical Perspective

Xiaoying Zhang, Ricardo S. Vieira-Pires, and Long Xu

Abstract

IgY Technology is a scientific field classically related to the generation, evaluation and use of avian egg yolk IgY antibodies to develop biomedical and biotechnological solutions and applications. The field is more than 120 years old. However, in the last decades, major advances in research and development areas such as Genetics, Biochemistry, Bioengineering and Bioprocessing, have prompted new approaches to this old technology. In this chapter, a retrospective summary is given on the major milestones, key events and developments of IgY technology. We also bring an overview of the main topics and themes gathered and further developed within this book edition. The discussion on the perspective and future trend analysis of IgY technology is given in Chap. 18.

Keywords

IgY technology · Egg yolk antibody (IgY) · History · Animal welfare · IgY product

X. Zhang (✉)
College of Biological Science and Engineering, Shaanxi University of Technology, Hanzhong, China
e-mail: zhxying@snut.edu.cn; xzhang67@uoguelph.ca; zhang@bio.uminho.pt

R. S. Vieira-Pires
CNC – Center for Neuroscience and Cell Biology, University of Coimbra, Coimbra, Portugal
e-mail: ricardo.pires@biocant.pt

L. Xu
College of Biological Science and Engineering, Shaanxi University of Technology, Hanzhong, China

CBMA, University of Minho, Braga, Portugal

X.-Y. Zhang et al. (eds.), *IgY-Technology: Production and Application of Egg Yolk Antibodies*, https://doi.org/10.1007/978-3-030-72688-1_1

1.1 Introduction

Hens produce an immunoglobulin class, immunoglobulin Y (IgY) whose natural role is the protection of the chicken and the developing chick in the egg. However, the immunization of hens results in the production of specific IgY against the antigen and the process of immunization and recovery of the IgY from the egg yolk, is termed IgY technology. The highly specific polyclonal antibodies are obtained by using different extraction and purification methods. This technology has developed rapidly in recent years and has good potentials in the field of medicine.

In the past two decades, biomedical research has been challenged by the growing public interest in the welfare of animals used in research; resulting in the search for ways to Refine, Reduce or Replace painful animal experiments. This also applies to the production of polyclonal and monoclonal antibodies in animals. Antibodies are important tools in biomedical research and are essential components of diagnostic methods for quantitative and qualitative determinations of a wide variety of biological molecules. Historically, polyclonal antibodies have been produced in rabbits and other rodents like mice, rats and guinea-pigs or in farm animals such as horses, sheep and goats. The production of polyclonal antibodies involves two invasive procedures, firstly the immunization procedure involving several injections and secondly, harvesting of the antibodies from the blood which involves bleeding the animals on several occasions.

1.2 Development of IgY Technology

The first experiments to demonstrate that a serum product could provide protection against a toxin were performed as early as 1890. The German scholar Emil von Behring and the Japanese scholar Kitasato Shibasaburo injected guinea pigs with heat-treated diphtheria toxin and found that the serum could protect animals from what would be a lethal dose of the live toxin. Emil von Behring was subsequently awarded the first Nobel Prize (1901) in Physiological Medicine for his work on diphtheria. Kitasato was experienced in the culture of the anaerobic tetanus bacilli, and worked on the determination of the toxin-destroying activity in serum; his research and the tetanus antitoxin saved the lives of many soldiers during the First World War (Kantha 1991). Started from the aforementioned antiserum work, antibody engineering and antibody based immune therapy have been well developed and have contributed to important developments in biotechnology and therapy.

Inspired by the antiserum study of the early 1890s, the German medical doctor Felix Klemperer did a classical experiment in 1892 (Fig. 1.1): laying hens were subjected to tetanus-causing bacilli (*Clostridium tetani*) by intraperitoneal injection every 5–15 days, for a total of five times. The egg yolk was first dissolved and mixed with a NaCl–Na_2SO_4 solution, and injected into mice at three different concentrations: high, medium and low doses. On the next day, these different groups were infected with 1.5 times the concentration of what would be a lethal dose (LD_{50}) of tetanus. Mice pre-exposed to high doses of the yolk-based solution all survived

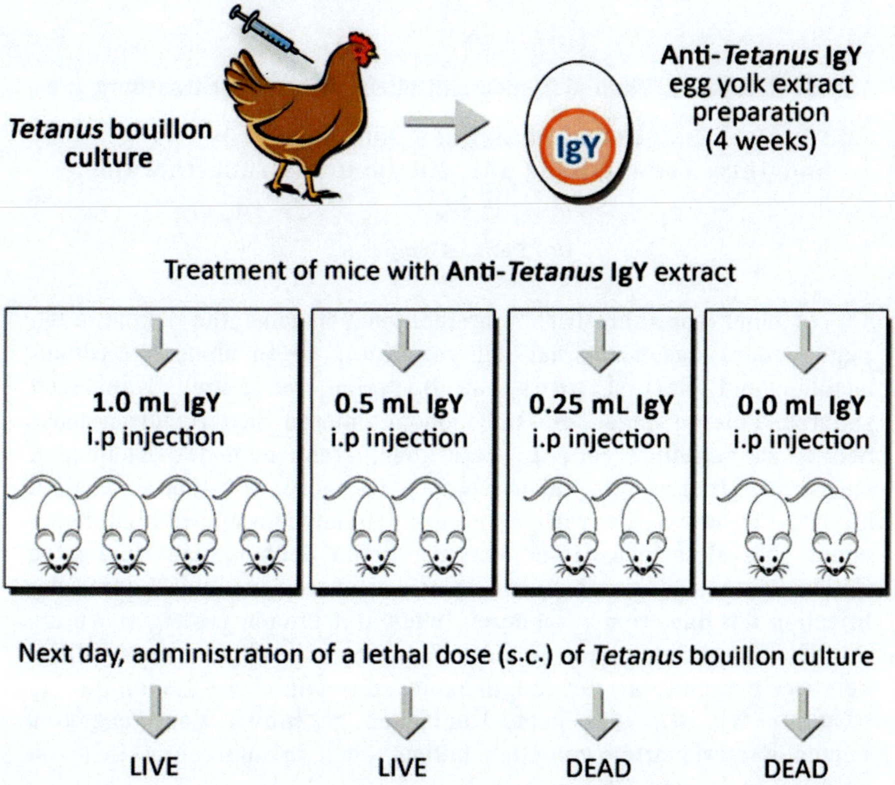

Fig. 1.1 Historical experiment of Dr. Felix Klemperer. The first relevant experiment in the field of IgY antibodies was performed by the German researcher Felix Klemperer in 1893. He showed for the first time that laying hens immunized with tetanus toxin could generate immunity and produce egg yolk extracts with protective characteristics. Adapted from IgYTechnology.com (www.igytechnology.com)

and those exposed to either the low dose or a control group with no pre-treatment died. This study proved for the first-time that immunized laying hens can produce specific antibodies with neutralizing ability, although Klemperer used the term "immunity" ("Immunität" in German) to refer to the "antibody". He claimed "extremely high immunity" obtained from the immunized hen egg yolk (Klemperer 1893). This discovery of a relatively easy way to obtain the specific antibody from laying hens represented a fundamental breakthrough in medical science, especially when considering that it was only 2 years after the discovery of the first mammalian antiserum therapy of Behring. Unfortunately, Klemperer's work and observations were very much undervalued, and its revival only happened after several decades (Fig. 1.2).

The chicken egg yolk antibody was called IgG for many decades since this immunoglobulin isotype exhibits properties quite similar to mammalian IgG (Chap. 5). The molecule was first termed "IgY" by Gerrie A. Leslie and L. W.

XV.

Aus dem Laboratorium der medicinischen Klinik zu Strassburg i. E.

Ueber natürliche Immunität und ihre Verwerthung für die Immunisirungstherapie.

Von

Dr. **Felix Klemperer**,
Assistent der Klinik.

In einer vor Jahresfrist mitgetheilten Versuchsreihe [1]) konnte ich zeigen, dass das Blutserum der von Natur gegen Mäusesepticämiebacillen und Friedländer'sche Bacterien refractären Kaninchen anderen Thieren gegen die Infection mit diesen Bacterienarten nicht Schutz zu verleihen vermag; dass aber, wenn man den Kaninchen steigende Mengen der Mäusesepticämie- resp. Pneumoniebacillen injicirt, ihr Serum die vorher vermisste immunisirende Fähigkeit gewinnt. Es liess sich dieser Vorgang dahin deuten, dass in beiden Fällen die von Natur vorhandene, relativ schwache Immunität durch die Injection der Bacterien zu höheren Immunitätsgraden gesteigert wurde.

Die Frage lag nahe, ob dieser Vorgang eine allgemeinere Gültigkeit besitze und ob die Möglichkeit, die natürliche Immunität zu steigern, wie wir dies beim Kaninchen gegenüber den genannten beiden Bacterienarten gefunden hatten, auch in anderen Fällen von natürlicher Immunität bestände.

Die Beantwortung dieser Frage schien mir nicht nur für das Verständniss der angeborenen Immunität von Bedeutung zu sein, sondern mehr noch für den weiteren experimentellen Ausbau der sogenannten Serumtherapie. Die Untersuchungen der letzten Jahre (Ehrlich) haben gezeigt, und Behring [2]) hat es jüngst noch mit besonderer Schärfe ausgesprochen, dass die Vorbereitung eines Thieres zur Lieferung eines Serums von starker immunisirender Fähigkeit zwei Acte umfasst: Einmal muss das Thier überhaupt erst immunisirt,

1) Berliner Klinische Wochenschrift 1892. Nr. 13.
2) Die Blutserumtherapie I. S. 40 u. ff. (Leipzig, Georg Thieme. 1892.)

Fig. 1.2 First IgY publication by Felix Klemperer in 1893

Clem in 1969 "based upon physical-chemical and antigenic characteristics", in order to better distinguish the yolk antibody which showed "insufficient correlation with any of the known human immunoglobulins" (Leslie and Clem 1969).

Since the middle of the twentieth century, as animal ethics and welfare has gained more attention in academia, the research results of Klemperer have gained more prominence and this was advanced by the animal welfare research of Russel and Burch (Russell and Burch 1959).

Since the 1980s, IgY has been widely used owing to the availability of standard commercial reagents such as IgY-specific purification kits (Chap. 11) and tagged secondary antibodies against IgY.

The term "IgY technology" was first used in 1995 and defined as "IgY production and application" in 1996 and recognized by the European Centre for the Validation of Alternative Method (ECVAM), as an international standardized technology, recommended to replace the use of mammalian IgG for animal welfare purpose (Schade et al. 1996) (Fig. 1.3).

In 2001, the world's first IgY laboratory manual was published in Germany, which standardized the laboratory practices of IgY technology (Schade et al. 2001).

In 2004–2007, an interdisciplinary study of the comprehensive utilization of eggs was launched through a Cooperative Organization Science and Technology action (COST 923) in European Union framework, with the aim to improve the comprehensive utilization of egg ingredients with new functions and usage, and to encourage the non-food applications of eggs. The biomedical use of IgY became the focus of the action plan, as emphasized in the project summary publications (Huopalahti et al. 2007).

In 2011, an IgY monograph in Chinese was published.

An analysis of the publications on avian IgY antibodies and IgY Technology performed on NCBI database (time window 1950–2020) with different search terms, namely "*IgY Technology*", "*IgY Antibodies*" and "*IgY*", shows a prominent increase of the numbers of published works since the 1990s. The publications cover a broad range of research fields including immunology, zoology, ornithology and poultry science, veterinary medicine and animal husbandry science, food engineering, medical sciences (e.g., Gastroenterology, Stomatology), demonstrating the breadth of application of IgY Technology research and development (Fig. 1.4).

1.3 Application of IgY Products at a Glance

IgY has the advantage of stable chemical properties, high yield, and low cost for its production. Hens are phylogenetically distant from mammals and thus can be used to raise antibodies against conserved mammalian proteins. Its passive immunity function can be used to study how to resist viral and bacterial diseases. Therefore, it has the potential of developing as a functional food and drugs for human use, and as a feed additive and veterinary drug to animals. For example, one of the research focuses of IgY technology has been treating gastrointestinal diseases and diarrhoea with specific IgY, which have been used more intensively in veterinary medicine, but also in human health. These will be further developed in Chaps. 15 and 16.

IgY-based products in the pipeline involve specific antibodies for the treatment of several pathogens such as *Helicobacter pylori*, *Streptococcus mutans*, *Candida*

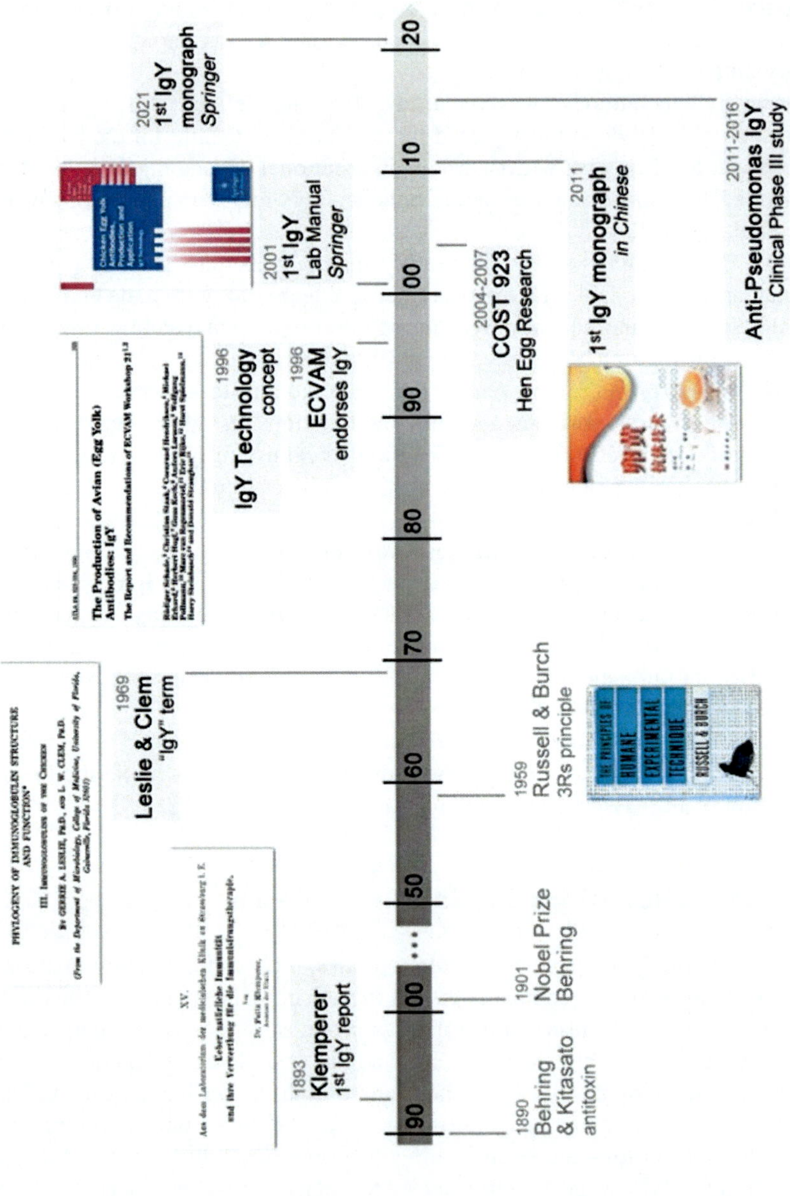

Fig. 1.3 Historical timeline of IgY technology

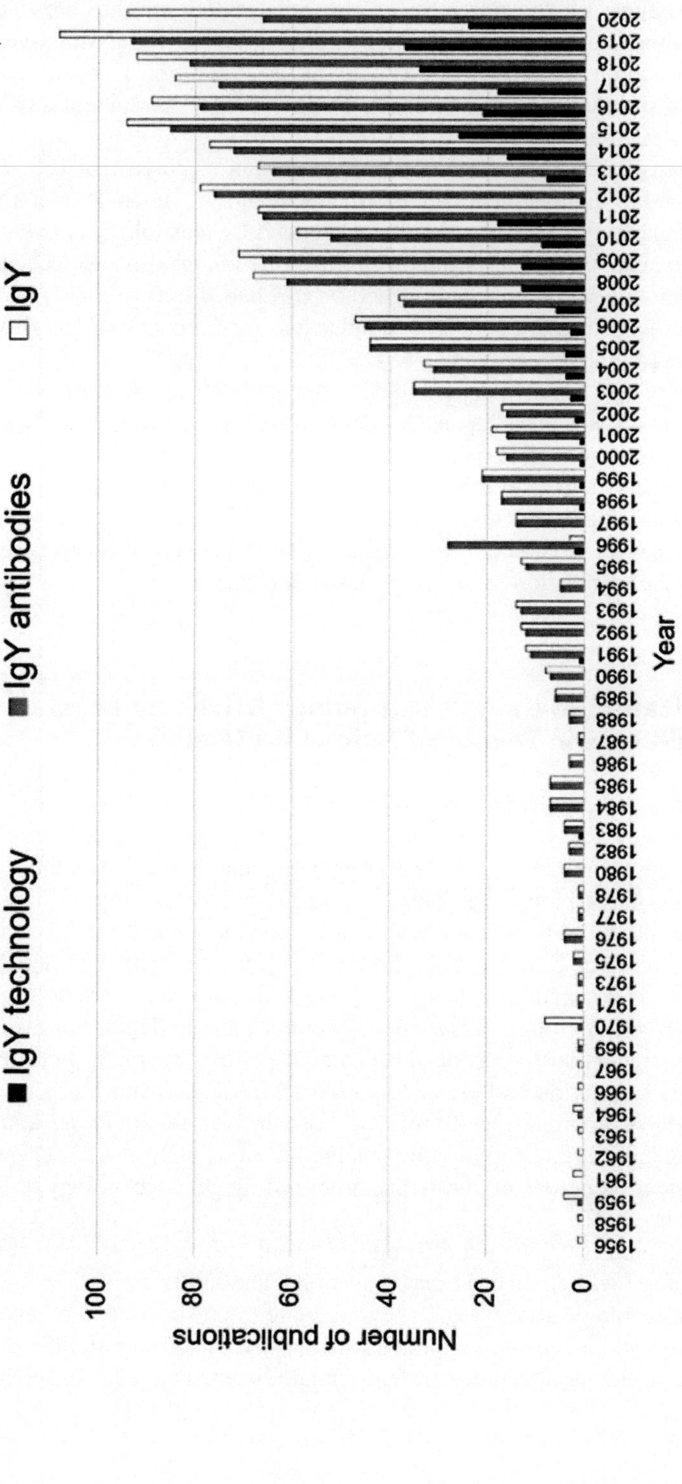

Fig. 1.4 Evolution of publications on avian IgY antibodies. A search of publications on avian IgY antibodies and IgY Technology was performed on NCBI database with different search terms on a time window from 1850 to 2020. The search terms were: "IgY Technology", "IgY Antibodies" and "IgY" and the total number of publications for each term were 268, 1366, and 1491, respectively (December 07, 2020)

albicans, *Porphyromonas gingivalis*, *Salmonella*, *Escherichia coli* and *Clostridium difficile*. IgY products for the treatment of clostridial enteric disease are under a phase II trial for the short-term evaluation of its efficacy and dosage. Moreover, the use of IgY in periodontitis, gingivitis, dental caries, gastric ulcer, dysbiosis, toxins, emerging viral diseases, nutritional diseases, chronic diseases and neoplasms have been presented in numerous patents (Leiva et al. 2020).

Traditionally, IgY technology refers to the generation of polyclonal IgY. However, since the end of the twentieth century, monoclonal IgY, monoclonal antibody fragment, particularly IgY-scFv can be also generated by technology convergence (Chap. 13). Also avian transgenesis and the production of biopharmaceuticals in egg white are relevant and will be briefly touched on in Chap. 14. This converges with the overall concept of using chickens as bioreactors for production of high-value proteins including biopharmaceuticals.

IgY technology was rapidly adopted in developing countries and regions such as South America, India, as well as Japan, Germany and other developed countries. The poultry industry is very well developed in China and the technology has broad application prospects there also. IgY technology research and application is broadly applied to veterinary medicine and functional food development (Chaps. 15 and 16). Many countries pay more attention to product exploitation and commercialization (Chap. 17), but the application to medicine requires further research.

1.4 Application of the IgY Technology Driven by Legal Regulations in Favour of Animal Protection

The implementation of the EU Directive 86/609/EEC (European Union 1986) for the protection of animals used for experiments and other purposes has stimulated efforts to increasingly apply alternatives to animal experimentation. Consequently, hens rather than rabbits may be immunized, and in more general terms scientists may even be asked to produce antibodies without using animals. In the longer run, the application of molecular genetics may allow the in vitro production of antibodies and as this stage is reached, classical polyclonal and monoclonal antibodies produced in animals will no longer be required. There is a strong debate on this issue, since many researchers still question the potential of fully synthetic libraries for in vitro antibody generation and selection; there are intrinsic features arising from natural evolution of immune repertoires (e.g., certain bias for preferred antibody structural arrangements) that can only be obtained from animal sources. Therefore, the immunological properties of antibodies produced by gene-technology will first have to be validated. The first production of monoclonal antibodies was welcomed with high expectations (Köhler and Milstein 1975), all of which were not met when the new technology was applied. In fact, polyclonal antibodies are still in use; and due to their immunological properties, they are quite often superior to monoclonal antibodies. Based on the reasons explained we are convinced that under the present legal environment the immunization of hens should be encouraged. Consequently, the scope of argumentation put forward by scientists who are not convinced of the use IgY technology would be reduced.

1.5 Advantages and Limitations Driving IgY Technology

A common misperception when claiming the main advantages and limitations driving IgY Technology results from the fact that biological and technical aspects are normally brought together in a random manner. Indeed, these are highly correlated hence the difficulty in distinguishing them. Nevertheless, we try to summarize the different dimensions of using avian hosts and avian IgY antibodies, hoping to clearly distinguish each of the features in separately and show how they are compounded to leverage the potential or emphasize the limitations for a given outcome.

Overall, there are five main dimensions grounded on which avian IgY antibody-based technologies have evolved as promising alternatives to conventional research, diagnostic and therapeutic approaches (Fig. 1.5). The five main dimensions are as follow: (1) Animal ethics and welfare—the method is non-invasive as the antibody is collected from the egg yolk not blood, greatly reducing animal harm and distress. (2) Egg crude source—the avian IgY antibody naturally accumulates in high amounts in the yolk of eggs (–100 mg of IgY per egg). As hens lay on average one egg per day, a continuous collection and processing of eggs results in a scalable, cost-effective and highly sustainable way of antibody production. (3) Phylogenetic distance of hosts—birds diverged from mammals more than 300 million years ago, while mammals hold a 98 million years old common ancestor (Chap. 4); this results in lower homology and thus higher immunogenicity between the two groups. It is possible to obtain antibodies against highly conserved mammalian proteins or against proteins that usually evade the immune system in mammals. (4) Genetic organization of avian immune repertoires—birds hold unique antibody repertoires, resulting from diversification based on gene conversion (Chap. 3); this organization allows simplified in vitro molecular cloning of e.g., phage-display libraries (Smith 1985), and similar antibody engineering possibilities as for mammalian antibodies. (5) Molecular structure of IgY—the avian IgY antibody molecule (M.W. 180 kDa) holds unique molecular features in comparison with its mammalian IgG counterpart (M.W. 150 kDa), namely no binding to mammalian Fc-receptor, mammalian rheumatoid factor or mammalian complement proteins (Chaps. 5 and 6). This favours its use in human therapeutics as secondary immunoreactions are less prone to occur.

The limitations to the use of IgY antibodies and related technologies arise naturally from one or several of the inherent features described above. Hence for example a therapeutic IgY, will have an intrinsic variability due to the use of multiple birds, or simply the collection of hyperimmune eggs at different time points with the same birds; changing the birds in the flock will directly compromise the consistency and the reproducibility of the final IgY product, leading to titre and reactivities that are not entirely predictable (Chap. 10). If such features need to be

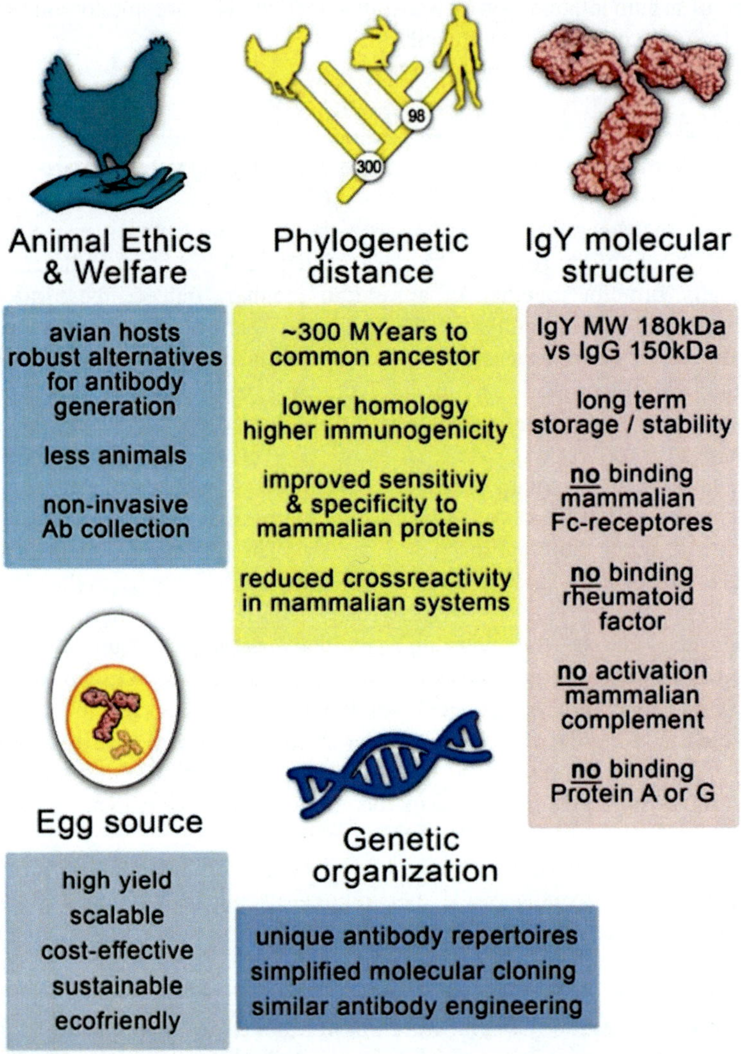

Fig. 1.5 Overall dimensions explored with IgY Technology. Animal ethics and welfare, phylogenetic distance of hosts, antibody crude source, molecular structure of IgY and genetic organization of avian immune repertoires, are the main dimensions grounded on which avian IgY antibody-based technologies have been evolving and growing as promising alternatives to conventional research, diagnostic and therapeutic approaches. Courtesy of IgYTechnology.com (www.igytechnology.com)

tightly controlled, as for e.g., human therapies, one as to account for such technical limitations inherent to any antiserum-like therapies derived from natural sources.

As mentioned above, the IgY molecule, namely its Fc domain (responsible for Fc-receptor binding and signalling functions) does not retain efficient binding capabilities to Protein A, G or other known microbial-binding proteins used to purify mammalian immunoglobulins. This has been a major limitation in the field for many decades and has driven development and optimization of novel IgY purification methods and approaches. IgY purification is particularly critical for large scale production of functional and therapeutic IgY. Having a simplified method for IgY purification even if only for purification of polyclonal mixtures, would definitely be a major breakthrough and driver for the field (Chap. 11).

The exploitation of avian antibody repertories towards development of novel monoclonal antibodies, faced an initial developmental stage where it was very much undervalued and even neglected. Interestingly, intrinsic technical limitations in translating the hybridoma technology to birds (Nishinaka et al. 1989), namely the absence of an efficient avian tumour cell line, made the approach lag behind in comparison to mammals. Even after the consolidation of phage-display technologies, as birds namely chickens, had become unpopular for monoclonal antibody development, it took some decades for birds to regain attention due to the potential of their distant phylogeny. Currently both phage-display and B-cell sorting technologies enable full exploitation of avian antibody repertories (Chap. 13). On the other hand, the intrinsic avian origin of the selected monoclonal antibodies typically demands further molecular engineering to obtain chimeric molecules with a structural background common to mammalian immunoglobulins.

Implications of all the features, advantages and limitations described above, will be further discussed and detailed in different chapters of the book and will hopefully become clear to the reader.

References

European Union (1986) Council Directive 86/609/EEC of 24 November 1986 on the approximation of laws, regulations and administrative provisions of the Member States regarding the protection of animals used for experimental and other scientific purposes. Official journal of the European Union. L&C, Legislation, Information and notices/EURLex

Huopalahti R, López-Fandiño R, Anton M, Schade R (2007) Bioactive egg compounds. Springer, Berlin

Kantha SS (1991) A centennial review; the 1890 tetanus antitoxin paper of von Behring and Kitasato and the related developments. Keio J Med 40(1):35–39

Klemperer F (1893) Ueber natürliche Immunität und ihre Verwerthung für die Immunisirungstherapie. Arch Exp Pathol Pharmakol 31(4–5):356–382

Köhler G, Milstein C (1975) Continuous cultures of fused cells secreting antibody of predefined specificity. Nature 256(5517):495–497

Leiva CL, Gallardo MJ, Casanova N, Terzolo H, Chacana P (2020) IgY-technology (egg yolk antibodies) in human medicine: a review of patents and clinical trials. Int Immunopharmacol 81:106269

Leslie GA, Clem LW (1969) Phylogen of immunoglobulin structure and function. 3. Immunoglobulins of the chicken. J Exp Med 130(6):1337–1352

Nishinaka S, Matsuda H, Murata M (1989) Establishment of a chicken X chicken hybridoma secreting specific antibody. Int Arch Allergy Appl Immunol 89(4):416–419

Russell WMS, Burch RL (1959) The principles of humane experimental technique. Methuen, London

Schade R, Staak C, Hendriksen C, Erhard MH, Straughan D (1996) The production of avian (egg yolk) antibodies: IgY. The report and recommendations of ECVAM workshop 21. Altern Lab Anim 24(6):925–934

Schade R, Behn I, Erhard M, Staak C, Hommel U, Hlinak A, Schwarzkopf C (2001) Isolation of IgY from yolk. In: Schade R, Behn I, Erhard M, Hlinak A, Staak C (eds) Chicken egg yolk antibodies, production and application: IgY-technology. Springer, Berlin, pp 65–107

Smith GP (1985) Filamentous fusion phage: novel expression vectors that display cloned antigens on the virion surface. Science 228(4705):1315–1317

2 The Domestic Hen

Shikun Ge, Álvaro Ferreira Júnior, Xiaoying Zhang, and
Patricia M. Morgan

Abstract

The egg-laying hen is one of the most common domestic animals. Birds are phylogenetically distinct from mammals and have several unique features: morphological differences such as the presence of air sacs, the medullary bone and cloacal chamber; distinct lymphoid tissues (bursa of Fabricius); and the presence of specific class of immunoglobulin (Ig) termed IgY in egg yolk. Genetic selection of particular traits for laying hens has improved their performance. Currently there are a large number of breeds of hen with different genotypes used for commercial and research purposes. The development of the oocyte and subsequent egg laying is hormonally controlled and formation of the egg as it passes through the oviduct has several defined stages. Hormonal factors affecting the initiation of egg laying and the components of the egg are detailed.

S. Ge · X. Zhang
College of Biological Science and Engineering, Shaanxi University of Technology, Hanzhong, China

CMBA, University of Minho, Braga, Portugal
e-mail: zhxying@snut.edu.cn; xzhang67@uoguelph.ca; zhang@bio.uminho.pt

Á. Ferreira Júnior
Escola de Veterinária e Zootecnia, Universidade Federal de Goiás, Goiânia, Brazil
e-mail: alvaro.ferreira@ufg.br

P. M. Morgan (✉)
Emerita Lecturer, School of Natural Sciences, National University of Ireland Galway, Galway, Ireland
e-mail: pat.morgan@nuigalway.ie

X.-Y. Zhang et al. (eds.), *IgY-Technology: Production and Application of Egg Yolk Antibodies*, https://doi.org/10.1007/978-3-030-72688-1_2

Keywords

Hen · Avian immune system · Reproductive system · Egg

2.1 Introduction

Birds exhibit unique biological features including the ability to fly, the presence of feathers, the ability to maintain a relatively uniform body temperature, migration in response to food sources and climate, and laying eggs. There are almost 10,000 species of birds (Prinzinger et al. 1991), with diverse living spaces, and diverse body structures, such as beak, foot and wing shapes. Avian physiology has two distinct branches; namely the physiology of wild birds (e.g., flight, migration and seasonal breeding) and the physiology of poultry (domesticated birds consisting of predominantly hens but also includes domestic turkeys, quail, ducks, geese, and the more primitive ratites such as ostriches which are discussed in Chap. 9.

2.2 Genetic Selection for Domestic Breeds of Chicken

Within domestic breeds of chicken there are enormous differences in both their genetic makeup and phenotypes (Tixier-Boichard et al. 2012). The differences in modern breeds (*Gallus gallus domesticus*) are probably as a result of their ancestral development from geographically separated subspecies of red jungle fowl (*Gallus gallus gallus*; *Gallus gallus spadiceus*, *Gallus gallus jaboullei*; *Gallus gallus murgha*) and a second species, grey jungle fowl *Gallus sonneratii* (Rubin et al. 2010; Tixier-Boichard et al. 2012) and cross breeding during several rounds of domestication. Whole-genome sequencing shows that several loci have been subjected to strong selection during domestication of chickens and specialization into meat producing (broiler) and egg producing (layer) chickens which are the subject of this chapter (Preisinger 2018; Qanbari et al. 2019). The genetic selection of laying hens resulted in reduced body weight, a shift to earlier sexual maturity, an increase in the quantity and quality of egg production, and decreases in the feed necessary to produce the same number of eggs (Preisinger 2018). Recent genetic trends in egg output, show an increase from 313 to 325 eggs per year from 2010 to 2015 and selection for this trait is likely to result in a further increase of two to three eggs per year (Preisinger 2018). Laying hens still retain some of the biological characteristics of birds, such as the habit of foraging and constant movement, acute hearing, acute daytime vision, neurotic characteristics, utilizing a wide range of food by eating gravel to grind it (Thiruvenkadan et al. 2010); traits which need to be considered in keeping hens (Chap. 8).

2.3 Breeds of Chicken

It is important to note at this juncture that genetic differences exist between different breeds of *Gallus gallus domesticus*: these result in different immunological adaptations as determined in whole blood (Bílková et al. 2017); nest use patterns (Villanueva et al. 2017); excitability and preference for enrichment elements (Kozak et al. 2019); timing of egg laying (Tůmová et al. 2017) and egg laying capacity (Mack et al. 2013; Table 2.1). The most common commercial laying hens include Romain Brown Shell Layer, Hyland Brown Shell Layer, Fresh Brown Shell Layer, Hay Sykes Brown Shell Layer, Elsa Brown Layer, Golden Brown Shell Layer, White Leghorns, Rhode Island Red, Golden Comet, Ameraucana, Barred Plymouth Rock, Golden Laced Wyandottes and New Hampshire Red (Agritech 2009; Ruppenthal 2012). Other breeds of chicken utilized in research in addition to the more commonly used White Leghorn and Rhode Island Red are the following: Anraja, Gramapriya, BlackRock, KalingaBrown (Agrawal et al. 2016); Araucana, Booted bantam, Czech, Minorca and Rosecomb bantam (Bílková et al. 2017); Greenlegged Partridge and Polbar (Kozak et al. 2019); Hy-Line Brown, Bovans Brown, DeKalb White, and Hy-Line W36 (Ali et al. 2016); Silkie and Dongxiang blue-shell (Sun et al. 2013); Erlang Mountainous (Liu et al. 2015); Bovans Sperwer, Isa Sussex, Moravia Barred and Moravia BSL (Tůmová et al. 2017).

Layers produce the most eggs in the first year, with 15%–20% decrease in egg productivity and 15% increase in mortality in the second year (Parkhurst and Mountney 1988). Peak production now approaches the biological limit of one egg per day, but in early production (at sexual maturity) and late production (persistency) genetic variation is still high (Thiruvenkadan et al. 2010).

Apart from breed difference, egg production is closely related to the physical fitness, including physical function, cognitive ability and living habits of hens. Hens with high annual egg production tend to live longer (Villanueva et al. 2017; Dudde et al. 2018). Many regions have rich varieties of laying hens' resources. Compared with the major commercial hens in the world, these hen breeds have received less attention, and they are marginalized by the extensive cultivation of commercial hens (Yang et al. 1996). However, such local layer resources are conducive to the preservation of biodiversity the genome and physiology of the hen and may be important resources for IgY technology.

2.4 Physiology of Chickens

The general physiological parameters of domestic chickens will now be discussed.

2.4.1 Body Temperature

Many birds undergo long distance migration and can adapt to extreme environments (e.g., hot, cold, and desert areas). Despite this, most birds maintain a relatively

Table 2.1 Strains of laying hens and hen laying capacity

Breed	First egg[a,b,c,d,e] (week)	Feed conversion ratio[a, b, e]	Mean egg production in the first laying year[b]	Mean egg weight[b, e] (g)	Peak egg production rate[d,e] (%)	Survivability[a, c] (%)
White Leghorn	18	NA	280–320	55–67	NA	91.6
Romain Brown Shell layer	18	2.0–2.2	300–305	63.5–65.5	92–94	94.6
Hyland Brown Shell layer	18	2.2–2.3	344	66.9	94–96	95
Fresh brown shell layer	20	2.25–2.4	287–296	63.5	90–94	93–97
Hay Sykes Brown Shell layer	17	2.24	324	63.2	NA	94.2
ILI-80	18	2.1	280	54	92	94
Golden-92	19	2.2	265	54	90	94
Priya	18	2.1	290	57	92	94
Sonali	19	2.2	275	54	90	94
Devendra	19	2.5	200	50	90	94

Notes: Data according to [a]Bradley (1960), Shi et al. (2020), [b]North and Bell (1990), [c]Parkhurst and Mountney (1988), [d]Mack et al. (2013), [e]Agritech (2009), NA: Data not available constant body temperature even when exposed to temperatures well below thermoneutrality (Hislop 1964). Their ability to do so depends on a combination of behavioural, insulative, metabolic, and other physiological responses (Chaffee and Roberts 1971). However, the average body temperature of all birds is higher than that of mammals. Birds have mean body temperatures of 38.54 °C ± 0.96 °C, 41.02 °C ± 1.29 °C and 43.85 °C ± 0.94 °C in resting, active phase and high activity periods, respectively; during the active phase birds maintain a relatively stable temperature despite varying environmental conditions (Prinzinger et al. 1991). The higher temperatures of birds should be taken into consideration in IgY-related technology and engineering; for instance, in generating monoclonal IgY using hybridoma technique (Chap. 14), the best temperature for cell culture might be higher than 37 °C (Stifani et al. 1990).

2.4.2 Anatomical Structures

Birds possess some unique organs, or, organs which have undergone special evolutionary pathways (Bradley 1960). The **Lungs** of avian species have a very different structure, as birds have a higher oxygen consumption at rest, compared to mammals. The lungs of mammals are suspended in the left and right thorax, which are completely closed and separated. In birds, about a third of the lung is buried deep in the intercostal spaces and limited by external stents and thus it does not dilate much. However, all parts of the avian lungs are directly connected to the **air sacs** which are distributed throughout the body, doubling the volume of the respiratory system compared to equally-sized mammals (King and Payne 1962). An air sac is a derivative organ of the lung with multiple functions. It can enhance the exchange of air in the lungs, reduce body weight, balance posture, strengthen vocal airflow, dissipate body heat and regulate body temperature. Thus, inflammation in the bird's lungs tends to be more severe than in mammals when it occurs (West et al. 2007).

2.4.3 Crop

The Crop is an expanded muscular pouch near the gullet or throat and is a place where food is temporarily stored in some birds. The duck has no real crop, forming only a fusiform bulge in the oesophagus and neck. The crop may also play a role in the immune response as plasma cells and B-lymphocytes have been isolated in the crop (Bartosz et al. 2016).

2.4.4 Cloacal Chamber

The Cloacal chamber is a dilated duct at the end of the intestine, which is the common channel where the digestive tract and the urinary and reproductive systems empty. The cloaca of the chicken appears to be a sudden dilation at the end of the rectum, opening to the outside with a drainage hole. In adult hens, the position of the connection between the rectum and the cloaca is changed due to the swelling of the end of the fallopian tube. The **bursa of Fabricius** is located at the top of the post-cloacal area. It consists of more than 10,000 follicles surrounded by connective tissue. This organ plays a central role in the development of the antibody-producing B-lymphocyte lineage in birds, particularly in early life stages (Ratcliffe 2006) (Chap. 3).

2.4.5 Medullary Bone

The **Medullary bone** is a bone tissue unique to birds. It is present only in females about to lay eggs and forms in the empty spaces within the skeleton (Wang et al. 2019). This bone tissue serves as a reservoir for calcium and phosphorus needed to form the eggshell. Medullary bone has been reported in a variety of non-avian dinosaurs including *Tyrannosaurus rex*, ornithopod dinosaurs like *Tenontosaurus*, and several sauropods (huge long-necked dinosaurs) including *Mussasaurus* (Wang et al. 2019). It has also been identified in pterosaurs, which are flying reptiles closely related to dinosaurs. Since the first report of medullary bone in a Mesozoic fossil, this tissue has attracted great interest because it links birds and dinosaurs (Taylor and Moore 1953).

2.5 Immune System of the Hen

The immune system of the hens differs from that of mammals in various ways as outlined in Chap. 3. Primarily, the existence of the bursa of Fabricius in birds but not in mammals and that the major serum antibody in birds is IgY instead of IgG in mammals.

2.5.1 Immune Tissues

The immune system of the hen and mammals is organized by groups of immune cells such as T cells and B cells and their homing into the organized lymphoid tissues which are strategically positioned to protect the host (Boehm et al. 2012; Rostami et al. 2018).

2.5.1.1 Thymus and Bursa of Fabricius

The primary lymphoid tissues of the hen include the thymus and the bursa of Fabricius (Boehm et al. 2012). The thymus is located at the ventral neck region and the bursa of Fabricius is found at the top of the cloaca region (Sun et al. 2016; Ifrah et al. 2016). The primary lymphoid tissue are the sources of lymphocytes such as T cells (thymus-dependent cell) and B cells (bursa of Fabricius-selected cells) which are required to both mount an appropriate immune response and to avoid autoimmunity (Sun et al. 2016; Ifrah et al. 2016).

2.5.1.2 Spleen

The activated T and B cells leave the primary lymphoid tissue and are attracted to their defensive position in the secondary lymphoid tissue such as spleen and mucosa-associated lymphoid tissue (MALT) (Lanning and Knight 2015; Madej et al. 2015; Sepahi and Salinas 2015). The spleen is a capsulated tissue found in the abdominal cavity closely to stomach (Zhang et al. 2015), whilst the MALT is a lymphoid tissue scattered through the body surfaces such as the mucosal surfaces of the digestive system, eyes (Harderian glands) (Van Ginkel et al. 2011), respiratory system and skin (Smialek et al. 2011; Lanning and Knight 2015; Sepahi and Salinas 2015). Following contact with pathogens, the spleen enlarges by hyperplasia know as splenomegaly. Lymphocyte hyperplasia has also been found in MALT. The germinal centres, within secondary lymphoid organs, are the areas where B cells proliferate and differentiate in response to an infection (Chap. 3).

2.5.1.3 Defence System in the Reproductive Organs

Infection of the reproductive tract has consequences for the transmission of disease to progeny and infection of eggs for human consumption continues to be a public health concern. The infection of the reproductive by bacterial and viral pathogens has been reviewed (Wigley et al. 2014). The hen ovary and oviduct can be infected by various bacterial pathogenic microbes such as *Salmonella enteritidis*, *Salmonella typhimurium*, and *Mycoplasma melagridis* (Wales and Davies 2011). Viral pathogenic infections include avian infectious bronchitis virus and avian influenza viruses (Nii et al. 2015). Latent and/or persistent infection of the reproductive tract also occurs by avian leukosis virus, reticuloendotheliosis virus and chicken infectious anaemia virus (Wigley et al. 2014). The health of the reproductive tract is important for formation and production of high quality and hygienic eggs (Yoshimura and Barua 2017). The hen ovary and oviduct have a lymphoid tissue populated by immunocompetent cells such as macrophages and lymphocytes. The influx of immune cells increases with hen maturity and decreases with aging (Yoshimura and Barua 2017; Ferreira Júnior et al. 2018). The follicular tissue of the hen express TLR molecules, triggering a production of pro-inflammatory cytokines and chemokines, and producing defensin molecules (Yoshimura and Barua 2017).

2.5.1.4 Immune Response

The innate and adaptive immune response of the hen is described in detail in Chap. 3. Interestingly, it has been demonstrated that some differences occur in populations of immunocompetent cells among different hen breeds, e.g., Araucana, Booted bantam, Czech, Minorca and Rosecomb bantam as measured by whole blood flow cytometry (Bílková et al. 2017). Czech, a European breed, had the lowest heterophil/lymphocyte (H/L) ratio and a high proportion of $CD4^+$ cells while Araucana, a South American breed had the highest H/L ratio and a high monocyte frequency; such differences may have an evolutionary basis (Bílková et al. 2017).

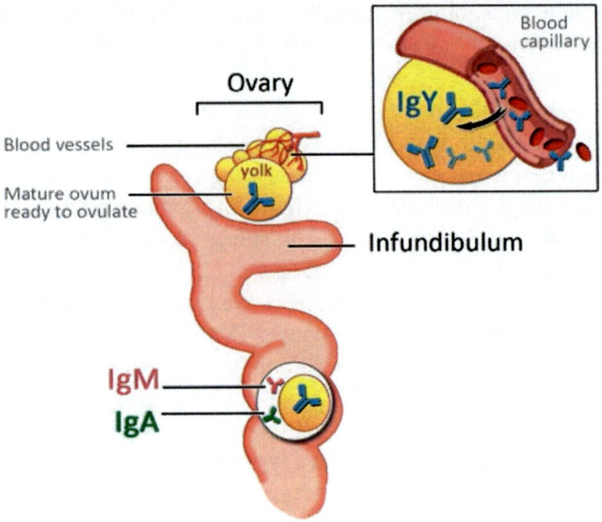

Fig. 2.1 IgY antibody transport from the blood vessels surrounding the developing follicle into the egg yolk and IgM and IgA deposition in egg white through the oviductal epithelium

The classes of chicken antibodies include IgM, IgA and IgY (Chap. 3). On a daily basis, the translocation of compounds, including IgY antibodies, from the blood of the hen to the ovarian follicle occurs (Fig. 2.1).

For detailed information on the immunology of the hen and biological basis of IgY, please refer to Chaps. 3 and 5, respectively.

2.6 Development of the Reproductive System and Egg Laying

Central to IgY technology is immunization of the hen (Chap. 10) and the production of eggs from which the IgY is extracted (Chap. 11). Therefore, the development of the reproductive system, initiation of laying, and the process of egg formation will be outlined in detail.

2.6.1 Embryology of the Chicken

When a fertilized egg is laid, it begins the 21-day period of embryonic development; all stages of development have been reviewed by Tong et al. and are outside the scope of this chapter (Tong et al. 2013). The avian embryo has no anatomical connection to the hen, all of its nutritive requirements and the maternally-derived antibodies must be contained within the egg (Chap. 11). The developing embryo requires a supply of oxygen and from very early on, the embryo develops special membranes external to its body to access the nutrients in the egg and to carry out essential bodily functions (Bradley 1960). The **Yolk sac** envelops the yolk and produces an enzyme that changes the yolk material into a form that can be used as a food source by the developing embryo (North and Bell 1990). The chicken FcRY receptor is located in the yolk sac (Chap. 6). The **Amnion** forms a fluid-filled sac in which the embryo floats and is protected from rocking of the egg. The **Allantois** develops an extensive circulatory system connected to that of the embryo and is driven by the new embryonic heart which is formed by 44 h of incubation (Tong et al. 2013). The allantois surrounds the embryo completely when it is fully developed. The **Chorion** fuses the inner shell membrane to the allantois and receives oxygen and eliminates carbon dioxide through the pores of the shell (Ribatti 2014). As early as day 5 or 6 of incubation, the reproductive organs differentiate, and the sex of the offspring is determined (Tong et al. 2013). In the final stages of development, the chick undergoes a series of internal and external pipping and hatches on day 21 of incubation (Tong et al. 2013).

2.6.2 The Ovary

2.6.2.1 Follicular Development and Oocyte Maturation

Eggs develop from the primordial follicles which are in the ovarian cortex (Pourquié 2018). In the developing follicles, the oocyte, containing a large amount of yolk, is surrounded by the perivitelline layer and the follicular wall, which consists of the granulosa layer, basal lamina, theca interna and externa, superficial tissue and epithelium (Yoshimura and Barua 2017).

The initial development of the primordial follicle pool occurs during the first 4 days post-hatching (Guo et al. 2019). It has been shown that Follicle-stimulating hormone (FSH) and Stem cell signalling are required for ovarian development in the chicken (Guo et al. 2019). A large number of genes have been identified, using in silico and microarray analysis, which are differentially expressed during final oocyte maturation in the hen (Elis et al. 2008). The oocyte may also influence the development of the follicle through the Kit system; an oocyte factor regulates Kit ligand expression which decreases with follicular development (Kundu et al. 2012).

2.6.3 Initiation of Laying

The control of sexual maturation in the hen has been reviewed by Bédécarrats et al. and is controlled by the hypthalmo-pituitary-gonadal axis, in common with other vertebrates (Bédécarrats et al. 2009). It is well documented that changes in light intensity impact on sexual maturation in hens (Shi et al. 2020). Photostimulation has been shown to induce an increase in hypothalamic gonadotropin releasing hormone (GnRH-1) which then increases FSH and Luteinizing Hormone (LH) release from the pituitary gland. However, the action of GnRH-1 is antagonized by Gonadotropin Inhibitory Hormone (GnIH) which was discovered in 2000 (Tsutsui et al. 2000). In summary, the following signalling events predominate; in young hens, not exposed to a stimulatory photoperiod, melatonin stimulates GnIH which interacts with its receptors in the pituitary gland and maintains the bird in its juvenile status; exposure to increasing light levels, decreases the level of GnIH and simultaneously increases the production of GnRH; the resulting release of LH and FSH from the pituitary gland, starts follicular growth and release of oestradiol which in turn inhibits GnIH synthesis; when hens are fully mature, high levels of oestradiol and progesterone inhibit the production of GnIH and gonadotropin release (Bédécarrats et al. 2009). The onset of laying ranges from week 17 to 20 depending on the breed of hen (Table 2.1).

2.6.4 Reproductive System of the Hen

The female reproductive system in domestic fowl consists of the **ovary** and the accompanying **oviduct**. While the female embryo in chicken has two ovaries, only the left ovary survives and reaches maturity to produce eggs. Anti-Mullerian hormone has a critical role in the regression of the right ovary and may play a role in follicular development (Johnson 2012). The left ovary is located just in front of the kidneys in the abdominal cavity and is firmly attached to the wall of the cavity. The ovary is well endowed with blood vessels to ensure effective transport of nutrients to the developing yolk. The ovary of hatched chickens contains more than 12,000 oocytes of which only a small proportion will develop into mature ovulatory oocytes (Nishio et al. 2018). With the onset of sexual maturity, follicles of different sizes develop by an accumulation of lipids (Johnson 2000). There are five to eight large, yolk filled follicles arranged in a hierarchy according to size, with the largest (F1) next to ovulate, followed by F2, and F3 (Fig. 2.2). A single follicle per day is selected from the pool of 6–8 mm follicles to enter the pre-ovulatory hierarchy (Woods and Johnson 2005). As the follicles develop, they protrude from the ovary (Fig. 2.2) and are categorised by size or colour as small white follicles (1–5 mm), small yellow follicles (6–8 mm) or preovulatory follicles (9–40 mm) in diameter

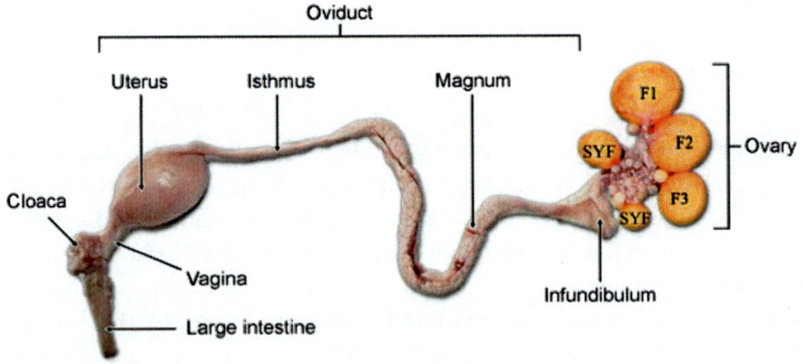

Fig. 2.2 Reproductive structures of the hen ovary and oviduct. Follicles labelled F1–F3 are the three largest pre-ovulatory follicles in decreasing diameter (9–40 mm). Small yellow follicles (SYF), which have accumulated lipids are <9 mm. The oviduct consists of the infundibulum, magnum, isthmus, uterus and vagina. The utero-vaginal junction (UVJ) is located between the uterus and vagina. Figure adapted from Apperson et al. (2017)

Table 2.2 Overview of the egg laying process

Parts of the oviduct	Duration of process (h)	Function
Infundibulum	< 0.5	Formation of vitelline membrane
Magnum	3–4	Albumen formation
Isthmus	1–2	Egg shell membranes formed
Shell gland (uterus)	19	Egg shell formation and cuticle deposition
Vagina	1	Oviposition
Cloaca	NA	Orifice through which egg is laid

Note: NA: Data not available

(Apperson et al. 2017) and weighing from –150 mg to –230 mg (Johnson 2012, 2015).

2.6.4.1 Egg Laying and the Oviduct

The oviduct of a laying hen contributes to three functions, namely the passage of the ovum from the ovary, formation of the egg membranes and egg shell and transport of the developing egg to the cloaca (Apperson et al. 2017). The oviduct consists of five segments including the infundibulum, magnum, isthmus, uterus (shell gland) and vagina (Fig. 2.2; Table 2.2; (Yoshimura and Barua 2017)).

Fertilization, if the hen is bred, occurs in the infundibulum before the outer perivitelline layer is formed. The vagina is at the caudal end of the oviduct which opens to the cloaca. The part between the uterus and vagina is called the UVJ, where sperm are kept alive for a few weeks (Yoshimura and Barua 2017). The egg shell is largely made from calcium carbonate. The calcium is obtained from the feed and also comes from medullary bone (Taylor and Moore 1953). If there is any pigment to be added to the egg, it happens in the shell gland. The egg remains in the shell gland for around 19 h.

The vagina is important in getting the egg laid properly and the egg is turned in a process called 'oviposition'. The egg enters the vagina narrow end first but is laid 'blunt' end first. Finally, the egg reaches the cloaca which is the exit. Once an egg has been laid, within 30 or so minutes another will start on its journey. A comparison of oviposition in brown and white egg layers demonstrated that brown-egg layers laid eggs more than 1 h earlier after lights came compared to white-egg hens (Tůmová et al. 2017). The sequence of egg laying is such that subsequent eggs are laid later in the day, starting from the morning, and as it is rare for a hen to start ovulation after 3 pm, so once every 6–7 days she will miss a day of production.

2.7 The Egg

Understanding of the composition of the egg is essential for the separation and purification of IgY. Yolk proteins and lipids, with the exception of immunoglobulins (Ig), are made in the liver and transferred via the blood stream to the developing

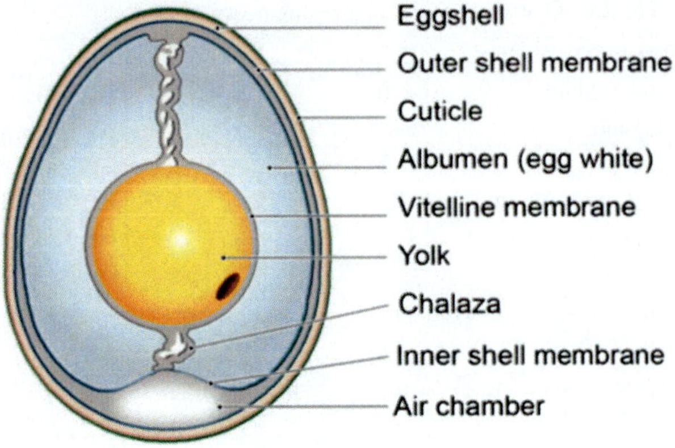

Fig. 2.3 The structure of a hen egg

follicle. The principle of IgY extraction is initially to remove the large amount of lipids, and to obtain water-soluble components for further extraction (Chap. 11).

2.7.1 Structure of Eggs

The principal components of eggs include the central egg yolk surrounded by the vitelline membrane, the albumen (egg white), egg shell membranes, the calcified egg shell and cuticle (Fig. 2.3). The egg shell is about 9.5% of the total egg volume. The external shell membranes are composed of inner and outer sheets that enclose the egg albumen; an outer membrane functions to avoid egg evaporation; the inner lower shell membrane allows air to passes freely; the air chamber is the space between two layers of shell membrane; the chalaza projects from the vituline membrane and holds the yolk in steady position in the egg (Hincke et al. 2012). The egg white is a semi-flowing gelatinous material inside the shell, accounting for 63% of egg in volume, and which can provide nutrients and protection for the development of fertilized eggs. Its main components are proteins, including ovalbumin, oval transferrin, ovomucoid, lysozyme, ovalbumin, cysteine protease inhibitors and antibiotic proteins, but it also contains a small amount of carbohydrates and trace amounts of fat (Stadelman et al. 1995).

2.7.2 Composition of Egg Yolk

The composition of the yolk is approximately 48% water, 33% lipid and 17% protein (Burley and Vadehra 1989). Importantly, for the extraction of IgY (Chap. 11) yolk can be separated into two parts: the major one is aqueous plasma consisting of 86% low density lipoproteins and the remainder α-livetin (serum albumin), β-livetin (α2-glycoprotein), and γ-livetin (IgY); the second portion contains granules of varying sizes composed of 70% high-density lipoproteins (α- and β-lipovitellins), 16% phosvitin (glycophospoprotein), and 12% low-density lipoproteins (LDL) (Burley

Table 2.3 Concentration of Immunoglobins in layer serum and egg

	IgY (mg/mL)	IgM (mg/mL)	IgA (mg/mL)
Serum[a]	5–7	1–2	0.33–0.65
Yolk[b]	15–25	<0.02	<0.03
Egg white[c]	<0.03	–0.15	–0.7

Notes: [a]Data according to Van Meter et al. (1969), Leslie and Clem (1970), Lebacq-Verheyden et al. (1974), Leslie et al. (1976), Härtle et al. (2014). [b],[c]Data according to Rose et al. (1974), Akita and Nakai (1993)

and Cook 1961; Chay Pak Ting et al. 2010; Anton 2013). The concentration and distribution of Ig in serum and eggs are shown in Table 2.3.

2.8 Concluding Remarks

The physiological structure and immune system of hens have been demonstrated to be competent to produce robust innate and acquired immune responses against different pathogens. In this context, IgY is a crucial molecule in the immune response of hens in terms of their specific antigenic binding properties and the wide range applications as an immunobiological reagent.

Avian and hen physiology still poses a number of significant questions. For example, how extreme environments (cold, hot, water, and hypoxia) impact on their physiological conditions and what are the unique regulatory mechanisms that allows them to control the response to these conditions? How is the hen's reproductive system controlled including the recruitment and hierarchy of follicular development, the circadian patterns of ovulation and oviposition, cross-talk between the ovary and oviduct and neural networks? What impact has the genetic selection into layers or broilers has on their immune and reproductive systems? A comprehensive understanding of the physiology of the domestic hen is essential to underpin the development of IgY technology.

References

Agrawal R, Hirpurkar SD, Sannat C, Gupta AK (2016) Comparative study on immunoglobulin Y transfer from breeding hens to egg yolk and progeny chicks in different breeds of poultry. Vet World 9(4):425–431

Agritech (2009). http://www.agritech.tnau.ac.in/animal_husbandry/ani_chik_breeds%20of%20chicken.html?tdsourcetag=s_pcqq_aiomsg. Accessed 2009

Akita EM, Nakai S (1993) Comparison of four purification methods for the production of immunoglobulins from eggs laid by hens immunized with an enterotoxigenic *E. coli* strain. J Immunol Methods 160(2):207–214

Ali AB, Campbell DL, Karcher DM, Siegford JM (2016) Influence of genetic strain and access to litter on spatial distribution of 4 strains of laying hens in an aviary system. Poult Sci 95 (11):2489–2502

Anton M (2013) Egg yolk: structures, functionalities and processes. J Sci Food Agric 93 (12):2871–2880. https://doi.org/10.1002/jsfa.6247
Apperson KD, Bird KE, Cherian G, Löhr CV (2017) Histology of the ovary of the laying hen (*Gallus domesticus*). Vet Sci 4(4):66
Bartosz K, Mateusz R, Jakub D, Sylwester Ś, Damian J (2016) Avian crop function – a review. Ann Anim Sci 16(3):653–678
Bédécarrats GY, McFarlane H, Maddineni SR, Ramachandran R (2009) Gonadotropin-inhibitory hormone receptor signaling and its impact on reproduction in chickens. Gen Comp Endocrinol 163(1–2):7–11
Bílková B, Swiderska Z, Janda J, Zita L, Vinkler M (2017) Different breeds, different blood: cytometric analysis of whole blood cellular composition in chicken breeds. Vet Immunol Immunopathol 188:71–77
Boehm T, Hess I, Swann J (2012) Evolution of lymphoid tissues. Trends Immunol 33(6):315–321
Bradley O (1960) The structure of the fowl. In: Grahame T (ed) 4th ed. Oliver and Boyd, London
Burley RW, Cook WH (1961) Isolation and composition of avian egg yolk granules and their constituent alpha- and beta-lipovitellins. Can J Biochem Physiol 39:1295–1307
Burley RW, Vadehra DV (1989) Egg yolk: structure and properties. In: Burley RW, Vadehra DV (eds) The avian egg: chemistry and biology. Wiley, New York
Chaffee R, Roberts J (1971) Temperature acclimation in birds and mammals. Annu Rev Physiol 33:155–202
Chay Pak Ting BP, Pouliot Y, Gauthier SF Mine Y (2010) Fractionation of egg proteins and peptides for nutraceutical applications. In: Separation, extraction and concentration processes in the food, beverage and nutraceutical industries, pp 595–618
Dudde A, Krause ET, Matthews L, Schrader L (2018) More than eggs – relationship between productivity and learning in laying hens. Front Psychol 9:2000
Elis S, Batellier F, Couty I, Balzergue S, Martin-Magniette ML, Monget P, Blesbois E, Govoroun MS (2008) Search for the genes involved in oocyte maturation and early embryo development in the hen. BMC Genomics 9:110
Ferreira Júnior Á, Santos JPD, Sousa IDO, Martin I, Alves EGL, Rosado IR (2018) *Gallus gallus domesticus*: immune system and its potential for generation of immunobiologics. Ciência Rural 48(8):e20180250
Guo C, Liu G, Zhao D, Mi Y, Zhang C, Li J (2019) Interaction of follicle-stimulating hormone and stem cell factor to promote primordial follicle assembly in the chicken. Front Endocrinol 10:91
Härtle S, Magor K, Göbel T, Davison F, Kaspers B (2014) Structure and evolution of avian immunoglobulins. In: Schat KA, Kaspers B, Kaiser P (eds), 2nd ed. [S.l.]. Elsevier, London, pp 103–114
Hincke MT, Nys Y, Gautron J, Mann K, Rodriguez-Navarro AB, McKee MD (2012) The eggshell: structure, composition and mineralization. Front Biosci (Landmark edition) 17:1266–1280. https://doi.org/10.2741/3985
Hislop HJ (1964) Handbook of physiology. Section 4. Adaptation to the environment. Am J Phys Anthropol 44(12):1121–1122
Ifrah M, Perelman B, Finger A, Uni Z (2016) The role of the bursa of Fabricius in the immune response to vaccinal antigens and the development of immune tolerance in chicks (*Gallus domesticus*) vaccinated at a very young age. Poult Sci 96(1):51–57
Johnson AL (2000) Granulosa cell apoptosis: conservation of cell signaling in an avian ovarian model system. Biol Signals Recept 9(2):96–101
Johnson P (2012) Follicle selection in the avian ovary. Reprod Domest Anim 4:283–287
Johnson AL (2015) Ovarian follicle selection and granulosa cell differentiation. Poult Sci 94 (4):781–785
King AS, Payne DC (1962) The maximum capacities of the lungs and air sacs of *Gallus domesticus*. J Anat 96(4):495–503
Kozak A, Kasperek K, Zięba G, Rozempolska-Rucińska I (2019) Variability of laying hen behaviour depending on the breed. Asian Australas J Anim Sci 32(7):1062–1068

Kundu MC, Wojtusik J, Johnson PA (2012) Expression and regulation of kit ligand in the ovary of the hen. Gen Comp Endocrinol 179(1):47–52

Lanning D, Knight K (2015) Diversification of the primary antibody repertoire by AID-mediated gene conversion. Results Probl Cell Differ 57:279–293

Lebacq-Verheyden AM, Vaerman JP, Heremans JF (1974) Quantification and distribution of chicken immunoglobulins IgA, IgM and IgG in serum and secretions. Immunology 27 (4):683–692

Leslie GA, Clem LW (1970) Chicken immunoglobulins: biological half-lives and normal adult serum concentrations of IgM and IgY. Proc Soc Exp Biol Med (New York, NY) 134 (1):195–198. https://doi.org/10.3181/00379727-134-34758

Leslie GA, Stankus RP, Martin LN (1976) Secretory immunological system of fowl. V. The gallbladder: an integral part of the secretory immunological system of fowl. Int Arch Allergy Appl Immunol 51(2):175–185. https://doi.org/10.1159/000231590

Liu L, Li D, Gilbert ER, Xiao Q, Zhao X, Wang Y, Yin H, Zhu Q (2015) Effect of monochromatic light on expression of estrogen receptor (ER) and progesterone receptor (PR) in ovarian follicles of chicken. PLoS One 10(12):e0144102

Mack LA, Felver-gant JN, Dennis RL, Cheng HW (2013) Genetic variations alter production and behavioral responses following heat stress in 2 strains of laying hens. Poult Sci 92(2):285–294

Madej J, Stefaniak T, Bednarczyk M (2015) Effect of in ovo-delivered prebiotics and synbiotics on lymphoid-organs' morphology in chickens. Poult Sci 94(6):1209–1219

Nii T, Isobe N, Yoshimura Y (2015) The effect of estrogen on the early cytotoxic response to IB virus infection in hen oviduct. Vet Immunol Immunopathol 164(1–2):56–66

Nishio S, Okumura H, Matsuda T (2018) Egg-coat and zona pellucida proteins of chicken as a typical species of Aves. Curr Top Dev Biol 130:307–329

North MO, Bell D (1990) Commercial chicken production manual, 4th edn. AVI, Westport

Parkhurst CR, Mountney CJ (1988) Poultry meat and egg production. Von Nostrand Reinbold, Melbourne

Pourquié O (2018) Somite formation in the chicken embryo. Int J Dev Biol 62(1-2-3):57–62

Preisinger R (2018) Innovative layer genetics to handle global challenges in egg production. Br Poult Sci 59(1):1–6

Prinzinger R, Preßmar AN, Schleucher E (1991) Body temperature in birds. Comp Biochem Physiol A Physiol 99(4):499–506

Qanbari S, Rubin CJ, Maqbool K, Weigend S, Weigend A, Geibel J, Kerje S, Wurmser C, Peterson AT, Brisbin IL Jr, Preisinger R, Fries R, Simianer H, Andersson L (2019) Genetics of adaptation in modern chicken. PLoS Genet 15(4):e1007989

Ratcliffe MJ (2006) Antibodies, immunoglobulin genes and the bursa of Fabricius in chicken B cell development. Dev Comp Immunol 30(1–2):101–118

Ribatti D (2014) The chick embryo chorioallantoic membrane as a model for tumor biology. Exp Cell Res 328(2):314–324

Rose ME, Orlans E, Buttress N (1974) Immunoglobulin classes in the hen's egg: their segregation in yolk and white. Eur J Immunol 4(7):521–523

Rostami H, Seidavi A, Dadashbeiki M, Asadpour Y, Simões J, Assar Ali S, Laudadio V, Losacco C, Perillo A, Tufarelli V (2018) Supplementing dietary rosemary (*Rosmarinus officinalis* L.) powder and vitamin E in broiler chickens: evaluation of humoral immune response, lymphoid organs, and blood proteins. Environ Sci Pollut Res 25(9):8836–8842

Rubin CJ, Zody MC, Eriksson J, Meadows JR, Sherwood E, Webster MT, Jiang L, Ingman M, Sharpe T, Ka S, Hallböök F, Besnier F, Carlborg O, Bed'hom B, Tixier-Boichard M, Jensen P, Siegel P, Lindblad-Toh K, Andersson L (2010) Whole-genome resequencing reveals loci under selection during chicken domestication. Nature 464(7288):587–591

Ruppenthal RJ (2012) Best chicken breeds: 12 types of hens that lay lots of eggs, make good pets, and fit in small yards. CreateSpace Independent Publishing Platform. ISBN 13: 9781480012813

Sepahi A, Salinas I (2015) The evolution of nasal immune systems in vertebrates. Mol Immunol 69:131–138

Shi L, Sun Y, Xu H, Liu Y, Li Y, Huang Z, Ni A, Chen C, Li D, Wang P, Fan J, Ma H, Chen J (2020) Effect of age at photostimulation on sexual maturation and egg-laying performance of layer breeders. Poult Sci 99(2):812–819

Smialek M, Tykałowski B, Stenzel T, Koncicki A (2011) Local immunity of the respiratory mucosal system in chickens and turkeys. Pol J Vet Sci 14(2):291–297

Stadelman WJ, Newkirk D, Newby L (1995) Egg science and technology, 4th edn. CRC, Boca Raton

Stifani S, Barber D, Nimpf J, Schneider W (1990) A single chicken oocyte plasma membrane protein mediates uptake of VLDL and VTG. Proc Natl Acad Sci USA 87(5):1955–1959

Sun H, Chen S, Cai X, Qu L (2013) Correlation analysis of the total IgY level in hen serum, egg yolk and offspring serum. J Anim Sci Biotechnol 4(1):10

Sun H, Liu P, Nolan L, Lamont S (2016) Thymus transcriptome reveals novel pathways in response to avian pathogenic *Escherichia coli* infection. Poult Sci 95(12):2803–2814

Taylor T, Moore J (1953) Avian medullary bone. Nature 172(4376):504–505

Thiruvenkadan AK, Panneerselvam S, Prabakaran R (2010) Layer breeding strategies: an overview. Worlds Poult Sci J 66:477–502

Tixier-Boichard M, Leenstra F, Flock DK, Hocking PM, Weigend S (2012) A century of poultry genetics. Worlds Poult Sci J 68(02):307–321

Tong Q, Romanini CE, Exadaktylos V, Bahr C, Berckmans D, Bergoug H, Eterradossi N, Roulston N, Verhelst R, McGonnell IM, Demmers T (2013) Embryonic development and the physiological factors that coordinate hatching in domestic chickens. Poult Sci 92(3):620–628

Tsutsui K, Saigoh E, Ukena K (2000) A novel avian hypothalamic peptide inhibiting gonadotropin release. Biochem Biophys Res Commun 275(2):661–667

Tůmová E, Uhlířová L, Tůma R, Chodová D, Máchal L (2017) Age related changes in laying pattern and egg weight of different laying hen genotypes. Anim Reprod Sci 183:21–26

Van Ginkel F, Gulley S, Lammers A, Hoerr F, Gurjar R, Toro H (2011) Conjunctiva-associated lymphoid tissue in avian mucosal immunity. Dev Comp Immunol 36(2):289–297

Van Meter R, Good RA, Cooper MD (1969) Ontogeny of circulating immunoglobulin in normal, bursectomized and irradiated chickens. J Immunol 102(2):370–374

Villanueva S, Ali A, Campbell D, Siegford J (2017) Nest use and patterns of egg laying and damage by 4 strains of laying hens in an aviary system1. Poult Sci 96(9):3011–3020

Wales A, Davies R (2011) A critical review of *Salmonella typhimurium* infection in laying hens. Avian Pathol 40(5):429–436

Wang M, O'Connor JK, Bailleul AM, Li Z (2019) Evolution and distribution of medullary bone: evidence from a new early cretaceous enantiornithine bird. Natl Sci Rev 7:1068–1078

West JB, Watson RR, Fu Z (2007) Major differences in the pulmonary circulation between birds and mammals. Respir Physiol Neurobiol 157(2–3):382–390

Wigley P, Barrow P, Schat K (2014) The avian reproductive immune system, pp 265–274. https://doi.org/10.1016/B978-0-12-396965-1.00015-7

Woods D, Johnson AL (2005) Regulation of follicle-stimulating hormone-receptor messenger RNA in hen granulosa cells relative to follicle selection. Biol Reprod 72(3):643–650

Yang N, Ning Z, Wang Q, Li J, Wu C (1996) Application of the sex-linked dwarf gene to improving feed efficiency of layers. In: Proceedings of the 20th World's Poultry Congress, New Delhi, pp 447–451

Yoshimura Y, Barua A (2017) Female reproductive system and immunology. In: Sasanami T (ed) Avian reproduction. Advances in experimental medicine and biology, vol 1001. Springer, Singapore

Zhang Q, Chen B, Yang P, Zhang L, Liu Y, Ullah S, Wu L, Waqas Y, Le Y, Chen W, Chen Q (2015) Identification and structural composition of the blood–spleen barrier in chickens. Vet J 204(1):110–116

Immune Response in Mammals and Chickens

3

Patricia M. Morgan

Abstract

The ability to distinguish between self and potential pathogens is a key feature of the immune response. There are many common mechanisms of action of the innate and adaptive immune systems shared across species as evolutionarily distinct as chickens and mammals. Toll-like receptors which detect pathogens are present in both species, but differ in number and function. The central roles of dendritic cells, Helper T cells and B cells as well as the cytokines released in response to pathogens are also shared. Lymphoid tissues in chickens are distinct from those in mammals and the maturation of B cells occurs in the bursa of Fabricius, an organ specific to avian species. The generation of antibody diversity occurs by gene rearrangement in mammals and by gene conversion in chickens. Chickens have three classes of immunoglobulins while mammals have five. The main serum immunoglobulin in mammals is IgG while that in chickens is IgY. The structural differences in these molecules is outlined.

Keywords

Innate · Adaptive · B cell · T cell · Chicken · Mammal

3.1 Evolutionary Context

The environment harbours a wide range of potential pathogens and all living organisms have of necessity developed defences against microbes and other toxic substances. An important parameter in this fight for survival is the recognition of

P. M. Morgan (✉)
Emerita Lecturer, School of Natural Sciences, National University of Ireland Galway, Galway, Ireland
e-mail: pat.morgan@nuigalway.ie

X.-Y. Zhang et al. (eds.), *IgY-Technology: Production and Application of Egg Yolk Antibodies*, https://doi.org/10.1007/978-3-030-72688-1_3

non-self. The basis of this recognition, in both unicellular and multicellular organisms, is cellular receptors that detect the molecular motifs that are conserved among pathogenic microorganisms (Buchmann 2014). It is also essential that the immune system does not attack its own tissues otherwise this would result in self-destruction such as in autoimmune disorders (Goodnow et al. 2005). During the development of immune cells, there is a selection process in which those immune cells that react strongly with self-antigens are killed or suppressed which is referred to as self-tolerance. As many proteins are conserved across mammalian species, this underlies the difficulty in producing antibodies in closely-related species and is one of the major advantages of antibody generation in chickens due to their evolutionary divergence from mammals (Chap. 4).

Primitive animals relied on a basic innate defence mechanism (Yatim and Lakkis 2015), but as species evolved and developed more complex organs the immune system needed to respond and the adaptive immune system appeared about 450 million years ago (Kasahara et al. 2004). However, many of the basic mechanisms used in innate immunity have been retained in adaptive immunity. Antibody production for research or biomedical applications has traditionally been in mammalian species (mice and rabbits). This chapter will focus on the general mechanisms involved in antibody production in mammalian systems with the differences between mammals and chickens being highlighted. Due to constraints of space, the scope of the chapter is limited to the cellular and molecular signals involved in the production of immunoglobulins (antibodies) with some reference to cytotoxic T cells. The physical barriers to pathogens and roles of other classes of leukocytes involved in the immune response in mammals (Chaplin 2010) and during early development of the chick (Hincke et al. 2019) will not be discussed here. The other arm of the immune system which is activated in innate and adaptive immunity is complement (Zhang et al. 2017) is also beyond the scope of this chapter.

3.2 Innate and Adaptive Immune Response

A fully effective immune response relies on the interplay of the innate and adaptive immune systems; the innate immune system provides some immediate non-specific protection against invaders and primes the adaptive immune system. The adaptive immune system exhibits high specificity for a given antigen, albeit with a delay for a number of days before full activation, and memory which ensures a swift response on re-exposure to the same antigen. As a general introduction, many features of the innate and adaptive immune systems of mammals and chickens are listed in Table 3.1 and will be discussed in more detail. It is worth noting that with increasing complexity and body size during evolution, secondary lymphoid organs assumed importance (Yatim and Lakkis 2015) and cytokine and chemokine production by dendritic cells results in the necessary interaction between the components of the innate and adaptive immune system.

Table 3.1 General features of the innate and adaptive immune systems of mammals and chickens

Immune system	Innate	Adaptive
Mode of action	Non-specific	Very high specificity
	No memory	Memory
Time of action	Immediate	Delay of 4–7 days
Cells involved	Macrophages, Dendritic Cells, Neutrophils[a] Heterophils[b]	Macrophages, Dendritic Cells, Neutrophils[a] Heterophils[b]
	Natural Killer (NK) cells	Natural Killer (NK)cells
	γδ T cells	γδ T cells
		B cells
		T cells
Receptor types	Toll Like Receptors (TLRs)	MHC Class I, MHC Class II
		T cell receptors
		B cell antigen receptors
Receptor Genes	Germ-line gene elements	Gene re-arrangement[c] Gene conversion[d]
Co-stimulatory molecules (activating)	Dendritic cell: CD40, CD80/86	Dendritic cell: CD40, CD80/86 T cell: CD28, CD40L B cell: CD80, CD40
Cytokines released	IL-1, IL-6, IL-8, and TNF, IL-12, Type I Interferons	IL-2, IL-4, IL-5, IFNγ, TNF, IL-6,
Other immune systems activated	Complement	Complement
	Defensins	Defensins
	Cathelicidins	Cathelicidins

[a]Absent in chickens
[b]Heterophils may play a similar role to neutrophils in chickens
[c]Generation of antibody diversity in most mammals
[d]Generation of antibody diversity in chickens

3.3 Innate Immune Response in Mammals and Chickens

The innate immune responses in mammalian and avian species have many common features but also some differences in the receptors and cellular responses which will be discussed in more detail. During the primary innate immune response, macrophages, dendritic and NK cells recognize the pathogens by their pathogen-associated molecular patterns (PAMPs) that bind to pattern-recognizing receptors (PRRs), such as Toll like receptors which then elicit an appropriate response (Takeda et al. 2003) (Fig. 3.1). PAMPs include the lipopolysaccharide from Gram-negative bacteria, peptidoglycan, bacterial flagellar proteins, and viral genomic material including DNA and RNA. The range and action of the chicken Toll like receptors will now be explored.

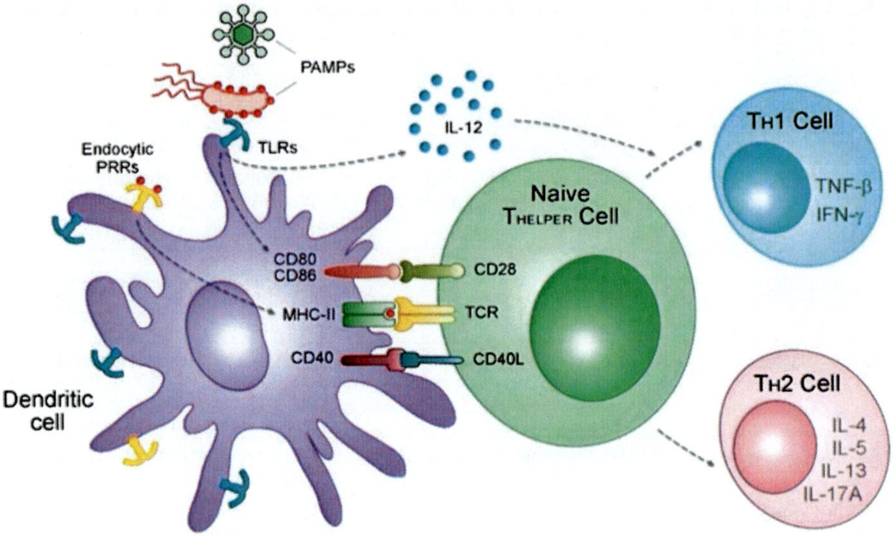

Fig. 3.1 Mature dendritic cells present antigens to naive T cells. Toll like receptors (TLRs) present in dendritic cells bind PAMP molecules present on pathogens (e.g., microbes, viruses) or even defective (e.g., tumour) cells. The antigen is processed and presented attached to MHC Class II molecules (MHC-II). Binding to the TLR upregulates the expression of co-stimulatory molecules such as CD80 and CD86 and IL-12 is released. A naïve T cell with the appropriate T cell receptor then binds to the mature dendritic cell, gets activated and the adaptive immune system is engaged. Endocytic **PRRs**, pattern-recognizing receptors. Figure courtesy of IgYTechnology.com (www.igytechnology.com)

3.3.1 Toll Like Receptors (TLRs)

Birds have ten TLRs while 13 mammalian TLRs have been identified (Boudinot et al. 2014) however, chicken TLRs differ to some extent from those expressed in mammals (Brownlie and Allan 2011). Each of the TLRs detects a distinct repertoire of highly conserved pathogen molecules. TLRs are also expressed on non-immune cells such as epithelial and endothelial cells, and fibroblasts and this distribution helps to promote a rapid immune response (McClure and Massari 2014). For a more detailed review of PRRs and TLRs in a number of other avian species, the reader is directed to a review by Chen et al. (2013).

Structurally, TLRs are integral membrane glycoproteins consisting of an extracellular ligand binding domain with leucine-rich repeats, a transmembrane domain and an intracellular Toll/interleukin 1 receptor (Chen et al. 2013). Chicken Toll-like receptor 3 (TLR-3) contributes to the induction of the innate immune response against viruses and is constitutively expressed in all chicken tissues; it is present in intracellular vesicles (endosome and endoplasmic reticulum) where viruses undergo uncoating during infection; it upregulates Type I interferon and inflammatory cytokine production (Table 3.1) (Chen et al. 2013). Chicken TLR-7 is located in intracellular vesicles and is involved in the interaction with RNA viruses; TLR

15 located on the plasma membrane and TLR21 in intracellular vesicles are involved in sensing both viruses and bacteria. Chickens, unlike ducks, lack the retinoic acid inducible gene 1 (RIG-1) however, this is compensated by the presence of melanoma differentiation-associated gene 5 (MDA5) which is necessary for the induction of type I interferon; the lack of RIG-I necessary for Interferon-ß production may explain the susceptibility of chickens to highly pathogenic influenza (Chen et al. 2013).

3.3.2 Cytokine Production and Host Defence Peptides

Comparative studies of cytokines in chickens and mammals suggest close functional capabilities although at the level of the genome and primary protein structures they are very different (Giansanti et al. 2007). The range of pro-inflammatory cytokines produced in response to binding to TLRs include interleukin-1 (IL-1), IL-6, IL-8 and tumour necrosis factor-alpha (TNF-α) (Table 3.1) (Gao et al. 2016; Sun et al. 2019; Alkie et al. 2019; Awad et al. 2019). The presence of TNF-α which was long thought to be absent in chickens has recently been confirmed (Rohde et al. 2018). Transcription factors, such as interferon regulatory factor-7 (IRF-7) and nuclear factor kappa B (NFκB), drive the transcription of genes encoding these cytokines (Goossens et al. 2013; Karnati et al. 2014; Markowska et al. 2017; Hoang et al. 2017; Liu et al. 2018). In addition, the maturation of antigen-presenting cells is promoted and the co-stimulatory molecules CD80 and CD86 are up-regulated (Table 3.1) (Mogensen 2009). Simultaneously to the humoral response, a cellular innate response is elicited by activation of tissue-homing cells such as natural killer cells, macrophages, and dendritic cells (Igyártó et al. 2006; Jansen et al. 2010; de Geus et al. 2012; Neulen and Göbel 2012; Sun et al. 2017; Jahromi et al. 2018; Feng et al. 2019; Lin et al. 2019; Yasmin et al. 2019). The NK cells produce IFN at early stages of a viral infection and later target the virus-infected cells to destroy them by apoptosis (Song et al. 2017).

Host defence peptides (HDPs), which are families of effector molecules found in mammals, insects and plants are also present in chickens (Table 3.1); the readers are referred to a review of their role, evolution and biological activities (Cuperus et al. 2013). Avian host defence peptides are present in nearly every organ and tissue supporting their importance in the innate immune response. Within the defensin family, β-defensins are the only defensins found in avian species so far. HDPs, as well as exerting a direct antimicrobial action, also function in immunomodulation and wound healing (Cuperus et al. 2013).

There are a number of other cell types (NK cells, γδT, NK-T, and B1) which are active in both the mammalian innate and adaptive immune response (Getz 2005). Cell types confirmed to be present in the chicken are as follows: NK cells (Jansen et al. 2010); γδT (Pieper et al. 2011); and NK-T (Rogers et al. 2008); γδT cells occur at a much higher frequency in chickens compared to mammals (Sowder et al. 1988). The common characteristics of mammalian γδT, NK, NK-T cells is their ability to recognize stressed cells in the absence of antibodies and MHC protein interactions.

Response to lipid and polysaccharide antigens occurs in these cells in a pathway independent of MHC Class I and II proteins but by a similar mechanism (Chaplin 2010). B1 cells produce natural antibodies and can respond without T-cell activation and show no evidence of somatic hypermutation of their receptor genes (Yatim and Lakkis 2015).

3.4 Adaptive Immune Response

The adaptive immune system normally has a small number of cells with specificity for any individual pathogen, toxin or allergen. The responding cells (T and B cells) must proliferate after encountering the antigen in order to have sufficient cells to mount an effective immune response. An example of the interplay between the innate and adaptive immune response is the role of dendritic cells; these cells present the antigen bound to MHC Class II molecules which then activates a T cell which displays a TCR which can bind that antigen (Fig. 3.1). Pathogens are complex and are able to interact with a large number of lymphocytes; each B lymphocyte is activated to proliferate and differentiate into plasma cells; each plasma cell only produces a single type of antibody but as a number of plasma cells are stimulated the overall response is polyclonal. Any protein when degraded by the endocytic pathway (Fig. 3.1) will result in a number of short fragments of about 15 amino acids, so a single protein will also give rise to a polyclonal response. A comparison of the lymphoid tissues of mammals and chickens and the origin and roles of B and T cells will now be discussed.

3.4.1 Lymphoid Tissues in Mammals and Chickens

The major chicken lymphoid organs involved in the adaptive immune response are the bone marrow, spleen and the bursa of Fabricius which is found only in birds (Table 3.2) (Boehm et al. 2012).The chicken's immune system lacks lymph nodes but has mucosa associated lymphoid tissues (MALT) that are scattered in the subcutaneous tissue and through the mucosal epithelia, lamina propria and serosa

Table 3.2 Comparison of the lymphoid tissues of mammals and chickens

	Mammals	Chicken	Observations
Bursa of Fabricius	No	Yes	B cell selection and antibody generation
Thymus	Yes	Yes	T cell selection
Lymph nodes	Yes	No	Site of development of adaptive immune response
Spleen	Yes	Yes	Site of development of adaptive immune response
Bone marrow	Yes	Yes	Haematopoiesis
Mucosa-associated lymphoid tissue	Yes	Yes	Regulating mucosal immunity

membranes (Hammer 1974). The Harderian gland, located in the orbit of the eye, forms part of the head-associated lymphoid tissue, has high levels of B-dependent lymphoid cells and plasma cells and is a key immune organ in chickens (Saelao et al. 2018). Birds do not have neutrophils (Table 3.1) but heterophils function in a similar manner to mammalian neutrophils (Chen et al. 2013). The spleen, a capsulated tissue found close to the stomach, mounts the immune response to blood borne pathogens. Immune cells (dendritic cells, macrophages, heterophils and lymphocytes) homing into the parenchyma of lymphoid organs function by trapping pathogens and removing them from an infected host (Zhang et al. 2015; Silveira et al. 2019).

Overview of the immune system of mammals and chicken comparing relevant aspects of their lymphoid tissues and immune response (Keestra-Gounder et al. 2013; Kaufman 2015, 2018; Taha-Abdelaziz et al. 2018).

3.4.2 B Cells, T Cells and MHC Class II Proteins

The roles of avian B and T cells in the production of antibodies has been reviewed (Ratcliffe and Härtle 2014; Smith and Göbel 2014). The development of B and T cells in the chicken begins with the migration of multipotent progenitors from the bone marrow to the bursa of Fabricius (B cells) and the thymus (T cells). The bursa begins development about embryonic day 5 and is populated by pre-bursal cells between embryonic days 8–14 (Ko et al. 2018). Chickens have three distinct B cell stages: pre-bursal, bursal and post-bursal B cells (Ko et al. 2018) which differs from B cell development in mammals. Overall, the function of the bursa is to generate naïve B cells which are then exported to the periphery, probably the spleen (Lassila 1989). Gene conversion (Sect. 3.6) occurs only in the bursa of Fabricius. As sexual maturity approaches, the bursa begins to regress and the adult depends on the post-bursal stem cells in the bone marrow as the source of B cells (Davani et al. 2014). In mammals, B cells mature in the bone marrow into committed B cells.

The T cell repertoire of the chicken and mammals develop by similar mechanisms. T cells are present in the chicken thymus by embryonic day 15 (Bucy et al. 1990) and by the 18th day of embryonic development the chicken is capable of producing both an innate and adaptive response to pathogens (Schilling et al. 2018 and references therein). As an aside, this feature is exploited in large-scale poultry production with the administration of *in ova* vaccination for multiple pathogens. The population of T cells are divided in two subpopulations named Cytotoxic T cells (CTL or $CD8^{+}$) and Helper T cells (Th or $CD4^{+}$) (Erf 2004). Binding of a Cytotoxic T cells through its T cell receptor to a MHC Class I peptide complex induces death of a cell infected by endogenous pathogens such as viruses (Gimeno and Schat 2018). T cells, in a process called immunosurveillance, are also involved in the detection and elimination of tumour cells. There is a comprehensive review on this arm of the immune system (Erf 2004) but it will not be discussed further here. There is a further distinction within Helper T cell populations as naïve T helper cells can differentiate into Th1 or Th2 type cells. Th2 cells are associated with antibody production and synthesis of IL-4, IL-5, IL-13 and IL-17A (Degen et al.

2005; Walliser and Göbel 2018). Helper T cells recognize exogenous antigen presented on macrophages, dendritic cell or B cell MHC class II molecules.

In the humoral response, the major class of MHC molecules involved in the production of antibodies are Class II. MHC Class II molecules are capable of antigen internalization, processing and surface display (Fig. 3.1). Mammals have 260 genes for the MHC and this polymorphism gives rise to six distinct binding grooves each of which can bind a diverse range of antigens (Chaplin 2010). Birds have approximately 19 MHC genes (Kaufman et al. 1999).

3.4.3 Antibody Secretion by B Cells

Mammalian and chicken T and B cells have receptors which are highly diverse, however each individual cell has a receptor of a single specificity. T cells recognize the antigen displayed in the MHC complex through their T cell receptor (Fig. 3.1). In mammals, to adequately distinguish between self and non-self, two proteasome subtypes have evolved, the immunoproteasome and the thymoproteasome. In the chicken these proteasome subtypes are lacking, which raises the question whether these animals have evolved compensatory mechanism for presentation of antigens to T cells (Kasahara and Flajnik 2019).

The major class of mammalian T cells displays the αβTCR. Chickens on the other hand may have frequencies of up to 50% γδ T cells (Sowder et al. 1988); the functions of these cells have not yet been fully determined. The activation of mammalian and chicken B cells follow similar routes. B cells can undergo antigen-dependent proliferation and differentiation into plasma cells (secrete Ig) and memory cells (Fig. 3.2). They can also act as antigen-presenting cells: antigens upon binding to the B cell receptor are internalised and displayed by MHC Class II molecules which can interact with the T cell receptor (Fig. 3.2). There is an absolute requirement for interactions between co-stimulatory molecules and the release of cytokines before B-cells proliferate (Fig. 3.2). Proliferation induces an antibody class switch (Sect. 3.5) that ultimately leads to plasma cells secreting specific antibodies against the antigenic target. Memory B cells are also generated which is the basis of vaccination (Fig. 3.2).

A subset of the cytokines that are released during the innate and adaptive immune response are listed in Table 3.1 (Chaplin 2010). These soluble proteins ensure a coordinated interaction of the cells of the immune system (Turner et al. 2014). Type I interferons block the spread of viruses and stimulate NK cells. IL-12 produced by activated dendritic cells induces the expression of the CD80/86 co-stimulatory molecules. Tumour Necrosis Factor (TNF) is involved in cell migration to the lymphoid organs. An example of the activation of IL-2 and IL-4 during B cell T cell contact is illustrated in Fig. 3.2.

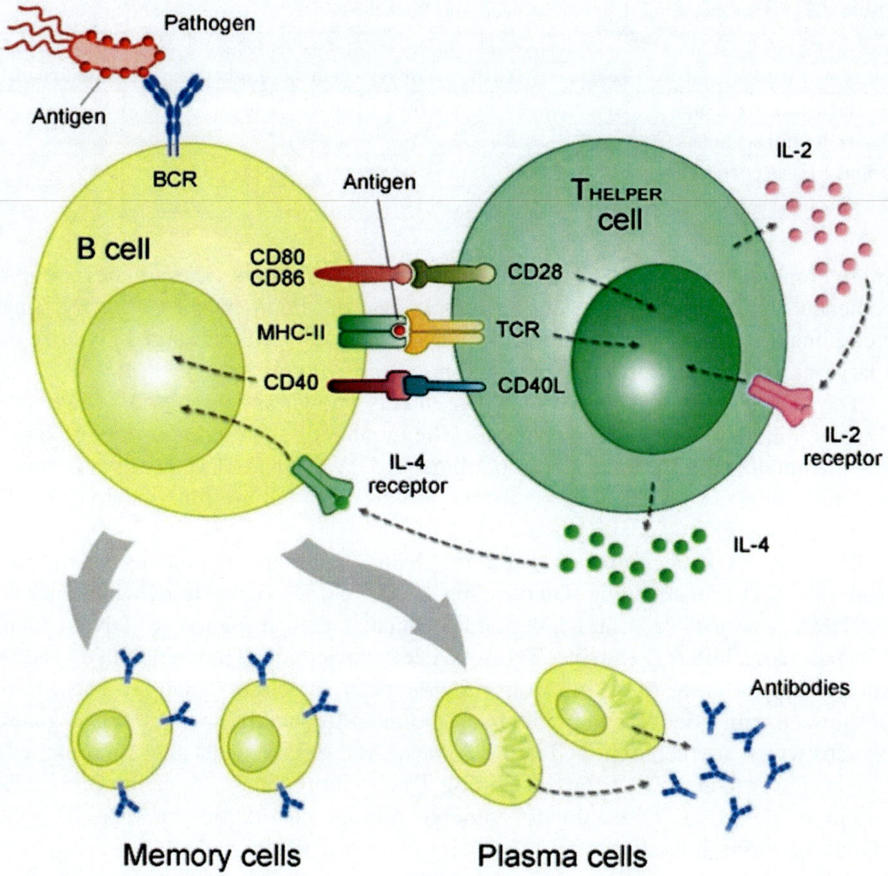

Fig. 3.2 Model of the interaction between B cells and Helper T cells. The B cell receptor (BCR) takes up the antigen and then displays it in the MHC Class II receptor (MHC-II) which reacts with a T cell receptor (TCR) which can interact with that specific antigen. Optimal T cell activation occurs when the costimulatory molecules CD80/CD86 interact with CD28 leading to the secretion of IL-2 and upregulation of the IL-2 receptor on the T cell. The activated T-helper cells upregulate CD40L and secrete IL-4 which binds to the receptor on the B cells. B cells differentiate into antibody secreting plasma cells and long-lived memory cells. Figure courtesy of IgYTechnology.com (www.igytechnology.com)

3.5 Genetic Basis for Diversity in Mammalian TCR and Immunoglobulins

Seminal events in the development of the adaptive immune system were two waves of extensive gene duplication which occurred as jawed vertebrates evolved and the acquisition of the recombination-activating genes (RAG) (Kasahara et al. 2004). RAG genes encode part of a protein complex which is involved in rearrangement

Table 3.3 Genes encoding mammalian TCR and Ig chains

	TCRα	TCRβ	Ig Heavy chain	IgGκ Light chain	Igλ Light chain
Variable (V) segments	70	67	65	44	38
Diversity (D) segments		2	27		
Joining (J) segments	61	13	6	5	5
Constant (C) segments	1	2	11	1	1

and recombination of genes encoding Ig and TCR molecules. These genes are not embedded in the genome but during lymphocyte development they arise by gene recombination where a fixed number of adjacent genes when rearranged give rise to a large number of unique antigen receptors, of the order of 10^{12}.

The genes for Ig and TCR contain segments known as variable (V), diversity (D) and joining (J) and constant regions. The number of gene segments reported for the mammalian light chain of Ig (Collins and Watson 2018) and heavy chain (Li et al. 2004) and TCR gene segments, not all of which are functional (Li et al. 2004), are shown in Table 3.3.

The main TCR is composed of the αβ chains which are somatically assembled from the V, D, J elements initiated by the RAG1 and RAG2 proteins which cleave the DNA near the V, D and J segments which is then re-joined by DNA repair enzymes. An additional enzyme, Terminal deoxynucleotidyl Transferase (TdT) adds nucleotides into some of the junctions providing extra diversity (Chap. 4). The action of these enzymes delivers huge diversity, but also frequently non-functional genes on cells which are then removed in the thymus. Autoreactive cells are also removed. In fact, fewer than 10% of the developing T cells survive these selection processes (Klein et al. 2014). TCRs do not undergo affinity maturation nor exhibit rapid evolution during infection.

In mammals, a similar process of somatic rearrangement of the gene elements occurs in Ig involving RAG genes and the Ig molecule is composed of two identical Heavy and Light chains. The combination of the two chains, imprecise joining of segments and somatic hypermutation all increase diversity (Mishra and Mariuzza 2018). In the initial response antibodies of the IgM class are expressed (Table 3.4) but during clonal expansion class switching occurs by recombination with other constant chain genes to give the following classes and subclasses: IgM, IgD, IgG (IgG_1, IgG_2, IgG_3 and IgG_4), IgA (IgA_1 and IgA_2) and IgE (Calonga-Solís et al. 2019). IgG_1 is the main immunoglobulin in serum having a plasma concentration of approximately 9 mg/mL and a half-life of 21 days (Almagro et al. 2017).

Affinity maturation occurs during the secondary immune response as IgM is switched to IgG; this results in Ig of higher affinity for the antigen (Mishra and Mariuzza 2018).

Table 3.4 Immunoglobulin classes in mammals

Immunoglobulin classes					
	IgM	IgG	IgA	IgE	IgD
Heavy chain	μ	γ	α	ε	δ
Antigen binding sites	10	2	4	2	2
% in serum	6%	80%	13%	0.002%	1%
Main functions	Main Ig of the primary immune response. Binds complement Monomer is a B cell receptor	Main serum Ig	In mucus, tears, saliva, colostrum	Ig related to allergic response	B cell receptor on naïve B cells

3.6 Genetic Basis for Diversity in Chicken TCR and Immunoglobulins

The generation of diversity of the TCR and Ig in the chicken occurs by two different mechanisms. The T cell repertoire is generated in a similar way to that of mammals and there is no evidence of somatic gene conversion mechanisms; a detailed review of the process can be accessed (Smith and Göbel 2014) and will not be discussed in detail here. In B cells, gene conversion is responsible for introducing sequence diversity in the Ig genes and appears to be as efficient as the gene rearrangement mechanism in mammals (Ratcliffe and Härtle 2014). The chicken contains only one light chain gene which is closely related to the mammalian λ light chain; it appears that the locus encoding the κ-type light chain was deleted during evolution (Reynaud et al. 1985); also, there is only one J segment. This contrasts with the situation in mammals (Table 3.3). RAG-1 and RAG-2 proteins involved in the deletion of coding nucleotides or the addition of palindromic (P) nucleotides induces diversity at the joining sites (Chap. 4); there is no evidence of the involvement of TdT in the chicken (Ratcliffe and Härtle 2014). There are three constant heavy chain segments encoding the μ, α and υ Ig heavy chains of IgM, IgA and IgY, respectively; this in contrast to the 11 heavy chain segments in mammals (Table 3.3). Variable, D and J segments encode the heavy chain domain; there is a single V and J segments and 15 D segments which however have low variability compared to those in mammals (Reynaud et al. 1991). Gene rearrangement in the heavy chain occurs in the same way as in the light chain. Additional diversity of the light chain is generated by gene conversion where a stretch of the rearranged L chain gene is replaced by a homologous sequence from one of 25 upstream pseudogenes; the length of the exchanged sequences can range from 10–300 nucleotides (Ratcliffe and Härtle 2014). Activation-induced cytidine deaminase (AID) is required to mediate repertoire diversity of B cell immunoglobulins genes by three closely associated pathways

Table 3.5 Immunoglobulin classes in chickens

Immunoglobulin classes			
	IgM	IgA	IgY
Heavy chain	μ	α	γ
Antigen binding sites	8 or 10	2, 4, 6	2
Serum concentrations (mg/mL)	1–2	0.35–0.65	5–7
Main functions	Main Ig of the primary immune response. Monomer is a B cell receptor	Mucosal protection (High in lachrymal fluid and cloaca)[a]	Main serum Ig; main maternal antibody in egg yolk

[a]Merino-Guzmán et al. (2017)

named somatic hypermutation, class switch recombination and gene conversion. Along with AID the transcription factor Bcl6 is a key to diversification by these three processes (Williams et al. 2016). Also, bursa of Fabricius-derived peptides are involved in local processes such as B cell differentiation and antibody production (Feng et al. 2019) Gene conversion in the heavy chain locus occurs by a similar mechanism. Greater detail of these gene rearrangements is given in Chap. 4 on the evolution of IgY. The contribution of affinity maturation to the development of specificity and effectiveness of chicken Ig is unclear (Ratcliffe and Härtle 2014). The characteristics of chicken Ig are given in Table 3.5 (Härtle et al. 2014).

3.7 Structure of IgG and IgY

IgG is the main serum antibody released during the mammalian humoral response. The relationship between the generation of antibody diversity and the structure of IgG is illustrated in Fig. 3.3a. The variable regions of the light and heavy chains (V_L and V_H) contain three sub-regions which are hypervariable and named the complementarity-determining regions (CDR1–3); these form the antigen binding domain (Fig. 3.3a). The three-dimensional alignment of the CDR of the light and heavy chains are shown in Fig. 3.3d. Each IgG has two identical antigen binding sites. The carboxy terminal region is constant for each class of antibody (fragment crystallizable, Fc) and is responsible for most of the effector functions including binding to Fc receptors and activating the complement system. The hinge region of the antibody, linking the antigen binding portions to the Fc region, is a flexible tether allowing some movement of the antigen binding arms. This is important for binding to antigens of different sizes.

The characteristics of the five classes of mammalian Ig: IgM, IgA, IgG, IgD and IgE (Table 3.4) are based on the structure of heavy chain (Schroeder and Cavacini 2010). Glycosylation sites are shown as green circles (Fig. 3.3a); the degree of

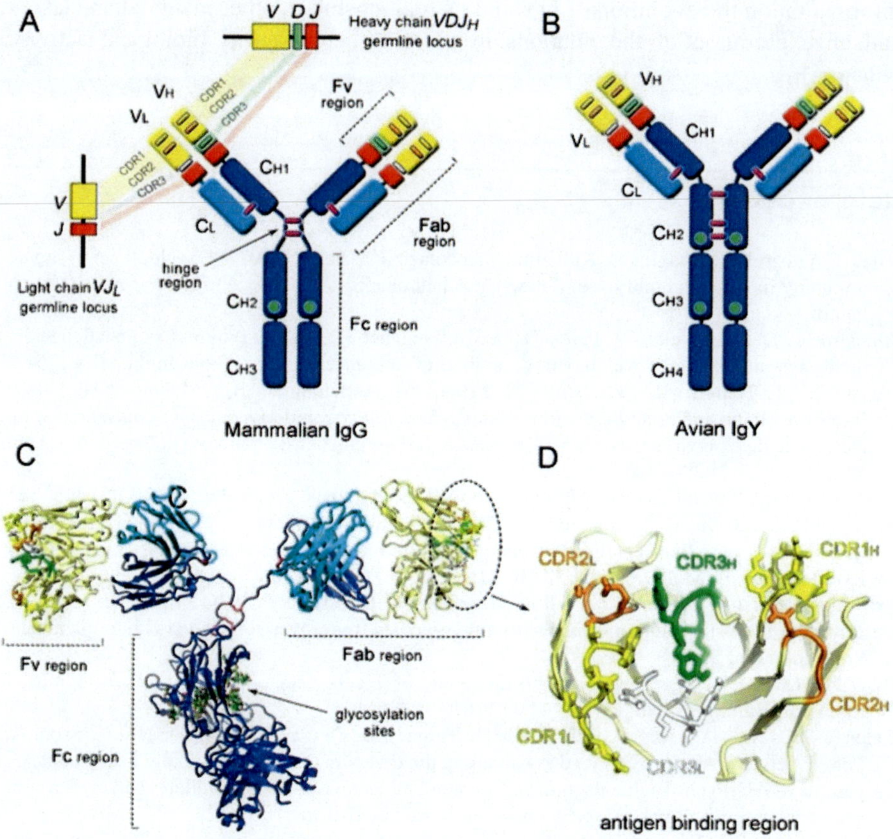

Fig. 3.3 Domain architecture and structure of mammalian IgG an avian IgY. (**a**) General domain architecture of mammalian IgG (MW: –150 kDa). Constant (C) and Variable (V) domains of the Heavy (H) and Light (L) chains of the antibodies are presented in different colours. Complementarity determining regions (CDR) are shown as yellow (CDR1), orange (CDR2) and white or green (CDR3) lines and interchain disulphide bonds as magenta lines; green circles indicate glycosylation sites. Fc (fragment crystallizable), Fab (fragment antigen-binding) and Fv (variable fragment) regions are indicated. (**b**) Similar representation of avian IgY (MW: –180 kDa), showing the additional CH domain (CH4). (**c**) Crystal structure of mammalian IgG (IgG2a) monoclonal antibody (PDB ID: 1IGT, (Harris et al. 1997)) in ribbon representation, following the same colouring scheme. (**d**) Close-up view of the antigen binding region showing the six CDRs, three from the light chain (L) and three from the heavy chain (H). Structure images prepared with PyMOL. Figure courtesy of IgYTechnology.com (www.igytechnology.com)

glycosylation may affect the activity of the molecule (Schroeder and Cavacini 2010) and circulatory half-life. IgG has been the antibody of choice for diagnostic and research tools.

IgY (Fig. 3.3b) has a similar structure to IgG with the following notable differences. It has an additional constant region (CH4), lacks a hinge region and has two glycosylation sites (Fig. 3.3b). The reader is referred to Chap. 4 for detailed information on the evolution of IgY and its relationship to other mammalian classes and an exploration of the relationship of its structure to its biological activity (Chap. 5).

References

Alkie T, Yitbarek A, Hodgins D, Kulkarni R, Abdelaziz K, Sharif S (2019) Development of innate immunity in chicken embryos and newly hatched chicks: a disease control perspective. Avian Pathol 48(4):288–310

Almagro JC, Daniels-Wells TR, Perez-Tapia SM, Penichet ML (2017) Progress and challenges in the design and clinical development of antibodies for cancer therapy. Front Immunol 8:1751

Awad NFS, El-Hamid MIA, Hashem YM, Erfan AM, Abdelrahman BA, Mahmoud HI (2019) Impact of single and mixed infections with *Escherichia coli* and *Mycoplasma gallisepticum* on Newcastle disease virus vaccine performance in broiler chickens: an in vivo perspective. J Appl Microbiol 127(2):396–405

Boehm T, Hess I, Swann JB (2012) Evolution of lymphoid tissues. Trends Immunol 33(6):315–321

Boudinot P, Zou J, Ota T, Buonocore F, Scapigliati G, Canapa A, Cannon J, Litman G, Hansen JD (2014) A tetrapod-like repertoire of innate immune receptors and effectors for coelacanths. J Exp Zool Part B, Mol Dev Evol 322(6):415–437

Brownlie R, Allan B (2011) Avian toll-like receptors. Cell Tissue Res 343(1):121–130

Buchmann K (2014) Evolution of innate immunity: clues from invertebrates via fish to mammals. Front Immunol 5:459

Bucy RP, Chen CH, Cooper MD (1990) Ontogeny of T cell receptors in the chicken thymus. J Immunol (Baltimore, Md: 1950) 144(4):1161–1168

Calonga-Solís V, Malheiros D, Beltrame MH, Vargas LB, Dourado RM, Issler HC, Wassem R, Petzl-Erler ML, Augusto DG (2019) Unveiling the diversity of immunoglobulin heavy constant gamma (IGHG) gene segments in Brazilian populations reveals 28 novel alleles and evidence of gene conversion and natural selection. Front Immunol 10:1161

Chaplin DD (2010) Overview of the immune response. J Allergy Clin Immunol 125(2):S3–S23

Chen S, Cheng A, Wang M (2013) Innate sensing of viruses by pattern recognition receptors in birds. Vet Res 44(1):82

Collins AM, Watson CT (2018) Immunoglobulin light chain gene rearrangements, receptor editing and the development of a self-tolerant antibody repertoire. Front Immunol 9:2249

Cuperus T, Coorens M, van Dijk A, Haagsman H (2013) Avian host defense peptides. Dev Comp Immunol 41(3):352–369

Davani D, Pancer Z, Ratcliffe MJ (2014) Ligation of surface Ig by gut-derived antigen positively selects chicken bursal and peripheral B cells. J Immunol (Baltimore, Md: 1950) 192 (7):3218–3227

de Geus ED, Jansen C, Vervelde L (2012) Uptake of particulate antigens in a nonmammalian lung: phenotypic and functional characterization of avian respiratory phagocytes using bacterial or viral antigens. J Immunol (Baltimore, Md: 1950) 188(9):4516–4526

Degen WG, Daal N, Rothwell L, Kaiser P, Schijns VE (2005) Th1/Th2 polarization by viral and helminth infection in birds. Vet Microbiol 105(3–4):163–167

Erf GF (2004) Cell-mediated immunity in poultry. Poult Sci 83(4):580–590

Feng XL, Zong MM, Zhou GF, Zheng Y, Yu YN, Cao RB, Chen PY, Yang M (2019) The functions and mechanism of a new oligopeptide BP9 from avian bursa on antibody responses, immature B cell, and autophagy. J Immunol Res 2019:1574383

Gao Y, Zhang Y, Yao Y, Guan X, Liu Y, Qi X, Wang Y, Liu C, Zhang Y, Gao H, Nair V, Wang X, Gao Y (2016) Avian leukosis virus subgroup J induces VEGF expression via NF-?B/PI3K-dependent IL-6 production. Oncotarget 7(49):80275–80287

Getz GS (2005) Thematic review series: the immune system and atherogenesis. Bridging the innate and adaptive immune systems. J Lipid Res 46(4):619–622

Giansanti F, Giardi MF, Botti D (2007) Avian cytokines—an overview. Curr Pharm Des 12 (24):3083–3099

Gimeno IM, Schat KA (2018) Virus-induced immunosuppression in chickens. Avian Dis 62 (3):272–285

Goodnow CC, Sprent J, Fazekas de St Groth B, Vinuesa CG (2005) Cellular and genetic mechanisms of self tolerance and autoimmunity. Nature 435(7042):590–597

Goossens K, Ward A, Lowenthal J, Bean A (2013) Chicken interferons, their receptors and interferon-stimulated genes. Dev Comp Immunol 41(3):470–476

Hammer DK (1974) The immune system in chickens. Avian Pathol 3(2):65–78

Harris LJ, Larson SB, Hasel KW, McPherson A (1997) Refined structure of an intact IgG2a monoclonal antibody. Biochemistry 36(7):1581–1597

Härtle S, Magor KE, Gobel TW, Davison F, Kaspers B (2014) Structure and evolution of avian immunoglobulins. In: Schat KA, Kaspers B, Kaiser P (eds) Avian immunology, 2nd edn. Elsevier, London, pp 103–114

Hincke MT, Da Silva M, Guyot N, Gautron J, McKee MD, Guabiraba-Brito R, Réhault-Godbert S (2019) Dynamics of structural barriers and innate immune components during incubation of the avian egg: critical interplay between autonomous embryonic development and maternal anticipation. J Innate Immun 11(2):111–124

Hoang CT, Truong AD, Lee J, Lee K, Hong Y (2017) Molecular cloning of chicken interleukin-17B, which induces proinflammatory cytokines through activation of the NF-κB signaling pathway. Dev Comp Immunol 74:40–48

Igyártó BZ, Lackó E, Oláh I, Magyar A (2006) Characterization of chicken epidermal dendritic cells. Immunology 119(2):278–288

Jahromi M, Bello MB, Abdolmaleki M, Yeap SK, Hair-Bejo M, Omar A (2018) Differential activation of intraepithelial lymphocyte-natural killer cells in chickens infected with very virulent and vaccine strains of infectious bursal disease virus. Dev Comp Immunol 87:116–123

Jansen C, Haar P, Haarlem D, Kooten P, De Wit J, van Eden W, Viertlböck B, Göbel T, Vervelde L (2010) Identification of new populations of chicken natural killer (NK) cells. Dev Comp Immunol 34(7):759–767

Karnati H, Pasupuleti S, Kandi R, Undi R, Sahu I, Ramasamy K, Subbiah M, Gutti R (2014) TLR-4 signalling pathway: MyD88 independent pathway up-regulation in chicken breeds upon LPS treatment. Vet Res Commun 39(1):73–78

Kasahara M, Flajnik M (2019) Origin and evolution of the specialized forms of proteasomes involved in antigen presentation. Immunogenetics 71(3):251–256

Kasahara M, Suzuki T, Pasquier LD (2004) On the origins of the adaptive immune system: novel insights from invertebrates and cold-blooded vertebrates. Trends Immunol 25(2):105–111

Kaufman J (2015) Co-evolution with chicken class I genes. Immunol Rev 267(1):56–71

Kaufman J (2018) Generalists and specialists: a new view of how MHC class I molecules fight infectious pathogens. Trends Immunol 39(5):367–379

Kaufman J, Milne S, Göbel TW, Walker BA, Jacob JP, Auffray C, Zoorob R, Beck S (1999) The chicken B locus is a minimal essential major histocompatibility complex. Nature 401 (6756):923–925

Keestra-Gounder A, de Zoete M, Bouwman L, Vaezirad M, Putten J (2013) Unique features of chicken toll-like receptors. Dev Comp Immunol 41(3):316–323

Klein L, Kyewski B, Allen PM, Hogquist KA (2014) Positive and negative selection of the T cell repertoire: what thymocytes see (and don't see). nature reviews. Immunology 14(6):377–391

Ko KH, Lee IK, Kim G, Gu MJ, Kim HY, Park BC, Park TS, Han SH, Yun CH (2018) Changes in bursal B cells in chicken during embryonic development and early life after hatching. Sci Rep 8 (1):16905

Lassila O (1989) Emigration of B cells from chicken bursa of Fabricius. Eur J Immunol 19 (5):955–958

Li A, Rue M, Zhou J, Wang H, Goldwasser MA, Neuberg D, Dalton V, Zuckerman D, Lyons C, Silverman LB, Sallan SE, Gribben JG (2004) Utilization of Ig heavy chain variable, diversity, and joining gene segments in children with B-lineage acute lymphoblastic leukemia: implications for the mechanisms of VDJ recombination and for pathogenesis. Blood 103 (12):4602–4609

Lin J, Wang Z, Wang J, Yang Q (2019) Microarray analysis of infectious bronchitis virus infection of chicken primary dendritic cells. BMC Genomics 20(1):557

Liu Y, Yuqiang C, Wenya S, Jingjiao M, Hengan W, Jianhe S, Y. Y. J. Developmental, C. Immunology (2018) Chicken interferon regulatory factor 1 (IRF1) involved in antiviral innate immunity via regulating IFN-β production. Dev Comp Immunol 88:77–82

Markowska M, Majewski PM, Skwarło-Sońta K (2017) Avian biological clock - immune system relationship. Dev Comp Immunol 66:130–138

McClure R, Massari P (2014) TLR-dependent human mucosal epithelial cell responses to microbial pathogens. Front Immunol 5:386

Merino-Guzmán R, Latorre JD, Delgado R, Hernandez-Velasco X, Wolfenden AD, Teague KD, Graham LE, Mahaffey BD, Baxter MFA, Hargis BM, Tellez G (2017) Comparison of total immunoglobulin A levels in different samples in Leghorn and broiler chickens. Asian Pac J Trop Biomed 7(2):116–120

Mishra AK, Mariuzza RA (2018) Insights into the structural basis of antibody affinity maturation from next-generation sequencing. Front Immunol 9:117

Mogensen TH (2009) Pathogen recognition and inflammatory signaling in innate immune defenses. Clin Microbiol Rev 22(2):240–273

Neulen ML, Göbel TW (2012) Chicken CD56 defines NK cell subsets in embryonic spleen and lung. Dev Comp Immunol 38(3):410–415

Pieper J, Methner U, Berndt A (2011) Characterization of avian γδ T-cell subsets after *Salmonella enterica* serovar Typhimurium infection of chicks. Infect Immun 79(2):822–829

Ratcliffe MJH, Härtle S (2014) B cells, the Bursa of Fabricius and the generation of antibody repertoires. In: Schat KA, Kaspers B, Kaiser P (eds) Avian immunology, 2nd edn. Elsevier, London, pp 65–89

Reynaud CA, Anquez V, Dahan A, Weill JC (1985) A single rearrangement event generates most of the chicken immunoglobulin light chain diversity. Cell 40(2):283–291

Reynaud CA, Anquez V, Weill JC (1991) The chicken D locus and its contribution to the immunoglobulin heavy chain repertoire. Eur J Immunol 21(11):2661–2670

Rogers SL, Viertlboeck BC, Göbel TW, Kaufman J (2008) Avian NK activities, cells and receptors. Semin Immunol 20(6):353–360

Rohde F, Schusser B, Hron T, Farkašová H, Plachý J, Härtle S, Hejnar J, Elleder D, Kaspers B (2018) Characterization of chicken tumor necrosis factor-α, a long missed cytokine in birds. Front Immunol 9:605

Saelao P, Wang Y, Gallardo RA, Lamont SJ, Dekkers JM, Kelly T, Zhou H (2018) Novel insights into the host immune response of chicken Harderian gland tissue during Newcastle disease virus infection and heat treatment. BMC Vet Res 14(1):280

Schilling MA, Katani R, Memari S, Cavanaugh M, Buza J, Radzio-Basu J, Mpenda FN, Deist MS, Lamont SJ, Kapur V (2018) Transcriptional innate immune response of the developing chicken embryo to Newcastle disease virus infection. Front Genet 9:61

Schroeder HW Jr, Cavacini L (2010) Structure and function of immunoglobulins. J Allergy Clin Immunol 125(2 Suppl 2):S41–S52

Silveira F, Felice V, Franzo G, Mescolini G, Catelli E, Cecchinato M, Berto G, Listorti V, Lupini C (2019) Inoculation of specific pathogen-free chickens with an infectious bursal disease virus of the ITA genotype (G6) leads to a high and persistent viral load in lymphoid tissues and to a delayed antiviral response. Vet Microbiol 235:136–142

Smith AL, Göbel TW (2014) Avian T cells: antigen recognition and lineages. In: Schat KA, Kaspers B, Kaiser P (eds) Avian immunology, 2nd edn. Elsevier, London

Song B, Li X, Ma J, Yu L, Feng Z, Liu Z, Cui Y (2017) Prokaryotic expression and anti-IBDV activity of chicken interleukin-18 and interferon-γ. Cytogenet Genome Res 153(1):36–45

Sowder JT, Chen CL, Ager LL, Chan MM, Cooper MD (1988) A large subpopulation of avian T cells express a homologue of the mammalian T gamma/delta receptor. J Exp Med 167 (2):315–322

Sun Z, Liu C, Pan T, Yao H, Shu L (2017) Selenium accelerates chicken dendritic cells differentiation and affects selenoproteins expression. Dev Comp Immunol 77:30–37

Sun GR, Zhou LY, Zhang YP, Zhang F, Yu ZH, Pan Q, Gao L, Li K, Wang YQ, Cui HY, Qi X, Gao YL, Wang XM, Liu CJ (2019) Differential expression of type I interferon mRNA and protein levels induced by virulent Marek's disease virus infection in chickens. Vet Immunol Immunopathol 212:15–22

Taha-Abdelaziz K, Hodgins DC, Lammers A, Alkie TN, Sharif S (2018) Effects of early feeding and dietary interventions on development of lymphoid organs and immune competence in neonatal chickens: a review. Vet Immunol Immunopathol 201:1–11

Takeda K, Kaisho T, Akira S (2003) Toll-like receptors. Annu Rev Immunol 21:335–376

Turner MD, Nedjai B, Hurst T, Pennington DJ (2014) Cytokines and chemokines: at the crossroads of cell signalling and inflammatory disease. Biochim Biophys Acta 1843(11):2563–2582

Walliser I, Göbel TW (2018) Chicken IL-17A is expressed in αβ and γδ T cell subsets and binds to a receptor present on macrophages, and T cells. Dev Comp Immunol 81:44–53

Williams AM, Maman Y, Alinikula J, Schatz DG (2016) Bcl6 is required for somatic hypermutation and gene conversion in chicken DT40 cells. PLoS One 11(2):e0149146

Yasmin AR, Omar AR, Farhanah MI, Hiscox AJ, Yeap SK (2019) Quantitative proteomics analysis revealed compromised chicken dendritic cells function at early stage of very virulent infectious bursal disease virus infection. Avian Dis 63(2):275–288

Yatim KM, Lakkis FG (2015) A brief journey through the immune system. Clin J Am Soc Nephrol 10(7):1274–1281

Zhang Q, Chen B, Yang P, Zhang L, Liu Y, Ullah S, Wu L, Waqas Y, Le Y, Chen W, Chen Q (2015) Identification and structural composition of the blood–spleen barrier in chickens. Vet J 204(1):110–116

Zhang W, Eijk M, Guo H, Dijk A, Bleijerveld OB, Verheije MH, Wang G, Haagsman HP, Veldhuizen EJA (2017) Expression and characterization of recombinant chicken mannose binding lectin. Immunobiology 222(3):518–528

Evolution of Immunoglobulins in Vertebrates

4

Jafar Salimian, Xiaoying Zhang, Ricardo S. Vieira-Pires, and Patricia M. Morgan

Abstract

Vertebrates evolved some 550 million years ago (Mya) and the vertebrate classes include three classes of fish (jawless, cartilaginous and bony) as well as amphibians, reptiles, birds and mammals. The adaptive immune system arose in cartilaginous fish about 500 Mya and involved the acquisition of the recombination-activating genes (RAG) and two waves of extensive gene duplication. As animals evolved the number of immunoglobulin classes increased and it is clear that IgY was central to antibody evolution. The role of RAG and gene organization is introduced.

Keywords

Evolution of immunoglobulins · Vertebrates

J. Salimian
Applied Microbiology Research Center, Chemical Injuries Research Center, Systems Biology and Poisoning Institute, Baqiyatallah University of Medical Sciences, Tehran, Iran
e-mail: jafar.salimian@bmsu.ac.ir

X. Zhang
College of Biological Science and Engineering, Shaanxi University of Technology, Hanzhong, China
e-mail: zhxying@snut.edu.cn; xzhang67@uoguelph.ca; zhang@bio.uminho.pt

R. S. Vieira-Pires
CNC – Center for Neuroscience and Cell Biology, University of Coimbra, Coimbra, Portugal
e-mail: ricardo.pires@biocant.pt

P. M. Morgan (✉)
Emerita Lecturer, School of Natural Sciences, National University of Ireland Galway, Galway, Ireland
e-mail: pat.morgan@nuigalway.ie

X.-Y. Zhang et al. (eds.), *IgY-Technology: Production and Application of Egg Yolk Antibodies*, https://doi.org/10.1007/978-3-030-72688-1_4

4.1 Introduction to Evolution

In evolutionary terms, life began in seas and ponds and was confined to those zones for some billions of years. The move onto land was initiated by cyanobacteria about 2 billion years ago. Eukaryotes formed about 2.7 billion years ago then split into the ancestors of modern, plants, fungi and animals. There are seven extant classes of vertebrates which can be traced back to about 550 million years ago (Mya) (Fig. 4.1). Three of these are fishes and include the following classes: Agnatha (jawless vertebrates e.g., hagfish and lampreys (Kasahara et al. 2004); Chondrichthyes (cartilaginous fish e.g., sharks and rays); and Osteichthyes (bony fish e.g., trout and zebrafish). The other four classes of vertebrates are called tetrapods and include the classes Amphibia (e.g., frogs and toads), Reptilia (e.g., snakes and crocodiles), Aves (e.g., ostrich and chickens) and Mammalia which includes monotremes (e.g., duck-billed platypus), marsupials (e.g., kangaroos) and placentals (e.g., mice and humans). The Devonian period (about 350–400 Mya) is known as the age of fishes as many new forms evolved in fresh and salt water. Reptiles, birds and mammals had to adapt to terrestrial life and are collectively called amniotes. This is based on the formation of an amniotic egg (Sect. 2.6.1) which is a shelled water-retaining structure which allows these classes to complete their life cycle on land. Some key features of this condition are retained in the mammals which give rise to live young; a monotreme, the duck-billed platypus continues to lay eggs.

The immune response is one of the many adaptive processes which enables animals to adjust to the adversarial nature of the environments in which they live.

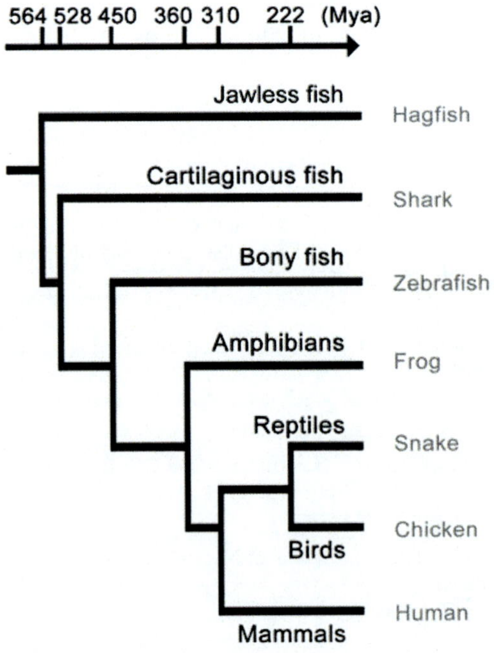

Fig. 4.1 Phylogeny of the vertebrates. Notes: Mya is million years ago. The time scale, shown on top is adapted from (Kumar and Hedges 1998)

As previously discussed (Chap. 3) primitive animals relied on an innate defence mechanism but as species evolved and developed more complex organs the immune system needed to respond and the adaptive immune system appeared about 500 million years ago (Sun et al. 2019) which corresponds to the period of evolution of cartilaginous fish (Fig. 4.1) The adaptive immune system (Chap. 3) includes the major histocompatibility complex proteins (MHC), immunoglobulins (Ig) and T-cell receptors (TCR); these underwent further development as new species evolved (Sun et al. 2019). The development of a process to rearrange genes in response to pathogenic infection is central to adaptive immunity (Chap. 3). Two major events in the development of the adaptive immune system were the acquisition of the recombination-activating genes (RAG); and two waves of extensive gene duplication which occurred as jawed vertebrates evolved (Kasahara et al. 2004). The RAG protein complex is involved in rearrangement and recombination of genes encoding Ig and TCR molecules (Sect. 4.5). There are an enormous diversity of organisms in each vertebrate class, and an ever increasing wealth of information on different immunoglobulin isotypes, however, for this chapter we will focus mainly on the classes of vertebrates rather than individual organisms. Insights into the particular structures and functions of Ig in a range of species has been recently reviewed (Sun et al. 2019; Olivieri et al. 2020) and is outside the scope of this chapter.

4.2 Immunoglobulin Classes in Vertebrates

In evolutionary terms cartilaginous fish were the first to be able to mount an adaptive immune response. As vertebrates evolved, there was diversification of the heavy chain gene resulting in an increased number of Ig classes (Fig. 4.2). There are five classes of mammalian Ig; IgM, IgA, IgG, IgD and IgE based on the structure of the heavy chain which are named μ, α, γ, δ and ε, respectively. IgM shows a remarkable consistency in structure and function from cartilaginous fish to mammals (Fig. 4.2). IgM antibodies are multivalent, and although they may be produced without an antibody challenge, they are the first antibodies produced in response to a new infection and are found predominantly in the blood and lymph fluid but are also involved in mucosal immunity (Zhang et al. 2017). The J chain is known to be involved, but not always required, in the formation of pentameric IgM and dimeric IgA (Fig. 4.2) and functions in the transport of the Igs to the mucosae (Koshland 1985).

Cartilaginous fish express a lineage-specific isotype IgNAR which is a heavy chain homodimer. In bony fish, in addition to IgM, IgZ (also known as IgT) is found. With the movement of vertebrates from water to land, there were genomic structural changes and an increase in number of Ig classes (Olivieri et al. 2020). Amphibians have both pentameric and hexameric IgM as well as IgX, IgY and IgF (Fig. 4.2). IgF, closely related to IgY, is present in only a few species and has only two heavy chain constant regions with a hinge between them (Zhang et al. 2017).

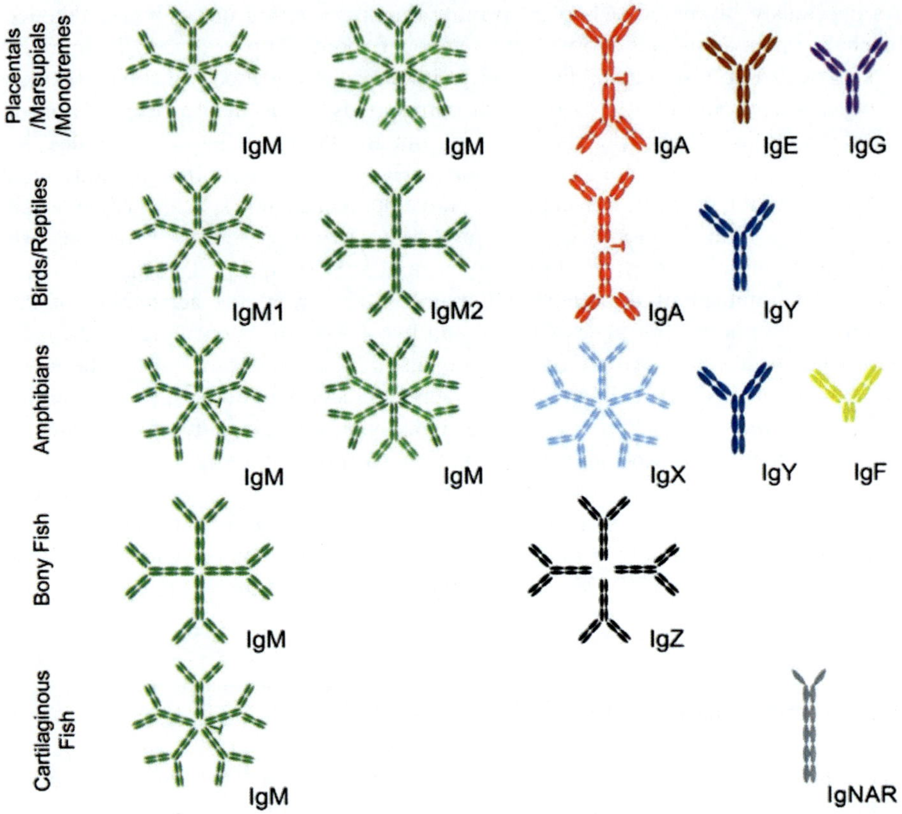

Fig. 4.2 Schematic representation of the domain structure of antibodies throughout jawed vertebrate evolution. The domain structure and polymeric state are indicated. The J chain which is part of pentameric IgM and IgA dimers is indicated by T-shaped symbols. IgD, IgH and IgW are not included in the figure. Figure adapted with publisher's permission from (Zhang et al. 2017)

4.3 Central Position of IgY in Antibody Evolution

The evolution of the difference classes of Ig appears to follow a quite complex pattern (Fig. 4.3). It is believed that IgY arose by a gene duplication event from IgM (Gregory et al. 1995) but through an intermediate step in which IgA evolved from IgM and then IgY evolved from both IgM and an ancestral IgA (Fig. 4.3) (Zhang et al. 2017). This hypothesis is supported by phylogenetic analyses (Sun et al. 2019). IgA is the principal mucosal Ig found in reptiles, birds and mammals; it shows a similar distribution to IgZ found in bony fish and IgX in amphibians. IgX and IgA are likely to have evolved from a common ancestor but the evolutionary relationship between IgZ and IgA is not clear (Mashoof et al. 2014). IgF found in amphibians, IgY in birds and reptiles and early mammals (Fig. 4.2) and IgE and IgG in mammals

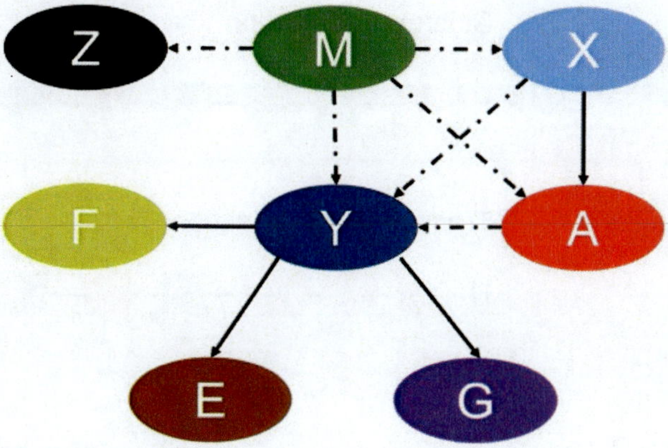

Fig. 4.3 The central position of IgY in immunoglobulin evolution. Solid arrows indicate an orthologous relationship between isotypes. Broken arrows connect isotypes which have a putative orthologous relationship, not yet verified. Colour coding for antibody isotypes follows that of Fig. 4.2. Figure adapted with the publisher's permission from Zhang et al. (2017)

evolved as serum proteins produced in response to infection: IgE on binding to its receptor is mainly present on tissue mast cells.

IgA shares some structural features with IgM (Fig. 4.3). Possibly the evolutionary sequence involved was IgM giving rise to IgX/IgA which then gave rise to IgY; it appears two domains of an ancestral IgA contributed to IgY (Zhang et al. 2017). IgY is thought to be the precursor of IgG and IgE (Gregory et al. 1995) based on analysis of shared exons and information from the genomic analysis of the duck-billed platypus. This monotreme has IgY instead of IgG, but it has a hinge region not present in avian IgY (Zhao et al. 2009). Furthermore, platypus IgE sequence displays features of avian IgY and mammalian IgE. These are good clues in speculating IgY's position in Ig evolution (Gambón-Deza et al. 2009). The discovery of IgY in amphibians (Zhao et al. 2006) and in 2009 in reptiles (Wei et al. 2009) and the duck-billed platypus (Gambón-Deza et al. 2009), as well as the emerging information on the crystal structure of part of the Fc molecule (Taylor et al. 2009) led to the acceptance that IgY is a distinct class of Ig and holds a key place in the evolution of Ig.

4.4 Gene Organisation Related to Ig Evolution

With the development of molecular genetics technology, researchers have been able to assess many genes coding for Ig in different species (Hsu 2009; Sun et al. 2013). In the germline, there a number of genetic elements which on recombination give rise to the variable domain involved in antigen binding. In the heavy chain, there are three elements, variable (VH), diversity (DH) and joining (JH) segment genes; the light chain is encoded by VL and JL gene segments. The order of these gene segments in cartilaginous fish is very different from later evolved species (Hsu 2016). Multiple individual loci in cartilaginous fish are able to encode Ig genes and IgNAR (new antigen receptor) is shown (Fig. 4.4). Each locus consists of constant regions and rearranging segmental elements. It is likely that another lineage

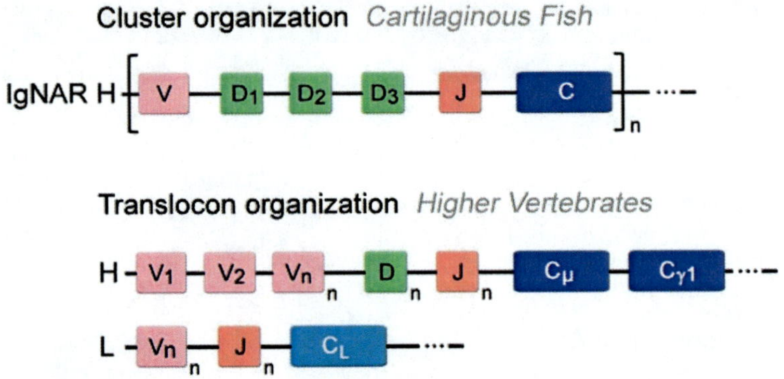

Fig. 4.4 Diagrammatic representation of the cluster and translocon organization of heavy chain genes in cartilaginous fish and higher vertebrates. V = variable gene, D = diversity gene, J = joining gene and C = constant gene while n represents a variable number of genes. Adapted from Das et al. (2012)

with the Ig in the translocon organization gave rise to Ig in all other vertebrates (Hsu et al. 2006).

4.5 Recombination-Activating Gene (RAG)-Mediated Rearrangement

Research has shown that the rearrangement of V, D and J segments with their neighbouring RSS exists in all types of jawed vertebrates (Hsu 2009). The enzymes RAG1/RAG2 and terminal deoxynucleotidyl transferase (TdT) play critical roles in DNA nicking and modification (Fig. 4.5). The study of these lymphocyte-specific enzymes has indicated that they are present in mice, and in their actions show a high degree of conservation in other animals (Krangel 2003). Initially, pair wise recognition of the RSS is needed and the RSS pair to be recombined must be composed of one RSS including a 12-bp spacer and the other a 23-bp spacer ("12/23 rule") (Krangel 2003). In all species of vertebrates where the genomic organization of the gene elements has been elucidated, the RSS that flank possibly recombinogenic gene elements show this pairing association (Hsu 2009).

Junctional diversity arises when the fusion of the V, D and J segments is not exact. This imprecise joining can result from the deletion or insertion of nucleotides (termed P and N). P nucleotides are so named as they make up palindromic sequences added to the end of gene segments. N-nucleotides are named as they are non-template-encoded but added by the enzyme TdT to the single-stranded ends of the coding DNA, after hairpin cleavage (Komori et al. 1993). RAG starts the process which results in breaking of double-stranded DNA at a gene segment and the coding ends being then linked by the cell's DNA repair pathways (Fig. 4.5).

TdT is a member of the X family of DNA polymerases whose expression is limited to lymphoid precursors of the B and T-cell lineage. Other X-family DNA

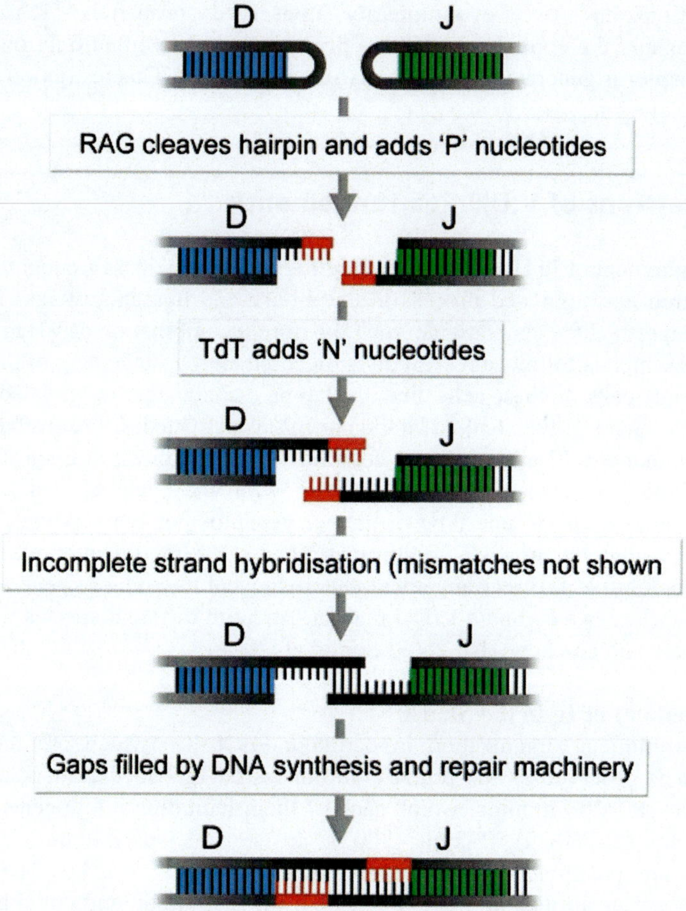

Fig. 4.5 Junctional diversity. Generation of junctional diversity through recombination illustrated between two gene segments: D (blue) and J (green). Sections highlighted in red show nucleotides added at each stage (Junctional Diversity by Robert Wilson, CC BY 3.0)

polymerases, Pol μ and Pol λ, have key roles at various points during coding end-processing (Bertocci et al. 2006). The early association of TdT in the evolution of V(D)J recombination reveals its key role in amplifying the selected features committed by gene rearrangement including variation of sequence and sequence length as a result of the breakage induced by RAG (Litman et al. 1999; Hsu 2009; Sun et al. 2013). It has been demonstrated that the highly diverse CDR3 region is the key determinant of specificity in antigen recognition (Xu and Davis 2000). In chickens TdT is not active and there is no evidence of N nucleotide additions.

The T cell receptors TCRαβ and TCRγδ (Chap. 3) have been conserved from all classes of jawed vertebrates, including mammals (i.e. marsupials, monotremes and placentals) (Sun et al. 2013). Multiple lines of evidence indicated that the V(D)J recombination has two evolutionarily conserved properties: RAG action mechanisms and the controlling of these processes to ensure that only one type of antigen receptor is generated by each cell (Sun et al. 2013). This restriction is known as allelic exclusion.

4.6 Patterns of V(D)J Rearrangement

The rearrangement of Ig Heavy (IgH) and Ig Light (IgL) chains is frequently defined as an ordered and regulated process in the mouse and human systems. The well-assessed stages include generation of the DJ before recombination of VH to the DJ in pro B-cells which is followed by cell division and then rearrangement of the L chain genes in pre B-cells. In these cells, first, the kappa L chain (Igκ) is activated and then the second L chain isotype (called lambda or Igλ) is activated. Chickens have only a single L chain locus. The controlled accessibility of RAG on various genes and gene segments leads to expression of one H chain with one allele of κ or λ L chain. Therefore, this results in one type of antigen receptor per lymphocyte. There are currently few data on Ig and TCR chromatin and DNA modification except in the mouse model, but V(D)J rearrangement patterns reveal the order of gene accessibility to RAG. Here, we compare V(D)J rearrangement in different species such as the shark, mouse and chicken (Hsu 2009; Sun et al. 2013).

Rearrangement of Ig in the Shark

The IgH minilocus organization in cartilaginous fishes which are the earliest vertebrates is observed as the primary and ancestral IgH locus. Sharks, rays and skates have 100–200 miniloci which each of them including a few gene segments (i.e., VH-D1-D2-JH-Cμ) (Flajnik 2002). In several species the rearranging sequences are positioned within an area with a length of 2 kb. The miniloci themselves are positioned at longer distance from each other, and could be located on various chromosomes (Litman et al. 1999; Lee et al. 2008). There is no stringent order of rearrangement for the VH, D1, D2 and JH gene segments. After activation of IgH gene in a precursor B-cell, all gene segments recombine at once and are followed by completion of gene rearrangement (Malecek et al. 2012).

In a study on a single B-cell it was observed that there were few genomic rearrangements and few Ig transcripts per lymphocyte (Mostoslavsky et al. 1998; Eason et al. 2004). There were 9–12 functional IgH genes and 1–3 VDJ genomic rearrangements in in each B-cell which only one emerged to make a viable receptor in nurse shark. Once started, recombination happens efficiently between the four gene segments. This information indicates that despite its unique IgH organization, H chain exclusion occurs in the shark.

Rearrangement of Ig in the Mouse

Information is presented on the mouse as it was the earliest animal used for the production of monoclonal antibodies by hybridoma technology. The IgH gene rearrangement is done in order and in a step-by-step process in mouse. Initially, in pro-B cells of the bone marrow the D and J gene segments are joined at both IgH alleles. Subsequently, the chromatin domain including the upstream VH genes become activated. There is a large space between the VH gene sequence and the DJ. Therefore, looping of the DNA and locus contraction are needed to bring them into close proximity for rearrangement. It has been indicated that exclusion of the H chain is a result of the staggering of the V to DJ phase among the two alleles.

Rearrangement of Ig in the Chicken

In birds, Ig genes exhibit several unusual characteristics in their genomic organisation and how they are expressed (Sun et al. 2013). In chickens, there is very little germline diversity in V, D and J gene segments and diversification occurs mainly by gene conversion. In this process, short sequences in the expressed rearranged V-region are replaced with sequences from the upstream V gene segment pseudogene. The enzyme activation-induced cytidine deaminase is essential for gene conversion. This is discussed in more detail in Sect. 5.3.

References

Bertocci B, Smet AD, Weill JC, Reynaud CA (2006) Nonoverlapping functions of DNA polymerases mu, lambda, and terminal deoxynucleotidyltransferase during immunoglobulin V (D)J recombination in vivo. Immunity 25(1):31–41

Das S, Hirano M, Tako R, McCallister C, Nikolaidis N (2012) Evolutionary genomics of immunoglobulin-encoding loci in vertebrates. Curr Genomics 13(2):95–102

Eason DD, Litman RT, Luer CA, Kerr W, Litman GW (2004) Expression of individual immunoglobulin genes occurs in an unusual system consisting of multiple independent loci. Eur J Immunol 34(9):2551–2558

Flajnik MF (2002) Comparative analyses of immunoglobulin genes: surprises and portents. Nat Rev Immunol 2(9):688–698

Gambón-Deza F, Sánchez-Espinel C, Magadán-Mompó S (2009) The immunoglobulin heavy chain locus in the platypus (*Ornithorhynchus anatinus*). Mol Immunol 46(13):2515–2523

Gregory WW, Katharine EM, David A (1995) IgY: clues to the origins of modern antibodies. Immunol Today 16(8):392–398

Hsu E (2009) V(D)J recombination: of mice and sharks. Adv Exp Med Biol 650:166–179

Hsu E (2016) Assembly and expression of shark Ig genes. J Immunol 196(9):3517–3523

Hsu E, Pulham N, Rumfelt LL, Flajnik MF (2006) The plasticity of immunoglobulin gene systems in evolution. Immunol Rev 210(1):8–26

Kasahara M, Suzuki T, Pasquier LD (2004) On the origins of the adaptive immune system: novel insights from invertebrates and cold-blooded vertebrates. Trends Immunol 25(2):105–111

Komori T, Okada A, Stewart V, Alt F (1993) Lack of N regions in antigen receptor variable region genes of TdT-deficient lymphocytes. Science 261(5125):1171–1175

Koshland ME (1985) The coming of age of the immunoglobulin J chain. Annu Rev Immunol 3 (1):425–453

Krangel MS (2003) Gene segment selection in V(D)J recombination: accessibility and beyond. Nat Immunol 4(7):624–630

Kumar S, Hedges SB (1998) A molecular timescale for vertebrate evolution. Nature 392 (6679):917–920

Lee V, Huang JL, Lui MF, Malecek K, Ohta Y, Mooers A, Hsu E (2008) The evolution of multiple Isotypic IgM heavy chain genes in the shark. J Immunol 181(4):2933–2934

Litman GW, Anderson MK, Rast JP (1999) Evolution of antigen binding receptors. Annu Rev Immunol 17(1):109–147

Malecek K, Lee V, Feng W, Huang JL, Flajnik MF, Ohta Y, Hsu E (2012) Immunoglobulin heavy chain exclusion in the shark. PLoS Biol 6(6):e157

Mashoof S, Pohlenz C, Chen PL, Deiss TC, Gatlin D, Buentello A, Criscitiello MF (2014) Expressed IgH μ and τ transcripts share diversity segment in ranched *Thunnus orientalis*. Dev Comp Immunol 43(1):76–86

Mostoslavsky R, Singh N, Kirillov A, Pelanda R, Cedar H, Chess A, Bergman Y (1998) Kappa chain monoallelic demethylation and the establishment of allelic? Exclusion. Genes Dev 12 (12):1801–1811

Olivieri DN, Mirete-Bachiller S, Gambón-Deza F (2020) Insights into the evolution of IG genes in amphibians and reptiles. Dev Comp Immunol 114:103868

Sun Y, Wei Z, Li N, Zhao Y (2013) A comparative overview of immunoglobulin genes and the generation of their diversity in tetrapods. Dev Comp Immunol 39(1–2):103–109

Sun Y, Huang T, Hammarström L, Zhao Y (2019) The immunoglobulins: new insights, implications, and applications. Annu Rev Anim Biosci 8:145–169

Taylor AI, Fabiane SM, Sutton BJ, Calvert RA (2009) The crystal structure of an avian IgY-Fc fragment reveals conservation with both mammalian IgG and IgE. Biochemistry 48(3):558–562

Wei Z, Wu Q, Ren L, Hu X, Guo Y, Warr GW, Hammarstrom L, Li N, Zhao Y (2009) Expression of IgM, IgD, and IgY in a reptile, *Anolis carolinensis*. J Immunol 183(6):3858–3864

Xu JL, Davis MM (2000) Diversity in the CDR3 region of VH is sufficient for Most antibody specificities. Immunity 13(1):37–45

Zhang X, Calvert RA, Sutton BJ, Doré KA (2017) IgY: a key isotype in antibody evolution. Biol Rev 92(4):2144–2156

Zhao Y, Qiang PH, Yu S, Wertz N, Zhang X, Li N, Butler JE, Hammarström L (2006) Identification of IgF, a hinge-region-containing Ig class, and IgD in Xenopus tropicalis. Proc Natl Acad Sci USA 103(32):12087–12092

Zhao Y, Cui H, Whittington CM, Wei Z, Zhang X, Zhang Z, Yu L, Ren L, Hu X, Zhang Y, Hellman L, Belov K, Li N, Hammarström L (2009) *Ornithorhynchus anatinus* (platypus) links the evolution of immunoglobulin genes in eutherian mammals and nonmammalian tetrapods. J Immunol (Baltimore, Md: 1950) 183(5):3285–3293

Biology and Molecular Structure of Avian IgY Antibody

5

Álvaro Ferreira Júnior, Patricia M. Morgan, Xiaoying Zhang, and Rüdiger Schade

Abstract

The source of IgY (egg yolk), the method of antibody collection, the monthly yield of antibody and the ability to raise antibodies to conserved mammalian antigens are some of the many advantages of IgY in comparison to immunoglobulin G (IgG). While the general structure of IgY is similar to immunoglobulin G (IgG) of mammals, there are several important differences. These include the presence of an additional heavy chain constant region (CH4), the absence of a flexible hinge region, differential glycosylation and effector functions. The generation of antibody diversity is also discussed. A truncated IgY, IgY(ΔFc), containing only CH1 and CH2 domains, is expressed in the serum of some birds and reptiles, such as duck and turtle. The diversity of IgY which is observed among different species has not yet been recognized as different isotypes.

Á. Ferreira Júnior
School of Veterinary and Zootechnics, Universidade Federal de Goiás, Goiânia, Brazil
e-mail: alvaro.ferreira@ufg.br

P. M. Morgan
Emerita Lecturer, School of Natural Sciences, National University of Ireland Galway, Galway, Ireland
e-mail: pat.morgan@nuigalway.ie

X. Zhang (✉)
College of Biological Science and Engineering, Shaanxi University of Technology, Hanzhong, China
e-mail: zhxying@snut.edu.cn; xzhang67@uoguelph.ca; zhang@bio.uminho.pt

R. Schade
Institute of Pharmacology, Charité – Universitätsmedizin Berlin, Berlin, Germany
e-mail: ruediger.schade@charite.de

X.-Y. Zhang et al. (eds.), *IgY-Technology: Production and Application of Egg Yolk Antibodies*, https://doi.org/10.1007/978-3-030-72688-1_5

Keywords

IgY structure · Truncated IgY (IgY(ΔFc)) · Avian · Vertebrates and reptile · Glycosylation

5.1 The Biology of IgY Production Compared to IgG

The most commonly used polyclonal antibody in research is immunoglobulin G (IgG). Rabbits are one of the primary species used in polyclonal antibody production due to their size, ease of housing, ease of vascular access and nature and robustness of their immune response (Lipman et al. 2005). Sheep and goats, due to their larger size, are used when larger quantities of antibody are required. Chickens offer a number of advantages compared to rabbits as outlined in Table 5.1. IgY can be easily sampled by a non-invasive method based on the simple action of egg collection, instead of the stressful bleeding of animals to obtain serum, which fulfils the requirement of 3R principle of animal welfare. The quantity of IgY harvested per month greatly exceeds that which can be recovered from a rabbit (Table 5.1). The antibody production of a chicken roughly corresponds to that of a large mammal, such as a sheep or a goat. There are no significant differences between the avidities (Woolley and Landon 1995; Svendsen et al. 1996) and affinities (Montes Pérez et al. 1994) of chicken IgY and rabbit IgG. IgY is relatively stable in various conditions (Table 5.1). Phylogenetic distance is the reason for the frequently described differences between the antibody specificities of mammals and chickens, even when identically immunised. In addition to differences between the mammalian and the avian immune system themselves, differences in phylogenetic development in these two animal classes contribute to the different antibody specificities. Chickens often raise antibodies against phylogenetically highly conserved mammalian antigens more efficiently than rabbits (Larsson and Sjöquist 1990; Furlong 2005), which provide a new strategy for the detection and diagnosis of these antigens as well as drug development. Furthermore, IgY has no cross-reactivity with rheumatoid factors and human anti-mouse antibodies (Table 5.1), and does not activate the complement system and hetero-agglutinins, which indicates that IgY could be used as an alternative in diagnostic and therapeutic applications. IgY-based immunoassays are being used to measure the content of proteins, peptides or small molecules via ELISA, RIA or other assays in clinical chemistry and basic research. However, there have recently been debates between IgY users as to whether IgY-based immunoprecipitation in some cases are rather diffuse and not as distinct as are IgG-based ones. This problem can be solved by using buffers with a high salt concentration (1.5 M NaCl with 5% PEG 6000). Most of the recent studies confirm that IgY can be used in the same way as IgG, but additionally offer advantages in terms of specificity, cross-reactivity and/or sensitivity. Systematic comparisons between IgY (chicken) and IgG (rabbit) in terms of their biological activities, molecular structure, characteristics and applications are given in Table 5.1.

Table 5.1 Overall comparison between IgY antibody (chicken) and IgG antibody (rabbit)

Characteristics	IgY (chicken)	IgG (rabbit)
Species	Birds, reptiles, amphibians and lungfish	Mammals
Sites of generation	Bursa of Fabricius, spleen and bone marrow	Lymph nodes, spleen and bone marrow
Antibody subclasses	Not for chicken[a]	IgG_1, IgG_2, IgG_3 and IgG_4
Source of antibodies	Egg	Blood
Antibody collection	Meets 3R principle of animal welfare	Can be painful
Average antibody levels per animal	50–100 mg/egg yolk	5 mg/mL of blood, blood collection up to 40 mL/month
Monthly antibody yield per animal	1000–2800 mg/chicken/month	200 mg/rabbit/month
Serum blood concentration (mg/mL)	6.9 ± 1.68	10–15
Egg yolk concentration (mg/mL)	8–10	Not applicable
Amount of antigen-specific antibodies	0.5%–10% total IgY	50–200 μg/mL serum
Phylogenetic distance to mammals	310 Mya	Near
Immune response to mammalian conserved antigens	Strong	Often weak
Affinity	Comparable to rabbit IgG	$1 \times 10^{-8} - 1 \times 10^{-10}$ mol/L
Antibody avidity	High, no significant difference to IgG	High
Molecular weight (kDa)	180	150
Isoelectric point	5.5–7.6[b]	to 8.5
pH stability	Stable at pH 3.5–11.0	Stable at pH 2.0–11.0
Extinction coefficient ($mL\ mg^{-1}\ cm^{-1}$ at 280 nm)	1.094, 1.33, 1.35[c]	1.4
Trypsin stability	Generally stable	Weaker than IgY
Thermo stability	Stable at 65 °C > 2 months; 100 °C > 6 min; 4 °C > 6 months	Generally higher than IgY[d]
Pepsin stability	At pH 4.0, 91% IgY activity for 1 h, 63% after 10 h	Higher than IgY
Fc chain (type of chain/number of domains/hinge region between Fab and Fc chains)	υ chain/3 constant domains/no	γ chain/2 constant domains/yes
Fab chain (type of chain/number of domains)	λ chain/1 variable domain and 1 constant domain	κ or λ chains/1 variable domain and 1 constant domain
Binding to mammalian complement or Fc receptors	No	Yes
Cross-reactivity with rheumatoid factor	No	Yes

(continued)

Table 5.1 (continued)

Characteristics	IgY (chicken)	IgG (rabbit)
Cross-reactivity with human anti-mouse antibody (HAMA)	No	Yes
Reaction to hetero-agglutinins (Coombs, blood group classification)	No	Yes
Non-affinity separation	Remove lipid mixture from egg yolk	Remove plasma mixture from blood
Purification by salting-out	Yes	Yes
Purification by affinity columns	Protein M[e]	Yes (protein A and G)
Conjugation with enzymes, fluorophores and colloidal gold	Yes	Yes
Immunoprecipitation	Relatively inefficient	Good
Immunosuppression applications	Can be used for xenotransplantation	Several products are under development
Diagnostic applications	IgY has been studied and applied for detection/ diagnosis	Widely used, especially monoclonal antibodies
Therapeutic applications	Under development	Well developed
Generation of full-length monoclonal antibody by hybridoma technology	Difficult, and lower yield of monoclonal IgY obtained	Routinely and commercially applied
Generation of scFv antibody fragment by phage display	Easily produced	Easily produced

Notes

[a]As described in this chapter, there is some limited diversity of IgY among species, particularly the IgY(ΔFc), however, it is not yet recognized as a sub-class of IgY. IgY(ΔFc) may exhibit some characteristics which are different to full length IgY (Sects. 5.4 and 5.5)

[b, c]For detailed analysis on isoelectric point and extinction coefficient please refer to Chap. 11

[d]For more detailed comparisons please refer to Shimizu et al. (1992)

[e]For more detailed information please refer to Chap. 11

Most data are direct reference from the review article (Schade et al. 2005), otherwise indicated as follows: Amount of antigen-specific antibodies (Lipman et al. 2005); Extinction coefficient (ml mg^{-1} cm^{-1} at 280 nm) (Pauly et al. 2011; Hode et al. 2013); Diagnostic applications (He et al. 2015); Therapeutic applications (Schade et al. 2007); Generation of full-length monoclonal antibody by hybridoma technology (Zhang et al. 2010); Generation of scFv antibody fragment by phage display (Barderas and Benito-Peña 2019); Protein M purification (Jiang et al. 2016)

5.2 Molecular Structure of IgY

The general structure of the IgY molecule is comparable to mammalian immunoglobulins (Fig. 5.1 and Table 5.2). It consists of two identical light and two identical heavy chains. The heavy chain is indicated by the Greek letter Y or υ and contains one variable domain and four constant domains. This is in contrast to the three constant domains of the IgG heavy chain which is indicated by the Greek letter γ. The molecular weight of IgY is 180 kDa (IgG 150 kDa); the heavy chains of approximately 65 kDa each and the light chains of approximately 27 kDa (Amro

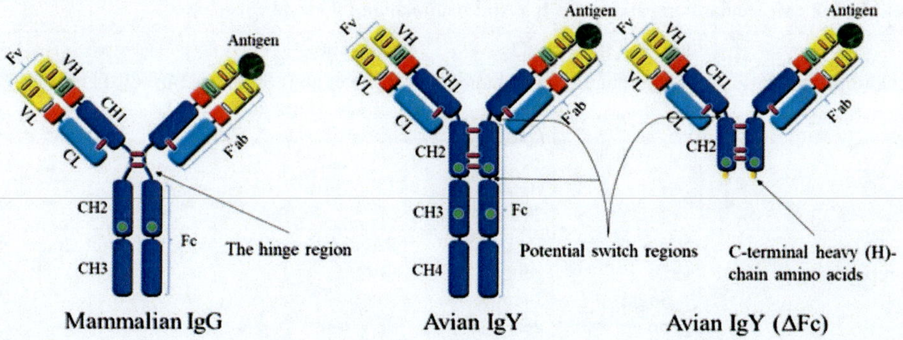

Fig. 5.1 The structure of IgG, full-length IgY and truncated IgY(ΔFc). Notes: IgG is composed of two identical heavy chains and two light chains. Each heavy chain is comprised of a variable domain (VH) and three constant domains (CH1, CH2, CH3) with a single carbohydrate site (green circle). Every light chain contains a variable domain (VL) and a constant domain (CL). VL and VH are the binding domain of antigen. Purple lines indicate the locations of interchain disulphide bonds. Bricks represent Complementarity Determining regions (CDRs). Compared to IgG, IgY is comprised of four constant domains per heavy chain (CH1–CH4), with two carbohydrate sites. The flexible hinge region found in IgG is absent in IgY and thus may restrict its flexibility in comparison to IgG. The truncated IgY (ΔFc) molecule does not possess CH3 or CH4 domains, and its two C-terminal heavy (H)-chain amino acids are not encoded within the CH2 or CH3 domains of the full-length isoform. Approx. 110–115 amino acid residues at each VL, VH, CL or constant domain, details in Table 5.2. (Source: XYZ-IgY lab)

et al. 2017). The additional CH4 region of IgY accounts for the higher molecular weight compared to IgG (Müller et al. 2015).

The antigen binding site of the Ig is composed of the variable regions of the heavy and light chains which are linked by disulphide bonds (Fig. 5.1). Within the variable regions, there are three hypervariable complementarity-determining regions (CDR) which interact with a specific antigen. The conserved regions next to the CDR are called framework regions which may also impact, but less directly, on antigen binding (Chap. 13). There is a certain genetic homology of the domains Cυ3 and Cυ4 with the domains Cγ2 and Cγ3 of mammalian IgG. The Cυ2 region forms a less flexible hinge region and is more comparable to mammalian IgE, rather than IgG (Chap. 4) (Müller et al. 2015). The limited flexibility of avian IgY may account for some of its unique biochemical properties, such as the inability to precipitate antigens at physiological salt concentrations, which is seen in chickens and ducks (Müller et al. 2015). For example, the two arms may be so closely aligned that they preclude cross-linking of epitopes on large antigens (Knight and Crane 1994). The Cυ2 domain may have been condensed in subsequent evolution to become the genetic hinge found in mammalian IgG (Chap. 4).

The Fc portion of IgY is the site of most biological effector functions as in the case of IgG. The Fc fragment of IgY includes two carbohydrate side chains in contrast to only one in IgG (Table 5.2). There are three interchain disulphide bonds between the heavy chains of IgY (Shimizu et al. 1992) and the number varies

Table 5.2 Molecular comparisons of IgY and mammalian IgG (rabbit)

	IgY	IgG	References
Chromosomes	λ light chain 15	λ light chain 21 κ light chain 2 Heavy chain 20	Zhao et al. (2000), Walport (2011), Schat (2013), Gertz et al. (2013)
VDJ gene repertoire	V-single gene J-single gene D-multigene	V-single gene J-single gene D-multigene	Reynaud et al. (1991), Knight and Crane (1994)
Pseudogene occupation	Yes	No	Schade et al. (2005)
Configuration	Monomer	Monomer	
Carbohydrate content (%)	4	2–4	Dávalos-Pantoja et al. (2000), Han et al. (2016)
Hinge region	No	Yes	
Antigen valency	2	2	
Number of amino acid residues	H-chain 599 L-chain 206	H-chain 440 L-chain 214	
Number of CH domains	4	3	
N-glycosylation site (H-chain)	-Asn (299)-Val-Ser -Asn (398)-Gly Thr-	-Asp (291)-Ser-Thr	Liu (2015), Gilgunn et al. (2016)

in IgG with two disulphide bonds between the heavy chains of IgG_1 and IgG_4, four for IgG_2 and 11 for IgG_3 (Liu 2012).

The role of the Fc region and its interaction with IgY receptors is reviewed in Chap. 6. Binding of IgY to these various receptors may also be implicated in leukocyte activation, and B cell activation and maturation (Lee et al. 2017).

Glycosylation is a vitally important post-translational modification of proteins that is frequent in the natural world (Oliveira-Ferrer et al. 2017; Sheng et al. 2018). According to the linkage structure between a sugar and an amino acid, there are four major forms of protein glycosylation: *N*-linked glycosylation, *O*-linked glycosylation, *C*-mannosylation and glycophosphatidylinositol (GPI) anchor attachments (Blom et al. 2004; Sheng et al. 2018). *N*-glycosylation is the main glycosylation structure found in immunoglobulins, and it has significant effects on protein stability, sensitivity to protease, immunogenicity and biological activity (Costa et al. 2013). A total of 217 *N*-glycosylation sites and 86 glycoproteins were identified in chicken egg yolk (CEY), and these glycoproteins are mainly involved in binding, biological regulation, catalytic activity, and metabolic processes (Geng et al. 2018). Among them, 15 were members of the complement and immune systems, which provide protection against potential threats during hatching (Geng et al. 2018).

IgY is more heavily glycosylated than its mammalian counterpart IgG as it contains two *N*-glycosylation sites in the Cυ3 domain, in comparison to the one on the CH2 (Cγ2) domain of IgG (Table 5.2 and Fig. 5.1). IgY contains mainly complex, bi-, tri- and tetra antennary glycans with or without core fucose and bisects, all with varying levels of galactosylation and sialylation, hybrid and high mannose glycans. Sialic acid content was high in IgY purified from serum (Gilgunn et al. 2016). Furthermore, no known non-human or immunogenic glycans were identified on IgY, as such antigenic structures can cause severe hypersensitivity reactions which are observed in many immunoglobulin-based bio-therapeutic agents (Gilgunn et al. 2016) (Chap. 13). The removal of *N*-glycan changed the conformation, storage stability, and resistance to guanidine hydrochloride and pepsin digestion of IgY; when IgY was digested by peptide-N4-(*N*-acetyl-beta-glucosaminyl) asparagine-amidase and *N*-glycan were removed, the IgY molecule became more flexible and disordered, with decreased storage stability and less resistance to guanidine hydrochloride induced unfolding, and more sensitive to pepsin digestion (Sheng et al. 2017).

From the glycobiology point of view, recombinant IgY antibody could be a potentially promising immune-therapeutic candidate after proper antibody engineering and expression.

5.3 Generation of Immunoglobulin Diversity

In humans and mice, immunoglobulins (Ig) genes are formed by a random assortment of variable (V), Diversity (D) and Joining (J) segments. After antigen stimulation this diversity is increased by somatic hypermutation and the Ig isotype can be changed by class switch recombination (Chap. 3). In the chicken, V(D)J recombination occurs to produce a single functional gene. Diversity is created mainly be gene conversion, where short sequences of the V-region gene are replaced by sequences from V gene segment pseudogenes. Similar modes of gene conversion occur in other species such as rabbit and farm animals (Flajnik 2002). The cluster of pseudo V genes is located upstream of the functional V gene segment in both the light and heavy chain loci. For the variable Light chain only the lambda (λ) light chain is expressed in avian species whereas mammals predominantly express the kappa (κ) light chain (Chap. 3).

There are 25 pseudo V genes at the light chain locus (Reynaud et al. 1987), and 80 pseudogenes at the heavy chain locus (Reynaud et al. 1989). During B-cell development in the bursa, segments of the pseudogenes appear in the V gene segments and multiple rounds of gene conversion can occur (Huang et al. 2012, 2016, 2018) (Chap. 4). This hyper-conversion process generates a large number of VH and VL repertoires, which when combined in the IgY molecule produce potentially –3 $\times$ 10^9 combinations (Hof et al. 2008). Additional diversity may be introduced by Activation-Induced-cytidine deaminase (AID) mediated point mutations.

The production of IgY against conserved mammalian proteins is possible due to the evolutionary distance (310 Mya) between chickens and mammals (Table 5.1). The complementarity-determining region (CDR) of the heavy chain (CDR-H3) of the chicken differs in both length and amino acid content from that of humans or mice (Finlay and Almagro 2012); in the chicken more smaller amino acids (e.g., glycine, serine, alanine) are present and a higher concentration of cysteine residues which may give rise to disulphide bonds which stabilize the longer CDR-H3 loops. These differences have been confirmed by high resolution crystal structure analyses (Zhang et al. 2017) and references therein. The diversity generated in the CDR1 and CDR2 of both heavy and light chains is mostly by gene conversion while somatic mutation is more important in the generation of diversity in CDR3 (Leighton et al. 2015). The chicken cell line DT40 is able to diversify Ig genes outside the bursal environment (Buerstedde et al. 1990) and human variable genes inserted in the chicken loci have been shown to be diversified using this mechanism (Schusser et al. 2013; Leighton et al. 2015). This was an important step in the development of transgenic hens expressing antibodies (Chap. 13).

5.4 Differential Expression of IgY Across Species

Although chicken IgY is the molecular model most usually adopted to represent IgY in general, there are IgY antibodies present as the predominant immunoglobulin in serum and yolk of other avian, amphibian and reptile species, and there are also secretory IgY and truncated IgY (IgY (ΔFc)) isotypes. While there is no formal definition of isotypes for IgY, a truncated IgY (IgY(ΔFc)), occurs containing only CH1 and CH2 domains (Fig. 5.1) and was first identified in ducks. It has a wide distribution in birds and has subsequently been identified in many other species, such as lizards (*Anolis carolinensis*) (Wei et al. 2009), turtles (*Pseudemys scripta* and *Trachemys scripta elegans*) (Leslie and Clem 1972; Li et al. 2012), snakes (*Elaphe taeniura*) (Wang et al. 2012), salamanders (*Andrias davidianus*), and geese (*Anser cygnoides orientalis*) (Sun et al. 2020).

5.5 Generation of IgY(ΔFc)

The generation of IgY(ΔFc) in different species occurs by different mechanisms. In ducks and geese, IgY and IgY(ΔFc) are generated by the alternative splicing of the gene for the full length IgY by using different transcriptional termination sites (Sun et al. 2020). In contrast, IgY and IgY(ΔFc) are encoded by separate genes in turtles (Li et al. 2012). A third mechanism occurs in some snakes: a single gene gives rise to identical transcripts which are then processed differently by posttranslational mechanisms to give rise to IgY and IgY(ΔFc) (Sun et al. 2020).

As shown in Fig. 5.1, neither IgY or IgY(ΔFc) have a hinge region which is present in IgG subclasses and which gives flexibility at that region. The switch regions (Fig. 5.1) of IgY and IgY(ΔFc) have been predicted to confer only limited flexibility. It not clear what advantages are conferred to the host by the presence of these two isoforms of IgY (Sun et al. 2020). Ducks on first exposure to an antigen mostly produce the full length IgY but repeated exposure shifts the production to IgY(ΔFc) (Grey 1967; Humphrey et al. 2004). It has been shown that the duck antibody response undergoes affinity maturation in response to an antigen (Higgins et al. 2001). The absence of the two constant regions in IgY(ΔFc) means that there is a loss of effector functions such as opsonization and Fc-mediated macrophage clearance of viruses. Antibodies intended to protect a host from viral infection may actually promote entry of the virus into the cell by the mechanism of Antibody-dependent enhancement of viral infection (ADE). This is dependent on the presence of the Fc portion of an antibody. Increasing levels of IgY(ΔFc) may result in limiting viral infections and represent an evolutionary selection pressure for this isoform (Meddings et al. 2014) although this has not yet been proven.

Secretory IgY is found expressed in the stomach and intestinal mucosae of young Axolotl (*Ambystoma mexicanum*) (Fellah et al. 1992). It has also been confirmed that different families of urodeles can express different immunoglobulin isotypes, including IgY (Schaerlinger and Frippiat 2008). Despite the limited investigations on isotypes of immunoglobins and IgY, particularly in amphibians and reptiles, it appears that there is more diversity of immunoglobins and IgY in these animals than anticipated, and they might be an important vestige of antibody evolution. It is also rational to speculate that IgY may have broader functions than our current understanding.

References

Amro W, Al-Qaisi W, Al-Razem F (2017) Production and purification of IgY antibodies from chicken egg yolk. J Genetic Eng Biotechnol 16(1):99–103

Barderas R, Benito-Peña E (2019) The 2018 Nobel Prize in chemistry: phage display of peptides and antibodies. Anal Bioanal Chem 411(12):2475–2479

Blom N, Sicheritz-Ponten T, Gupta R, Gammeltoft S, Brunak S (2004) Prediction of post-translational glycosylation and phosphorylation of proteins from the amino acid sequence. Proteomics 4:1633–1649

Buerstedde JM, Reynaud CA, Humphries E, Olson W, Ewert D, Weill J (1990) Light chain gene conversion continues at high rate in an ALV-induced cell line. EMBO J 9(3):921–927

Costa AR, Rodrigues M, Henriques M, Oliveira R, Azeredo J (2013) Glycosylation: impact, control and improvement during therapeutic protein production. Crit Rev Biotechnol 34(4):281–299

Dávalos-Pantoja L, Ortega-Vinuesa J, Bastos-González D, Hidalgo-Alvarez R (2000) A comparative study between the adsorption of IgY and IgG on latex particles. J Biomater Sci Polym Ed 11:657–673

Fellah J, Iscaki S, Vaerman J, Charlemagne J (1992) Transient developmental expression of IgY and secretory component like protein in the gut of the axolotl (*Ambystoma mexicanum*). Dev Immunol 2:181–190

Finlay W, Almagro J (2012) Natural and man-made V-gene repertoires for antibody discovery. Front Immunol 3:342

Flajnik MF (2002) Comparative analyses of immunoglobulin genes: surprises and portents. Nature reviews. Immunology 2(9):688–698

Furlong RF (2005) Insights into vertebrate evolution from the chicken genome sequence. Genome Biol 6(2):207

Geng F, Xie Y, Wang J, Majumder K, Qiu N, Ma M (2018) N-glycoproteomic analysis of chicken egg yolk. J Agric Food Chem 66(43):11510–11516

Gertz EM, Schäffer AA, Agarwala R, Bonnet-Garnier A, Rogel-Gaillard C, Hayes H, Mage RG (2013) Accuracy and coverage assessment of *Oryctolagus cuniculus* (rabbit) genes encoding immunoglobulins in the whole genome sequence assembly (OryCun2.0) and localization of the IGH locus to chromosome 20. Immunogenetics 65(10):749–762

Gilgunn S, Millan S, Wormald M, Zapatero Rodríguez J, Conroy P, O'Kennedy R, Rudd P, Saldova R (2016) Comprehensive N-glycan profiling of avian immunoglobulin Y. PLoS One 11:e0159859

Grey H (1967) Duck immunoglobulins. II. Biologic and immunochemical studies. J Immunol (Baltimore, Md: 1950) 98:820–826

Han B, Yuan H, Wang T, Li B, Ma L, Yu S, Huang T, Li Y, Dongming F, Chen X, Wang Y, Qiu S, Guo Y, Fei J, Ren L, Pan-Hammarström Q, Hammarström L, Wang J, Wang J-T, Zhao Y (2016) Multiple IgH isotypes including IgD, subclasses of IgM, and IgY are expressed in the common ancestors of modern birds. J Immunol (Baltimore, Md: 1950) 196

He J, Wang Y, Sun S, Zhang X (2015) Evaluation of chicken IgY generated against canine parvovirus viral-like particles and development of enzyme-linked immunosorbent assay and immunochromatographic assay for canine parvovirus detection. Viral Immunoll 28(9):489–494

Higgins D, Ko O, Chan S (2001) Duck antibody responses to keyhole limpet haemocyanin, human immunoglobulin G and the trinitrophenyl hapten. Evidence of affinity maturation. Avian Pathol 30:381–390

Hode P, Trefil P, Imnek J, Hudecek J, Stiborova M (2013) Optimized protocol of chicken antibody (IgY) purification providing electrophoretically homogenous preparations. Int J Biol Macromol 8(1):113–124

Hof D, Hoeke MO, Raats JM (2008) Multiple-antigen immunization of chickens facilitates the generation of recombinant antibodies to autoantigens. Clin Exp Immunol 151(2):367–377. https://doi.org/10.1111/j.1365-2249.2007.03569.x

Huang T, Zhang M, Wei Z, Wang P, Sun Y, Hu X, Ren L, Meng Q, Zhang R, Guo Y, Hammarstrom L, Li N, Zhao Y (2012) Analysis of immunoglobulin transcripts in the ostrich *Struthio camelus*, a primitive avian species. PLoS One 7(3):e34346

Huang T, Wu K, Yuan X, Shao S, Wang W, Wei S, Cao G (2016) Molecular analysis of the immunoglobulin genes in goose. Dev Comp Immunol 60:160–166

Huang T, Wang X, Si R, Chi H, Han B, Han H, Cao G, Zhao Y (2018) Identification of a transcriptionally forward alpha gene and two upsilon genes within the pigeon (*Columba livia*) IgH gene locus. J Immunol 200(11):3720–3728

Humphrey B, Calvert C, Klasing K (2004) The ratio of full length IgY to truncated IgY in immune complexes affects macrophage phagocytosis and the acute phase response of mallard ducks (*Anas platyrhynchos*). Dev Comp Immunol 28:665–672

Jiang X, Diraviyam T, Zhang X (2016) Affinity purification of egg yolk immunoglobulins (IgY) using a human mycoplasma protein. Chromatogr B Analyt Technol Biomed Life Sci 1012–1013:37–41

Knight KL, Crane MA (1994) Generating the antibody repertoire in rabbit. Adv Immunol 56:179–218

Larsson A, Sjöquist J (1990) Chicken IgY: utilizing the evolutionary difference. Comp Immunol Microbiol Infect Dis 13(4):199–201

Lee W, Syed Atif A, Tan SC, Leow CH (2017) Insights into the chicken IgY with emphasis on the generation and applications of chicken recombinant monoclonal antibodies. J Immunol Methods 447:71–85

Leighton PA, Schusser B, Yi H, Glanville J, Harriman W (2015) A diverse repertoire of human immunoglobulin variable genes in a chicken B cell line is generated by both gene conversion and somatic hypermutation. Front Immunol 6:126

Leslie GA, Clem LW (1972) Phylogeny of immunoglobulin structure and function. VI. 17S, 7.5S and 5.7S anti-DNP of the turtle, Pseudamys scripta. J Immunol 108(6):1656–1664

Li L, Wang T, Sun Y, Cheng G, Yang H, Wei Z, Wang P, Hu X, Ren L, Meng Q, Zhang R, Guo Y, Hammarström L, Li N, Zhao Y (2012) Extensive diversification of IgD-, IgY-, and truncated IgY(δFc)-encoding genes in the red-eared turtle (*Trachemys scripta elegans*). J Immunol 189 (8):3995–4004

Lipman NS, Jackson LR, Trudel LJ, Weis-Garcia F (2005) Monoclonal versus polyclonal antibodies: distinguishing characteristics, applications, and information resources. ILAR J 46 (3):258–268

Liu HK (2012) Disulfide bond structures of IgG molecules: structural variations, chemical modifications and possible impacts to stability and biological function. MAbs 4(1):17–23

Liu L (2015) Antibody glycosylation and its impact on the pharmacokinetics and pharmacodynamics of monoclonal antibodies and fc-fusion proteins. J Pharm Sci 104(6):1866–1884

Meddings JI, Owens L, Burgess G, Ariel E (2014) Revelations in reptilian and avian immunology: a proposed evolutionary selection pressure for truncated immunoglobulin-Y. Int J Immunol Stud 2(1):29

Montes Pérez RC, Murcia Mejía C, Zarco Quintero L (1994) Producción de anticuerpos antiprogesterona apartir de la yema de huevo de gallinas y del suero sanguíneo de conejos, para ser utilizados en radioinmunoanálisis. J Vet Med Ser A-Physiol Pathol Clin Med 25 (2):117–125

Müller S, Schubert A, Zajac J, Dyck T, Oelkrug C (2015) IgY antibodies in human nutrition for disease prevention. Nutr J 14:109

Oliveira-Ferrer L, Legler K, Milde-Langosch K (2017) Role of protein glycosylation in cancer metastasis. Semin Cancer Biol 44

Pauly D, Chacana PA, Calzado EG, Brembs B, Schade R (2011) IgY technology: extraction of chicken antibodies from egg yolk by polyethylene glycol (PEG) precipitation. Jove-J Visualized Exp 51

Reynaud CA, Anquez V, Grimal H, Weill JC (1987) A hyperconversion mechanism generates the chicken light chain preimmune repertoire. Cell 48(3):379–388

Reynaud CA, Dahan A, Anquez V, Weill JC (1989) Somatic hyperconversion diversifies the single Vh gene of the chicken with a high incidence in the D region. Cell 59(1):171–183

Reynaud CA, Anquez V, Weill JC (1991) The chicken D locus and its contribution to the immunoglobulin heavy chain repertoire. Eur J Immunol 21(11):2661–2670

Schade R, Calzado EG, Sarmiento R, Chacana PA, Porankiewicz-Asplund J, Terzolo HR (2005) Chicken egg yolk antibodies (IgY-technology): a review of progress in production and use in research and human and veterinary medicine. Atla-Alternat Lab Anim 33(2):129–154

Schade R, Zhang X-Y, Terzolo HR (2007) Use of IgY antibodies in human and veterinary medicine. In: Huopalahti R, López-Fandiño R, Anton M, Schade R (eds) Bioactive egg compounds. Springer, Berlin, pp 213–222

Schaerlinger B, Frippiat J-P (2008) IgX antibodies in the urodele amphibian *Ambystoma mexicanum*. Dev Comp Immunol 32:908–915

Schat BKK (2013) Avian immunology. Academic. Hardcover ISBN: 9780123969651

Schusser B, Yi H, Collarini E, Izquierdo S, Harriman W, Etches R, Leighton P (2013) Harnessing gene conversion in chicken B cells to create a human antibody sequence repertoire. PLoS One 8: e80108

Sheng L, He Z, Chen J, Liu Y, Ma M, Cai Z (2017) The impact of N-glycosylation on conformation and stability of immunoglobulin Y from egg yolk. Int J Biol Macromol 96:129–136

Sheng L, He Z, Liu Y, Ma M, Cai Z (2018) Mass spectrometry characterization for N-glycosylation of immunoglobulin Y from hen egg yolk. Int J Biol Macromol 108:277–283

Shimizu M, Nagashima H, Sano K, Hashimoto K, Ozeki M, Tsuda K, Hatta H (1992) Molecular stability of chicken and rabbit immunoglobulin G. Biosci Biotechnol Biochem 56(2):270–274

Sun Y, Huang T, Hammarstrom L, Zhao Y (2020) The immunoglobulins: new insights, implications, and applications. Annu Rev Anim Biosci 8:145–169

Svendsen BL, Crowley A, Stodulski G, Hau J (1996) Antibody production in rabbits and chickens immunized with human IgG. A comparison of titre and avidity development in rabbit serum, chicken serum and egg yolk using three different adjuvants. J Immunol Methods 191 (2):113–120

Walport KMMPTM (2011) Janeway's immunobiology. ISBN 13: 9780815342434

Wang T, Sun Y, Shao W, Cheng G, Li L, Cao Z, Yang Z, Zou H, Zhang W, Han B, Hu Y, Ren L, Hu X, Guo Y, Fei J, Hammarström L, Li N, Zhao Y (2012) Evidence of IgY subclass diversification in snakes: evolutionary implications. J Immunol 189(7):3557–3565

Wei Z, Wu Q, Ren L, Hu X, Guo Y, Warr GW, Hammarström L, Li N, Zhao Y (2009) Expression of IgM, IgD, and IgY in a reptile, *Anolis carolinensis*. J Immunol 183(6):3858–3864

Woolley JA, Landon J (1995) Comparison of antibody production to human interleukin-6 (IL-6) by sheep and chickens. J Immunol Methods 178(2):253–265

Zhang X, Chen H, Tian Z, Chen S, Schade R (2010) Chicken monoclonal IgY antibody: a novel antibody development strategy. Avian Biol Res 3(3):97–106

Zhang X, Calvert RA, Sutton BJ, Dore KA (2017) IgY: a key isotype in antibody evolution. Biol Rev 92(4):2144–2156

Zhao Y, Rabbani H, Shimizu A, Hammarström L (2000) Mapping of the chicken immunoglobulin heavy-chain constant region gene locus reveals an inverted alpha gene upstream of a condensed upsilon gene. Immunology 101(3):348–353

6 IgY Cell Receptors and Immunity Transfer

Xiaoying Zhang, Shikun Ge, and Patricia M. Morgan

Abstract

Receptors play important roles in the recognition, transportation and mediation of immunoglobulins. The Fc receptors on the cell surface bind to the constant region (Fc) of the Ig and modulate the immune response. In avian species, three receptors for IgY have been discovered: IgY's Fc receptor (FcRY), Chicken like Ig-like receptor AB1 (CHIR-AB1), and *Gallus gallus* receptor (ggFcR). FcRY, despite being considered as a functional homologue of the neonatal Fc receptor (FcRn) in mammals belongs to a different protein structural family (mannose family). CHIR-AB1 functions as a classical FcR expressed on chicken B cells, macrophages, monocytes and NK cells. ggFcR is found highly expressed in peripheral blood mononuclear cells. The transfer of IgY from the hen to the embryo is a two-step receptor mediated process: the transfer from the maternal bloodstream to the yolk sac; and secondly from the yolk sac to the embryo. FcRY mainly functions in the second process, while the receptor involved in the initial step, as well as the functions of CHIR-AB1 and ggFcR, remain largely unknown.

X. Zhang (✉)
College of Biological Science and Engineering, Shaanxi University of Technology, Hanzhong, China
e-mail: zhxying@snut.edu.cn; xzhang67@uoguelph.ca; zhang@bio.uminho.pt

S. Ge
College of Biological Science and Engineering, Shaanxi University of Technology, Hanzhong, China

CBMA, University of Minho, Braga, Portugal

P. M. Morgan
School of Natural Sciences, National University of Ireland Galway, Galway, Ireland
e-mail: pat.morgan@nuigalway.ie

X.-Y. Zhang et al. (eds.), *IgY-Technology: Production and Application of Egg Yolk Antibodies*, https://doi.org/10.1007/978-3-030-72688-1_6

A deeper understanding of IgY's receptors is crucial to a full understanding IgY's transportation, particularly in the unique process of "hen-egg-offspring", which is quite different to mammals. Knowledge of the receptors can also clarify the evolution of immunoglobulins and is important in the development of transgenic hens which could express complex human proteins.

Keywords

IgY transfer · IgY's neonatal Fc receptor (FcRY) · Chicken like Ig-like receptor AB1 (CHIR-AB1) · Gallus gallus receptor (ggFcR) · Immunoglobulin mediation

6.1 Constant Region of Immunoglobulins (Ig)

The structures and the generation of diversity in the antigen binding regions of mammalian immunoglobulins and IgY have been described in Chap. 3. The constant region of an antibody is termed the Fragment crystallizable (Fc) region can be recognized by various immune cells and regulate the immune response. These signalling pathways have been recently reviewed in detail (Koenderman 2019). In summary, the Fc region of an immunoglobulin specifies the different sub-classes and plays a role in mammalian complement activation (Chap. 3) and importantly binds to specific Fc receptors which are the focus of this chapter. In order to exert control of the immune response, it is necessary that the Fc receptors are tightly regulated; receptors therefore function as activating or inhibitory of cellular proliferation (Koenderman 2019).

6.2 Transfer of Maternal Immunoglobulins

Transportation of maternal immunoglobulins to offspring plays a pivotal role in neonatal humoral immunity (Brambell 1958). In mammals, delivery of maternal IgG can occur at both the pre- (placental or yolk sac barrier) and post-natal (milk/colostrum) stages (Ghetie and Ward 2000; Ward and Ober 2009) . In fact, Brambell (1958) was the first to propose that there was a receptor system which bound the Fc of Ig and that it was involved in the transfer of maternal antibodies to the foetus (Brambell 1958). The identity of the neonatal Fc receptor (FcRn) remained unknown for about 30 years (Pyzik et al. 2019). This receptor functions by bidirectionally transporting ligands across cellular membranes and is structurally related to MHC-I molecules (Pyzik et al. 2019).

6.3 Transfer of IgY

In avian species, the maternal immunoglobulin is transferred to the egg. The reproductive system of a hen, described in detailed in Chap. 2, is composed of two parts: the ovary and the oviduct. With the onset of sexual maturity in the chicken, follicles of different sizes develop by an accumulation of lipids and give rise to one pre-ovulatory follicle per day (Woods and Johnson 2005; Johnson 2015). The yolk contains approximately 33% lipids, 17% protein (predominantly Ig), other trace components (2% made up of carbohydrates, vitamins and organic compounds) and the remainder is water; formation of the yolk is completed 24 h before ovulation (Schade et al. 2001). After ovulation the follicle moves into the oviduct where the egg goes through its stages of development; formation of the vitelline membrane, albumen formation, formation of egg shell membranes and finally before laying the egg shell is completed the cuticle which is a protective layer is deposited (Chap. 2). Birds have a unique IgY transportation system, which is quite different to IgG's, and involves two major steps (Fig. 6.1 (West et al. 2004)).

6.3.1 Transfer from the Maternal Circulation to the Yolk of the Oocytes

Immunoglobulin (Ig)-secreting B cells of chick origin have been detected in circulation after 6 days post-hatch, meaning that during the first days of the post-hatching period, humoral immunity is totally dependent on the maternal transfer of Igs (Agrawal et al. 2016). With the onset of sexual maturity, and the growth of the ovarian follicles, the developing oocyte with yolk is concentrically surrounded by several layers of supporting tissue, including the theca layer, acellular layer and epithelial layer (Murai 2013). The transfer of IgY from the bloodstream to the follicle increases with follicular size, up to 2–3 days prior to ovulation; the increased

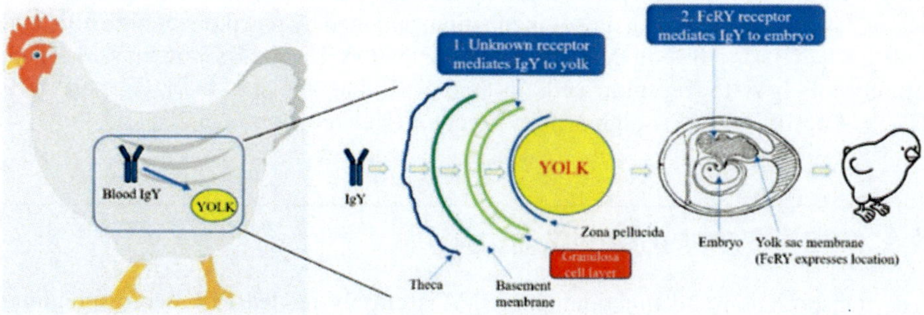

Fig. 6.1 Two-step transfer of maternal IgY to embryo. Notes: in the first step, involving transfer from the maternal circulation to the yolk of the oocyte, the transport receptor(s) are unknown but may be present in the granulosa cells. In the second step, involving transfer from the embryonic yolk sac to the embryonic blood stream the FcRY receptor is present in the yolk sac membrane. (Source: XYZ-IgY lab)

transfer rates are due modification of the follicular epithelium which is thinner in large follicles (Patterson et al. 1962). Most of the yolk components are derived from blood plasma yolk precursors that are mainly synthesized in the liver. Of the three avian Ig classes (IgA, IgM and IgY), only IgY is actively transferred into the egg yolk, suggesting the existence of a selective IgY transport system in the maternal ovary, which endocytoses IgY into the developing oocyte. However, the molecular basis of IgY transfer by oocyte membrane receptors is still largely unknown (Fig. 6.1).

6.3.2 Transfer from the Embryonic Yolk Sac to the Embryonic Blood Stream

The yolk sac starts to develop around day 2 of embryonic growth as a well vascularised membrane surrounding the egg yolk; however, the transfer of yolk nutrients is limited until embryonic day 11. After day 11, the yolk sac totally surrounds the yolk and the transfer of nutrients is accelerated and there is rapid growth of the embryo (Yadgary et al. 2014). Transfer of IgY from the embryonic yolk sac to the embryonic blood stream relies on the mediation of special IgY-Fc receptors called FcRY (Fig. 6.1), which are functionally equivalent to the mammalian FcRn. Unexpectedly, FcRY is not a structural homologue of MHC-I molecules like FcRn. It is a homologue of the mammalian phospholipase A2 receptor, a member of the mannose receptor family (West et al. 2004). FcRY is expressed in multiple tissues including the ovary and plays a critical role in IgY homeostasis like mammalian FcRn; but FcRY does not seem to be involved in IgY transportation from maternal circulation to the yolk of the oocytes. This speculation is based on the observation of human IgG uptake into avian egg yolk by uptake receptor (Morrison et al. 2002). Therefore, it has been suggested that there is a distinct receptor involved in IgY transport from maternal circulation to the yolk of the oocytes. This association of IgY-Fc and yolk sac receptors is first mediated by low affinity sac receptors with a K_D of 3.4×10^{-7} on the 8th day of incubation, and then by receptors of high affinity, with a K_D of 3.4×10^{-8} on the 18th day of incubation (He and Bjorkman 2011). The affinity of IgY to receptor cells is more similar to that of IgG than IgE ($K_D = 1 \times 10^{-10}$) for its high affinity receptor (He and Bjorkman 2011).

6.4 IgY Receptor Family

The transportation and functioning of IgY is closely related to its corresponding receptors. IgY has three characterized receptors in chicken; yolk sac IgY receptor (FcRY), Chicken like Ig-like receptor AB1 (CHIR-AB1) and *Gallus gallus* receptor (ggFcR). They bind to IgY-Fc at sites that are structural homologous to mammalian counterparts FcRn/IgG, FcαRI/IgA, as well as FcεRI/IgE and FcγR/IgG (Zhang et al. 2017).

Table 6.1 Major receptors for IgY

Receptor	FcRY	CHIR-AB1	ggFcR
Cellular effect	Activating	Activating and inhibitory	Activating
Affinity (L/mol)	pH-dependent	High	Low
Receptor-ligand binding ratio	2:1 (4:1)[a]	2:1	1:1
Binding site on ligand	Cυ3 and Cυ4 junction	Cυ3Cυ4	Cυ2-Cυ4
Confirmed function	Mediates IgY transportation from the embryonic yolk sac to the embryonic blood stream; protects IgY from catabolism	Activates calcium release in chicken B cell lines; activates IgY aggregation	Unknown

Note: [a]Predicted, still unclear or unproven

As stated earlier, FcRY located on the avian yolk sac membrane, is responsible for IgY transportation from the egg yolk to the embryonic bloodstream; this occurs by interacting with the Cυ3/Cυ4 interface of IgY-Fc (West et al. 2004; Murai 2013). In common with PLA2R and other MR family members, FcRY is a type I membrane glycoprotein with a large ectodomain comprising 10 domains of known structure: a N-terminal cysteine-rich (CysR) domain, a fibronectin type II (FNII) repeat, and eight C-type lectin-like domains (CTLDs) (West et al. 2004). A recombinant form of the FcRY ectodomain was shown to bind IgY and the IgY-Fc fragment of IgY with high affinity at acidic, but not basic, pH, and the full-length FcRY expressed in polarized mammalian epithelial cells is functional in endocytosis, bidirectional transcytosis, and recycling of chicken FcY/IgY, analogous to the functions of FcRn in epithelial and endothelial cells (Table 6.1; (Zhang et al. 2017)). IgY might bind to one surface of the epithelial cells via FcRY and the FcRY-IgY complexes are taken up by receptor-mediated endocytosis (He and Bjorkman 2011). The complexes might be then transcytosed across the cells and delivered via exocytosis at the basolateral surface of the cells (He and Bjorkman 2011). FcRY-IgY dissociation occurs at this site owing to the instability of the pH-dependent interaction (West et al. 2004). This model has formed the basis of our current understanding of the FcRY-IgY transfer mechanism (Fig. 6.2).

CHIRs correspond to the leukocyte receptor family (LRC) and include a large family of highly polymorphic genes predicted to encode activating (CHIR-A), inhibitory (CHIR-B), or bifunctional (CHIR-AB) receptors (Viertlboeck et al. 2004, 2005). CHIR-A is an activating receptor which contains a histidine residue in its trans-membrane segment and relatively short cytoplasmic tail and activates intracellular tyrosine kinases. CHIR-B on the other hand, has a non-polar transmembrane domain, and a relatively long cytoplasmic tail containing two immune-receptor tyrosine based inhibitory motifs. CHIRs are present on a variety of immune

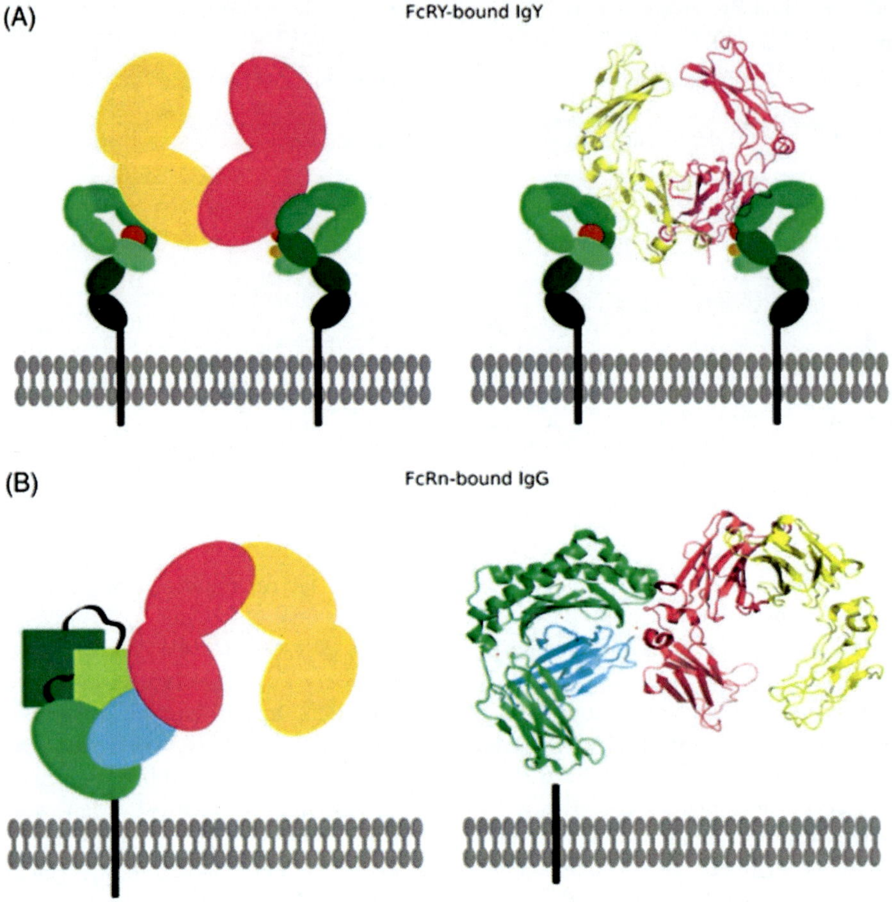

Fig. 6.2 Comparison of the binding of IgY to FcRY and the binding of IgG to FcRn. Notes: Schematic images on the left correspond to the structures on the right. (**a**) Fcυ3-4 (yellow and purple) interacting with the tandem array of domains in FcRY (from N-terminal cysteine-rich domain in red, fibronectin type II domain in tan, and tandem array of eight C-type lectin-like domains in light to dark green; from cryo-EM analysis; (He and Bjorkman 2011). (**b**) IgG-Fc (yellow and purple) bound to FcRn (green domains with β2-microglobulin in light blue). Reproduced with permission from Zhang et al. (2017)

cells, including those of the innate system, and may have a role in innate immune recognition (Viertlboeck et al. 2009) as well as regulation of adaptive immunity (Guselnikov et al. 2009). One of these receptors, CHIR-AB1, functions as a classical FcR expressed on chicken B cells, macrophages, monocytes and NK cells (Viertlboeck et al. 2007). The critical IgY residues interacting with CHIR-AB1 overlap partly with the residues that are critical for IgY transportation into egg yolk, however, the functions of the CHIR family, including CHIR-AB1, remain largely obscure. Although the potential relevance of CHIR family members on IgY transport into egg yolk remains unknown, the isolation of IgY receptor from the inner layer of ovarian follicles and a thorough binding study of isolated IgY receptor are necessary to identify a true receptor for maternal IgY transfer in the future (Murai 2013).

Chicken FcRs have been demonstrated on macrophages decades ago, and FcRs have multifaceted roles in the immune system; however, a novel FcR gene was identified in the chicken, designated *Gallus gallus* FcR (ggFcR) (accession no. FM200428). ggFcR was found to be highly expressed in peripheral blood mononuclear cells (PBMC), spleen cells, macrophages and thrombocytes (Viertlboeck et al. 2009). It consists of four extracellular C2-set Ig domains, where each Ig domain is encoded by a separate exon, followed by a transmembrane region containing arginine as a positively charged amino acid and a short cytoplasmic tail (Murai 2013). ggFcR is a potential activating receptor, which interacts with the common γ-chain, and selectively binds to chicken IgY instead of IgM or IgA (Viertlboeck et al. 2009). ggFcR mainly interacts with the Cυ2 domain of IgY-Fc (Schreiner et al. 2012), however, the Cυ2 domain is not essential to maintain IgY-Fc transportation into egg yolks (Bae et al. 2010), suggesting that the likelihood of ggFcR involvement in IgY transport from the maternal circulation to the yolk of the oocytes is low (Murai 2013).

6.5 Prospective

There are only limited studies on the major receptors for IgY, which have focused on the receptors' structure and the binding mode to IgY, while their functions remain largely unknown. Furthermore, there are a series of fundamental issues associated with IgY receptors which need to be revealed, for example, is there an evolutionary link among FcRY, CHIR-AB1 and ggFcR and IgG/IgA/IgE related receptors? In addition to transporting IgY in chickens, do they participate in broader physiological activities?

Knowledge on IgY's receptors is of particular value and is fundamental for the idea of the generation of transgenic hens which would express complex human proteins and transfer these to the eggs. The production of human proteins and particularly modified antibodies with biomedical value, depends on an in-depth understanding of receptor-mediated IgY transfer and enrichment in the egg yolk. Despite the scientific communities realising the value of such a system, the practice is still very limited, especially compared to expression of proteins in milk. Definitely, IgY receptors could be used as major targets for such avian manipulation, although we are still lacking relevant immunological information.

References

Agrawal R, Hirpurkar SD, Sannat C, Gupta AK (2016) Comparative study on immunoglobulin Y transfer from breeding hens to egg yolk and progeny chicks in different breeds of poultry. Vet World 9(4):425–431

Bae H-D, Honda H, Murota R, Kobayashi M, Horio F, Murai A (2010) Production of recombinant chicken IgY-Fc and evaluation of its transport ability into avian egg yolks. J Poultry Sci 47:256–261

Brambell F (1958) The passive immunity of the young mammal. Biol Rev 33:488–531

Ghetie V, Ward ES (2000) Multiple roles for the major histocompatibility complex class I- related receptor FcRn. Annu Rev Immunol 18:739–766

Guselnikov SV, Ramanayake T, Robert J, Taranin AV (2009) Diversity of the FcR- and KIR-related genes in an amphibian Xenopus. Front Biosci 14:130–140

He Y, Bjorkman PJ (2011) Structure of FcRY, an avian immunoglobulin receptor related to mammalian mannose receptors, and its complex with IgY. Proc Natl Acad Sci USA 108 (30):12431–12436

Johnson AL (2015) Reproduction in the female. In: Scanes CG (ed) Sturkie's avian physiology, 6th edn. Academic, San Diego, pp 635–665

Koenderman L (2019) Inside-out control of Fc-receptors. Front Immunol 10:544

Morrison SL, Mohammed MS, Wims LA, Trinh R, Etches R (2002) Sequences in antibody molecules important for receptor-mediated transport into the chicken egg yolk. Mol Immunol 38(8):619–625

Murai A (2013) Maternal transfer of immunoglobulins into egg yolks of birds. J Poult Sci 50:185–193

Patterson R, Youngner JS, Weigle WO, Dixon FJ (1962) Antibody production and transfer to egg yolk in chickens. J Immunol 89:272–278

Pyzik M, Sand KMK, Hubbard JJ, Andersen JT, Sandlie I, Blumberg RS (2019) The neonatal Fc receptor (FcRn): a misnomer? Front Immunol 10:1540

Schade R, Schwarzkopf C, Behn I, Hommel U, Hlinak A, Staak C, Erhard M (2001) Isolation of IgY from yolk. In: Chicken egg yolk antibodies, production and application, pp 65–107

Schreiner B, Viertlboeck BC, Göbel TW (2012) A striking example of convergent evolution observed for the ggFcR:IgY interaction closely resembling that of mammalian FcR:IgG. Dev Comp Immunol 36(3):566–571

Viertlboeck BC, Crooijmans RP, Groenen MA, Göbel TW (2004) Chicken Ig-like receptor B2, a member of a multigene family, is mainly expressed on B lymphocytes, recruits both Src homology 2 domain containing protein tyrosine phosphatase (SHP)-1 and SHP-2, and inhibits proliferation. J Immunol 173(12):7385–7393

Viertlboeck BC, Habermann FA, Schmitt R, Groenen MA, Du Pasquier L, Göbel TW (2005) The chicken leukocyte receptor complex: a highly diverse multigene family encoding at least six structurally distinct receptor types. J Immunol 175(1):385–393

Viertlboeck BC, Schweinsberg S, Hanczaruk MA, Schmitt R, Du Pasquier L, Herberg FW, Göbel TW (2007) The chicken leukocyte receptor complex encodes a primordial, activating, high-affinity IgY Fc receptor. Proc Natl Acad Sci USA 104(28):11718–11723

Viertlboeck BC, Schmitt R, Hanczaruk MA, Crooijmans RP, Groenen MA, Göbel TW (2009) A novel activating chicken IgY FcR is related to leukocyte receptor complex (LRC) genes but is located on a chromosomal region distinct from the LRC and FcR gene clusters. J Immunol 182 (3):1533–1540

Ward ES, Ober RJ (2009) Chapter 4: multitasking by exploitation of intracellular transport functions the many faces of FcRn. Adv Immunol 103:77–115

West AP Jr, Herr AB, Bjorkman PJ (2004) The chicken yolk sac IgY receptor, a functional equivalent of the mammalian MHC-related Fc receptor, is a phospholipase A2 receptor homolog. Immunity 20(5):601–610

Woods DC, Johnson AL (2005) Regulation of follicle-stimulating hormone-receptor messenger RNA in hen granulosa cells relative to follicle selection. Biol Reprod 72(3):643–650

Yadgary L, Wong EA, Uni Z (2014) Temporal transcriptome analysis of the chicken embryo yolk sac. BMC Genomics 15(1):690

Zhang X, Calvert RA, Sutton BJ, Doré KA (2017) IgY: a key isotype in antibody evolution. Biol Rev Camb Philos Soc 92(4):2144–2156

Biorhythms of Hens

7

Long Xu, Ben Li, Patricia M. Morgan, and Xiaoying Zhang

Abstract

The study of biorhythms, which are internal responses to external stimuli, is knows as chronobiology. Biorhythms, and their impact on physiological processes, exist across a wide range of different species. Hens exhibit strong biorhythms and knowledge of these biorhythms are necessary for understanding the range of IgY levels reported in the blood and yolk of laying hens. Infradian (60-day in serum and yolk) and circaseptan (7 day in yolk, 4 day in serum) rhythms were observed in both yolk IgY and serum IgY concentrations. Ultradian rhythms were observed in serum IgY concentrations, which reached a peak in the morning, decreased to a minimum during the daytime and increased again at night. Total and specific IgY increases with the age of a hen, leading to an increased yield per hen, despite decreased egg productivity over time. There is a correlation between concentrations of IgY in the serum and egg with some variation been reported for different breeds of hen. It appears that the

L. Xu
CMBA, University of Minho, Braga, Portugal

College of Biological Science and Engineering, Shaanxi University of Technology, Hanzhong, China

B. Li · X. Zhang (✉)
College of Biological Science and Engineering, Shaanxi University of Technology, Hanzhong, China
e-mail: zhxying@snut.edu.cn; xzhang67@uoguelph.ca; zhang@bio.uminho.pt

P. M. Morgan
Emerita Lecturer, School of Natural Sciences, National University of Ireland Galway, Galway, Ireland
e-mail: pat.morgan@nuigalway.ie

X.-Y. Zhang et al. (eds.), *IgY-Technology: Production and Application of Egg Yolk Antibodies*, https://doi.org/10.1007/978-3-030-72688-1_7

immunisation program, used to induce a specific antibody response, does not affect biorhythms in hens. However, the genetic basis and other factors influencing biorhythms in hens still remain largely unclear, and it is interesting to explore the mechanisms underlying the different rhythms of IgY in serum and yolk.

Keywords

Hen · Biorhythms · Infradian · Circaseptan · Ultradian · IgY concentration

7.1 Introduction

Interest in biological rhythms and their effect on physiological processes has been growing for many years. Three basic rhythms related to a day exist: ultradian when the period is shorter than 24 h and which may last from seconds to several hours; circadian rhythms of approximately 24 h; and infradian rhythms which are repeated every few days, months or even once per year (He et al. 2014).

Circadian rhythms were among those first demonstrated and are hypothesized to have evolved in aerobic organisms to anticipate changes in environmental oxygen levels and in mammals it has been estimated that approximately 10% of the genome is under circadian control (Scheiermann et al. 2013). Light patterns are the major external driver of an organism's circadian rhythm as it sets the rest-activity cycle and is linked indirectly to the timing of food intake which also influences the rhythm (Scheiermann et al. 2013). The key immune parameters in the blood which exhibit circadian rhythms are circulating haematopoietic cells, as well as the levels of hormones and cytokines (Haus and Smolensky 1999). Also, 7-day rhythms, known as circaseptan rhythms, are displayed in diverse species from sea algae, to plants, insects, fish, birds and mammals (Reinberg et al. 2016). Circaseptan rhythms have been noted in connection with cardiac mortality, blood pressure, excretion of steroids, as well as in the immune system (Cornelissen 1993; Labrecque et al. 1995; Halberg et al. 2003). In the course of many studies, the modifications of biorhythms due to ageing have also been considered (Yamazaki et al. 2002). Thus, knowledge of chronobiology impacts not only on basic research but also pharmacological, clinical, and therapeutic problems. In this chapter, the impact of rhythms on the levels of IgY in the blood and egg yolk will be reviewed as well as informing future research.

7.2 Observed Rhythms in Hens

IgY as a component of the immune system, and thus in common with all lymphoid organs and immune cell types studied to date, is controlled by the circadian clock. IgY is secreted in the serum and is transported into the yolk of the egg (Chap. 6) with a delay of a number of days. Additional rhythms appear to regulate IgY levels of the serum and yolk. A circadian rhythm was observed in hen serum IgY; with a peak in the morning, followed by decreasing values during the day which then increased overnight to reach the morning peak value (Fig. 7.1) (He et al. 2014). The total IgY concentration varied from 31.33 mg/egg to a peak level of 57.05 mg/egg (He et al. 2014). IgY is selectively transferred from serum to the yolk (Loeken and Roth 1983) and the concentration of total IgY in the yolk is directly related to the circulating IgY level in hen's blood (Hamal et al. 2006). There appears to be a 4 day rhythm (circaquattran) in serum IgY and a circaseptan rhythm in yolk IgY concentrations (Schade et al. 2001; He et al. 2014). The 7 day rhythm in the yolk may reflect both the 4 day rhythm in the serum and the delay in the appearance of IgY in the yolk as it is transported from the serum.

As stated earlier, light patterns are the major regulator of circadian rhythms; the pineal gland responds to the state of the light-dark cycle with the release of melatonin rhythmically. Melatonin levels increase during the night and serum IgY concentrations may be correlated with this increase (Skwarlo-Sonta et al. 2003). However, it is still necessary to explain the differences between the rhythms in the serum and yolk. Pauly interpreted an observed 60-day rhythm as an approximate multiple of seven, a finding that was also reported for several components of the immune system (Fig. 7.2) (He et al. 2014). The biological significance of the 7 day rhythm of IgY in the yolk is not fully understood. The export of IgY from the serum to the egg yolk is to ensure that each egg has sufficient IgY to protect the newly hatched chick before its own immune system develops. Very limited variations have been reported in the amount of IgY in individual eggs (Carlander et al. 2001). The total amount of IgY present in serum of the hatched chick has been reported to be about 2 mg; the fate of the 90% of the yolk IgY is uncertain but it may have a nutritional role (Kowalczyk et al. 1985). Given the large excess of IgY in the egg, it would appear that the biological rhythm would not pose risk to the developing chick. The exact mechanism for the rhythmic control of synthetic activity of B lymphocytes which impacts on the biorhythm of IgY in the serum and yolk could be further explored.

There are various reports on the delay between the appearance of IgY in serum and subsequently in the yolk: a delay of 5–6 days (Patterson et al. 1962) or 3–4 days (Woolley and Landon 1995; He et al. 2014). Factors affecting this delay, may be the number of large follicles being produced (Sect. 2.6.4), the serum IgY rhythm and the half-life of the circulating IgY. The half-life of the circulating IgY of a hen is approximately 36–65 h (Patterson et al. 1962) and thus the hen loses 30%–40% of the IgY due to normal turnover (Klasing 1998). The concentration of IgY in the yolk is constant throughout the maturation of the oocyte (8 mg/mL) and the rate of uptake is at a maximum during the last 3 days of follicular growth before ovulation and can be as high as 45 mg/day (Kowalczyk et al. 1985).

The correlation between serum and yolk levels of IgY has been extensively studied (Schade et al. 2005). The reported IgY concentrations in the yolk of individual birds range from 3 to 7 mg/mL (Carlander et al. 2001) and one study reported that IgY concentrations in yolk were 1.23 times higher than that in serum (Woolley and Landon 1995). There are varying reports on the differences of yolk IgY concentration in different breeds and the interplay of a number of factors, such

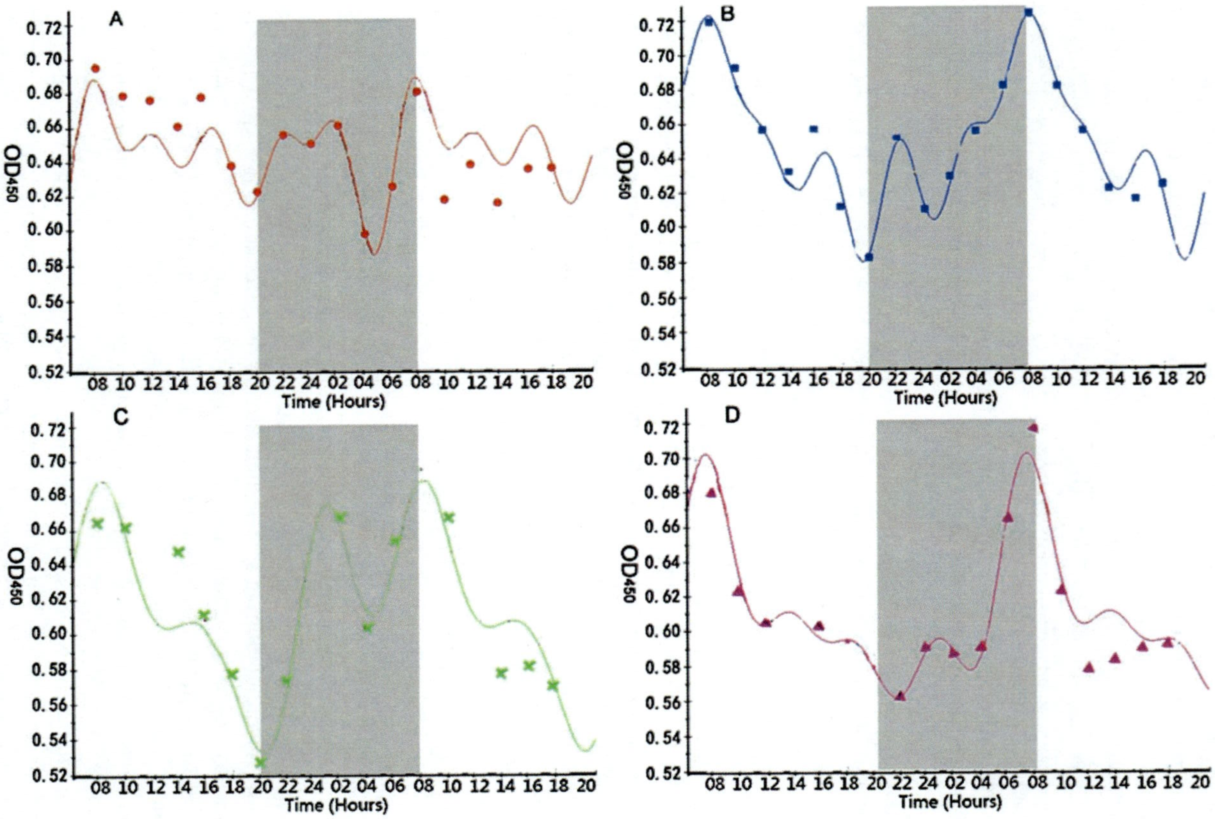

Fig. 7.1 Circadian rhythm of IgY in serum. Notes: A, B, C, D represented the circadian rhythm of IgY in a single hen for four consecutive 36 h, Gray zone = 12-h dark span. OD_{450} value represented IgY determined by ELISA at 450 nm absorbance. Figure with publisher's permission from He et al. (2014)

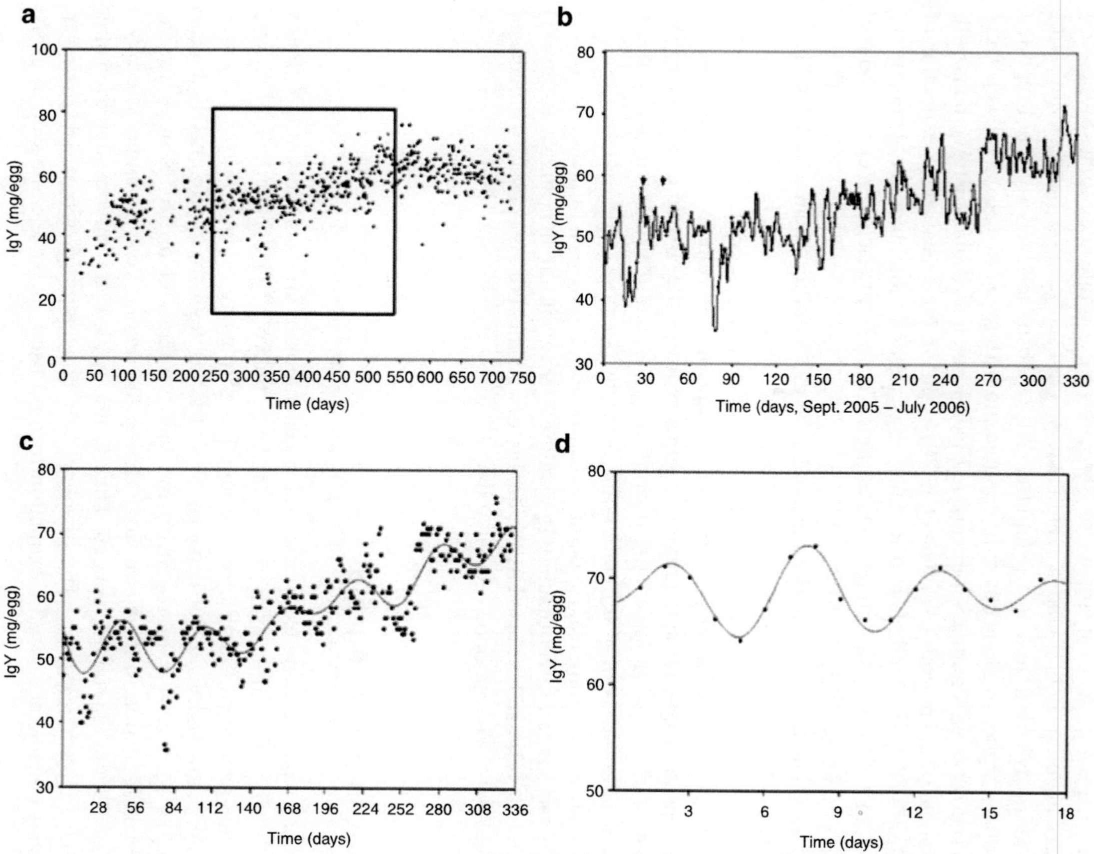

Fig. 7.2 IgY concentrations and biological rhythms as observed in hen eggs over various time periods. Notes: (**a**) Total IgY concentration in the eggs over a 2-year period. (**b**) Time-dependent oscillation of the daily mean IgY of hen eggs. The original data were processed by means of the moving average. The period

as the genetic lines or breeds of chickens, purification methods and measurement techniques, biological rhythms of serum and yolk IgY, may account for these differences. For instance, there was a reported significant difference between different breeds (Carlander et al. 2003) but mean levels of yolk IgY only ranged from 1.75 mg/mL for Rhode Island Red, 2.0 for SLU-1329 and 2.2 for Single Comb White Leghorn. A comparison of three breeds, White Leghorn, Silkie and Dongxiang blue-shell, found a significant positive correlation between levels of total serum IgY and yolk IgY (Sun et al. 2013); absolute levels were not reported but appear higher in Dongxiang. In contrast, no difference in yolk IgY levels were observed between Vanraja, Gramapriya, Black Rock and Kalinga Brown hens with an average of 2.5 mg/mL (Agrawal et al. 2016). In a more recent study, extraction of hen yolk IgY from Rhode Island Red hen eggs yielded more than double quantities of yolk IgY (8.37 mg/mL) compared to Single Comb White Leghorn eggs (3.66 mg/mL (Amro et al. 2017).

At this juncture, it is understandable that the significant circadian rhythm is superimposed by an ultradian rhythm. The concentration of yolk IgY level increases in both immunized and non-immunized hens (Pauly et al. 2009). There is a significant circaseptan rhythm in yolk IgY and circaquattran rhythm in serum IgY and in addition a noticeable ultradian rhythm is superimposing the circadian rhythm of IgY. On the whole, these biological oscillations of IgY are not really dependent upon the immunization protocol of different target antigens (He et al. 2014).

The observed biological rhythmic variation of IgY may also influence the observation vaccination effect (namely IgY antibody level) in the hen, higher (or lower) antibody levels can be caused by the quality of vaccine, but also possibly influenced by the rhythms of antibody level in a day.

7.3 Impact of Ageing

Hens can be used for up to 2 years for the generation of high amounts of IgY (Pauly et al. 2009). After immunisation of young hens, the yolk IgY concentration followed clear age-dependent kinetics: with around 38–45 mg/egg in the first 10 weeks after immunization and values of between 53 and 60 mg/egg for approximately 1 year; during a second year of the study the values increased up to 68 mg/egg. The observed reduction in laying capacity during the second year is compensated by a greater amount of total IgY per egg. Thus, antibody recovery remains high even after prolonged immunizations. The absolute IgY concentration recovered, however, depends considerably on the extraction method used, as well as the weight of the

Fig. 7.2 (continued) corresponds to the period of maximum egg-laying of the observed hen (weeks 36–82). (**c**) As in (**b**), but the graph was analysed using the Chronos-Fit program (significant period length around 60 days, 73% rhythm). (**d**) As in (**b**), but data were analysed in a more detailed period of time as indicated by arrowheads (significant period length around 7 days, 96% rhythm). Figure with the publisher's permission from Pauly et al. (2009)

egg, the breed of the hen, and the method used for quantification of the IgY (Chap. 11).

References

Agrawal R, Hirpurkar SD, Sannat C, Gupta AK (2016) Comparative study on immunoglobulin Y transfer from breeding hens to egg yolk and progeny chicks in different breeds of poultry. Vet World 9(4):425–431

Amro WA, Al-Qaisi W, Al-Razem F (2017) Production and purification of IgY antibodies from chicken egg yolk. J Genetic Eng Biotechnol 16(1):99–103

Carlander D, Wilhelmson M, Larsson A (2001) Limited day to day variation of IgY levels in eggs from individual laying hens. Food Agric Immunol 13(2):87–92

Carlander D, Wilhelmson M, Larsson A (2003) Immunoglobulin Y levels in egg yolk from three chicken genotypes. Food Agric Immunol 15(1):35–40

Cornelissen G (1993) Beyond circadian chronorisk: worldwide circaseptan-circasemiseptan patterns of myocardial infarctions, other vascular events, and emergencies. Chronobiologia 20 (1–2):87–115

Halberg F, Cornélissen G, Wang Z, Wan C, Ulmer W, Katinas G, Singh R, Singh RK, Singh RK, Gupta BD (2003) Chronomics: circadian and circaseptan timing of radiotherapy, drugs, calories, perhaps nutriceuticals and beyond. J Exp Ther Oncol 3(5):223–260

Hamal KR, Burgess SC, Pevzner IY, Erf GF (2006) Maternal antibody transfer from dams to their egg yolks, egg whites, and chicks in meat lines of chickens. Poult Sci 85(8):1364–1372

Haus E, Smolensky MH (1999) Biologic rhythms in the immune system. Chronobiol Int 16 (5):581–622

He JX, Thirumalai D, Schade R, Zhang X-Y (2014) Chronobiological studies of chicken IgY: monitoring of infradian, circadian and ultradian rhythms of IgY in blood and yolk of chickens. Vet Immunol Immunopathol 160(3–4):266–272

Klasing KC (1998) Nutritional modulation of resistance to infectious diseases. Poult Sci 77 (8):1119–1125

Kowalczyk K, Daiss J, Halpern JDJ, Roth TF (1985) Quantitation of maternal-fetal IgG transport in the chicken. Immunology 54(4):755–762

Labrecque G, Bureau JP, Reinberg AE (1995) Biological rhythms in the inflammatory response and in the effects of non-steroidal anti-inflammatory drugs. Pharmacol Ther 66(2):285–300

Loeken MR, Roth TF (1983) Analysis of maternal IgG subpopulations which are transported into the chicken oocyte. Immunology 49(1):21–28

Patterson R, Youngner JS, Weigle WO, Dixon FJ (1962) Antibody production and transfer to egg yolk in chickens. J Immunol 89(2):272–278

Pauly D, Dorner M, Zhang X, Hlinak A, Dorner B, Schade R (2009) Monitoring of laying capacity, immunoglobulin Y concentration, and antibody titer development in chickens immunized with ricin and botulinum toxins over a two-year period. Poult Sci 88(2):281–290

Reinberg AE, Dejardin L, Smolensky MH, Touitou Y (2016) Seven-day human biological rhythms: an expedition in search of their origin, synchronization, functional advantage, adaptive value and clinical relevance. Chronobiol Int 1–30

Schade R, Behn I, Erhard M, Hlinak A, Staak C (2001) Chicken egg yolk antibodies, production and application. Springer Nature. https://doi.org/10.1007/978-3-662-04488-9

Schade R, Calzado EG, Sarmiento R, Chacana PA, Porankiewicz-Asplund J, Terzolo HR (2005) Chicken egg yolk antibodies (IgY-technology): a review of progress in production and use in research and human and veterinary medicine. Altern Lab Anim 33(2):129–154

Scheiermann C, Kunisaki Y, Frenette PS (2013) Circadian control of the immune system. Nat Rev Immunol 13(3):190–198

Skwarlo-Sonta K, Majewski P, Markowska M, Oblap R, Olszanska B (2003) Bidirectional communication between the pineal gland and the immune system. Can J Physiol Pharmacol 81 (4):342–349

Sun H, Chen S, Cai X, Xu G, Qu L (2013) Correlation analysis of the total IgY level in hen serum, egg yolk and offspring serum. J Anim Sci Biotechnol 4(1):10

Woolley JA, Landon J (1995) Comparison of antibody production to human interleukin-6 (IL-6) by sheep and chickens. J Immunol Methods 178(2):253–265

Yamazaki S, Straume M, Tei H, Sakaki Y, Menaker M, Block GD (2002) Effects of aging on central and peripheral mammalian clocks. Proc Natl Acad Sci USA 99(16):10801–10806

Part II

Core Methods of IgY Technology

Check for updates

Keeping Laying Hens to Obtain Antibodies 8

Leonard M. Moran, Catherine D. Caulfield, Martin G. Murphy, and Patricia M. Morgan

Abstract

For the production of IgY, it is essential to maintain the laying hens in an environment which is conducive to egg production and which complies with the highest international animal welfare standards and legislation. The environment should allow the normal expression of patterns of behaviour and minimise stress. The legislative framework, as detailed in European Union Council directives, with regard to housing conditions for hens and the competence of staff responsible for their welfare is outlined. A particular focus is the sources of enrichment provided for the hens. Factors impacting IgY production are also discussed. An example of a conventional and a SPF housing system are described.

Keywords

Laying hens · Welfare · SPF housing system · IgY

8.1 General Introduction

The general physiology and anatomy of the domestic hen, their genetic selection, common breeds and the egg laying process are detailed in Chap. 2. There is an extensive range of reference material on the physiology and husbandry of laying

L. M. Moran · C. D. Caulfield · M. G. Murphy
Ovagen Group Ltd., Carrentrila, Ballina, Ireland
e-mail: leonard.moran@ovagen.com; Catherine.Caulfield@ovagen.com; martin.murphy@ovagen.com

P. M. Morgan (✉)
Emerita Lecturer, School of Natural Sciences, National University of Ireland Galway, Galway, Ireland
e-mail: pat.morgan@nuigalway.ie

X.-Y. Zhang et al. (eds.), *IgY-Technology: Production and Application of Egg Yolk Antibodies*, https://doi.org/10.1007/978-3-030-72688-1_8

hens, in both experimental and commercial settings, which provides an introduction to new researchers in the field (Hawkins et al. 2001; LayWel 2006; FASS 2010). In this chapter, we outline the central principles of husbandry, complying with relevant legislation, involved in keeping hens for the production of IgY antibodies. Domestic fowl are derived from the geographically separated subspecies of Red Jungle fowl (*Gallus gallus gallus*; *Gallus gallus spadiceus*, *Gallus gallus jaboullei*; *Gallus gallus murgha)* and a second species of Grey jungle fowl *Gallus sonneratii* (Rubin et al. 2010; Tixier-Boichard et al. 2012; Malomane et al. 2019). They retain much of the associated behaviour and biology of Jungle fowl (Hawkins et al. 2001). In the wild, birds spend up to75% of the day searching for seeds fruits, and insects. Nesting, perching, and scratching, pecking and dust-bathing in litter are the behaviours that are most important to domestic fowl (*Gallus gallus domesticus*; (FAWC 1997) It is important that birds should be housed in order to permit these natural behaviours, whenever possible. In the wild, the dominant social structure is of one male to about four females, however they may form greater groupings of around 20 birds in less confined situations (Hawkins et al. 2001) The optimal ratio of males to females in most egg-producing flocks is 1 male per 12 to 15 females (FASS 2010).

8.2 Legislation

Directive 2010/63/EU on the protection of animals used for scientific purposes, recognised to impose the world's strictest framework in the field, also governs the use of birds in research performed in Member States (2010/63/EU). An overarching principle is that the care and use of live animals for scientific purposes is governed by the internationally established principles of the Three Rs: Replacement, Reduction and Refinement (Russell and Burch 1959). The use of birds instead of mammals for the production of antibodies, is considered to be both a reduction and a refinement in animal use. It is a refinement in that antibody is extracted from the egg yolk and does not involve blood collection. It also results in the reduction in the number of animals needed since antibody production is much higher in chickens than laboratory animals (Schade et al. 2001).

The welfare of the birds is highly dependent on the quality and professional competence of all staff involved in their welfare. This principle is embedded in Article 23 of the EU Directive 2010/63/EU where it explicitly states that staff must be "adequately educated and trained" and that "they will be supervised in the performance of their tasks until they have demonstrated the requisite knowledge" (2010/63/EU). Authorisation requirements exist at three levels: the Establishment, the Project and each person responsible for care of the animals (LASA 2016). The holder of the Establishment authorisation ensures that there is a functional system of training and development in place, including supervision and assessment of competence. Direct responsibility for ensuring that a suitable level of supervision is provided for all holders of personal authorisations, lies with the Compliance officer. The personal authorisation holder is responsible for keeping records of supervision and competence. An expert working group has produced a document on an education and training framework to meet the requirements under the Directive (Framework 2014). The maintenance of biosecurity, defined here as the unintentional transfer of pathogens to animals and humans, is a further consideration (FASS 2010). Personal hygiene should be facilitated by the provision of accessible showering and washing facilities as well as frequent changes of clothing.

Various articles of the Directive 2010/63/EU concern the general physical design and function of the accommodation provided for animals used for scientific purposes. A number of extracts from Annex III illuminate these requirements.

- "All facilities shall be constructed so as to provide an environment which takes into account the physiological and ethological needs of the species kept in them. Facilities shall also be designed and managed to prevent access by unauthorised persons and the ingress or escape of animals." With regard to control of the environment, insulation, heating and ventilation shall ensure that air circulation, dust levels and gas concentrations are kept within non-harmful limits. "The temperature shall be measured and logged on a daily basis." "Controlled lighting shall be provided to satisfy the biological requirements of the animals and to provide a satisfactory working environment" and allow inspection of the animals.
- "Animals, except those which are naturally solitary, shall be socially housed in stable groups of compatible individuals." "All animals shall be provided with space of sufficient complexity to allow expression of a wide range of normal behaviour. They shall be given a degree of control and choice over their environment to reduce stress-induced behaviour. Establishments shall have appropriate enrichment techniques in place, to extend the range of activities available to the animals and increase their coping activities including physical exercise, foraging, manipulative and cognitive activities, as appropriate to the species." "The enrichment strategies in establishments shall be regularly reviewed and updated."

Specifically with regard to domestic fowl, the minimum standards with regard to housing are shown in Table 1 (Directive 2010/63/EU Table 8.1.). Member States may allow exemptions from the requirements of Article 33 for scientific, animal-welfare or animal-health reasons. "During agricultural research, when the aim of the project requires that the animals are kept under similar conditions to those under which commercial farm animals are kept, the keeping of the animals shall comply at least with the standards laid down in Directives 98/58/EC" and 1999/74/EC (98/58/EC 1998; 99/74/EC 1999).

"Where these minimum enclosure sizes cannot be provided for scientific reasons, the duration of the confinement shall be justified by the experimenter in consultation with veterinary staff" and the regulatory authority. "In such circumstances, birds can be housed in smaller enclosures containing appropriate enrichment and with a minimum floor area of 0.75 m^2."

Table 8.1 Minimum enclosure spaces and feed trough length for birds of increasing mass (Directive 2010/63/EU Table 8.1.)

Body mass (g)	Minimum enclosure size (m^2)	Minimum area per bird (m^2)	Minimum height (cm)	Minimum length of feed trough per bird (cm)	Date referred to in Article 33(2)
Up to 200	1.00	0.025	30	3	1 January, 2017
Over 200 to 300	1.00	0.03	30	3	
Over 300 to 600	1.00	0.05	40	7	
Over 600 to 1200	2.00	0.09	50	15	
Over 1200 to 1800	2.00	0.11	75	15	
Over 1800 to 2400	2.00	0.13	75	15	
Over 2400	2.00	0.21	75	15	

8.3 Climate, Air Quality, Light & Noise

The thermo-neutral zone for adult hens is from 12–24 °C (EFSA 2005). Hens cope with the effects of high temperature by panting which results in evaporative cooling. Therefore the relative humidity and temperature should be considered together. Adult birds should be maintained at a relative humidity of 30-70%. The bulk of cage dust consists of skin debris interspersed with food particles. Use of dust-free wood shavings will decrease the overall level of dust in the atmosphere. Ammonia concentrations rise with increasing temperatures and litter-moisture content and are normally in the range of 1-2 ppm. The health and productivity of hens may be impacted adversely by ammonia levels above –25 ppm (EFSA 2005).

There are a number of factors to be considered with regarding to lighting; including the period of light/darkness, light intensity and the light colour or wavelength (Hofmann et al. 2020). In the European Union, an uninterrupted period of darkness of 8 h for laying hens is mandatory (99/74/EC 1999). Keeping chickens with a least 6 hours of darkness could lead to a stronger response of the adaptive arm of immune response than those provided with longer light periods (Hofmann et al. 2020). The role of lighting on the initiation of lay is discussed in Chap. 2. For chickens up to 18 weeks old, 8-10 hours of light is suggested; the lighting period may be increased to 12–16 hours when the chickens start to lay. A normal laying rate is maintained by a light intensity of 20 to 30 lux. Chickens are sensitive to light at different wavelengths than humans and perceive light on the infrared and ultraviolet spectra (Hofmann et al. 2020). Serum IgY levels do not appear to be influenced by light colour (red, green or blue) although higher egg production was observed under monochromatic red light (Hassan et al. 2013). In a more recent study it has been demonstrated that the use of LED lighting that contains red light positively influences the productivity and welfare of chickens during the early lay period (Archer 2018).

Most birds have an audible range that includes 1 to 5 kHz (Hawkins et al. 2001). Birds are likely to be stressed by excessive noises, however, entering the housing system too quietly may also cause fright in the flock; hearing a staff member approach will reduce this stress to birds (Hawkins et al. 2001).

8.4 Enrichment and Nourishment

There are a number of reports which indicate that enrichment of the environment of hens has a positive impact on improving the immune system of hens under stress and that the autonomic responses to aversive events were also attenuated in hens housed in enriched cages (Matur et al. 2015; Ross et al. 2020).

8.4.1 Nest Box

The most important physical enrichment for laying hens is a nest box. Hens prepare to lay by searching for a nesting site, and then forming the substrate into a nest by pecking, treading and moulding it (FASS 2010). Laying hens prefer enclosed nest boxes and those with an appropriate substrate to allow for this nest-building behaviour (Hawkins et al. 2001). Before sexual maturity, birds should become familiar with nest boxes before the onset of lay (FASS 2010). There must be "at least one nest for every seven hens. If group nests are used, there must be at least 1 m^2 of nest space for a maximum of 120 hens" (99/74/EC 1999).

8.4.2 Perches

To improve laying-hen welfare, Council Directive 1999/74/EC sets forth the minimum standards that laying hens must have "adequate perches without sharp edges and providing at least 15 cm per hen. Perches must not be mounted above the litter and the horizontal distance between perches must be at least 30 cm and the horizontal distance between the perch and the wall must be at least 20 cm." In a recent study, (Liu et al. 2018) found that during the day laying hens spent about 10% of time on perches and at night over 75% of hens were on perches.

8.4.3 Flooring

Species that spend a significant proportion of their time walking should be maintained on solid floors with substrates (litter) which supports their natural behaviour of foraging and dust-bathing (Hawkins et al. 2001; FASS 2010). Birds will dust bathe in different types of loose material; with a minimum standard of "250 cm^2 of littered area per hen, the litter occupying at least one third of the ground surface"(99/74/EC 1999). A recent study showed that the addition of a pecking device made of stones on the cage floor lowered the agonistic behaviour of caged hens compared to controls (Moroki and Tanaka 2016).

8.4.4 Food and Water

Food and water should be available *ad libitum*. The standards for access to feeders are as follows: "linear feeders providing at least 10 cm per bird or circular feeders providing at least 4 cm per bird"(99/74/EC 1999). This will ensure that all birds get access to food as there are peaks in feeding behaviour, usually at the start and conclusion of the light period (Hawkins et al. 2001). Hens are fed a standard formulated diet for laying hens with the nutritional requirements as recommended. The basis for recommendations for energy sources are reviewed (Jeroch 2011). The recommended quantity of feed per bird per day is 100–200 g. Calcium requirements increase during egg laying. Calcium is supplemented by the addition of ground oyster shell and grit on a monthly basis which achieves excellent results in the quality of the egg shell. Investigations into the effect of the addition of oil seeds (Camelina or flax) showed an increase in IgY in Lohman brown hens and it was suggested that it warranted further research into the immunomodulation properties of this oil seed (Cherian and Quezada 2016). An investigation on the possible effects of the addition of quercetin to the diet of broilers found that it enhanced IgY antibody production (Hager-Theodorides et al. 2014). Clean drinking water should be provided *ad libitum* as "either continuous drinking troughs providing 2.5 cm per hen or circular drinking troughs providing 1 cm per hen. In addition, where nipple drinkers or cups are used, there shall be at least one nipple drinker or cup for every 10 hens. Where drinking points are plumbed in, at least two cups or two nipple drinkers shall be within reach of each hen"(99/74/EC 1999).

8.5 Factors Impacting on IgY Production

The annual total amount of IgY produced per laying hen has been reported to vary from 20 g to 100 g, as reviewed by (Marcq et al. 2013) and is influenced by the quantitative method used and the extraction procedure (Chap. 11). Indeed there is a complex interplay between a number of factors such as the housing of birds, stress levels, the circadian clock, genetic lines and the methods used to purify and assess the IgY levels. A number of individual studies have examined these factors. An early study, albeit at intensive stocking levels, involved keeping hens in cages, or on the floor. Hens in cages produced higher mean IgY concentrations and specific antibody titres compared to those kept on the floor (Erhard et al. 2000). It has also been shown that there are distinct biological fluctuations in serum and yolk IgY levels over a 60 day period and that IgY levels are under the control of the circadian clock (He et al. 2014) and Chap. 7. A number of reports have suggested that IgY levels are more closely correlated with the genetic background of the chickens. To date, five candidate genes related to IgY levels have been determined using a genome-wide association study (Zhang et al. 2015). A recent study showed that Rhode Island Red chicken eggs yielded higher quantities of IgY (8.37 mg/mL) compared to (3.66 mg/mL) by Single Comb White Leghorn chickens (Amro et al. 2018). IgY concentrations in the blood have been measured for three chicken strains used in organic production (Hellevad, Hisex and Bovans) with Bovans having the highest concentration (Kjærup et al. 2017). In contrast, Cook and Trott reported identical antibody content of eggs of different commercial lines (Cook and Trott 2010). They reported that egg size and the rate of lay were the dominant parameters governing the production of IgY over a year. The productivity of the chicken breed may therefore be the key parameter to consider when selecting birds intended for IgY production (Chap. 2). Pauly et al. have demonstrated that chickens can be used for up to two years: although egg production rate was reduced in the second year, the greater amount of IgY per egg compensates for this reduction (Pauly et al. 2009). In practice, the normal laying period is approximately 300 days.

8.6 Example of a Conventional Housing System

In compliance with the Directive 2010/63/EU, hens must be provided with enriched cages (2010/63/EU 2010). Individual hens may kept in a cage for experimental purposes or in small groups as illustrated in Fig. 8.1.

8.7 Specific Pathogen Free (SPF) Hens

As detailed in the European Pharmacopoeia, "a designated SPF flock is derived from chickens shown to be free from vertically-transmissible agents" listed in Table 8.2 below (European Pharmacopoeia 7.0 5.2.2. n.d. "This is achieved by the testing of two generations prior to the designated SPF flock". Routine testing must be performed as outlined in the European Pharmacopoeia and all manufacturers of medicines or substances for pharmaceutical use therefore must apply the European Pharmacopoeia quality standards in order to be able to market and use these products in Europe. Antibody production may be impacted by the occurrence of viral diseases in chickens. Thus, if the antibodies produced are to be used for therapeutic purposes, the use of SPF chickens is compulsory (Narat 2003). As previously discussed, the factors affecting IgY production (both total IgY and specific to an antigen) are dependent on a number of factors and there are conflicting reports on whether SPF

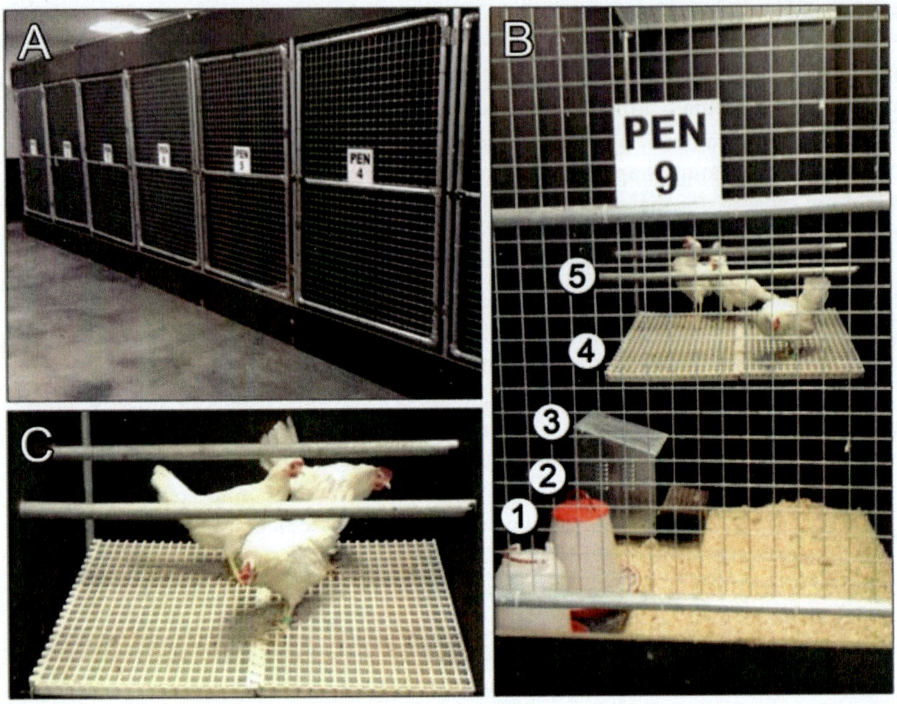

Fig. 8.1 Caging systems used for experiments. A number of individual experimental pens are shown in photo (**a**). Each pen is 2 m^2 and has the capacity to hold up to 20 chickens. In photo (**b**), 3 chickens are shown in a pen with dust-free wood shavings. Food and water are provided in removable containers (1 and 2) and are monitored daily. There is a semi-enclosed nest box with a step which doubles as an egg collection box (3). Eggs are removed daily. A platform for chickens to rest and walk on is shown at 4. Faeces collect below the platform. Dust-free wood shavings are replaced every two to three days, depending on the stocking density. Perches are shown at 5 and are clearly visible in photo (**c**). Photo courtesy of Ovagen Group Ltd. (www.ovagen.ie)

eggs give higher antibody titres. Schwarzkopf et al. researched the influence of housing conditions and the specificity of IgY produced and concluded that the use of SPF-hens does not lead to any improvement (Schwarzkopf et al. 2001). In a review by Schade et al. (2005) and references therein, Hommel & Behn showed that an identical immunisation scheme carried out in two White Leghorn hen groups, one kept under SPF conditions and the other under conventional conditions, resulted in comparable antibody titers in both groups (Schade et al. 2005). Currently, there are insufficient data to show the superiority of SPF conditions for IgY production, however for therapeutic uses antibodies must be derived from SPF chickens.

Table 8.2 A designated SPF flock is derived from chickens shown to be free from the vertical transmission of the listed agents (European Pharmacopoeia 7.0 n.d.)

Agent	Test to be used[a]	Vertical transmission	Rapid/Slow spread
Avian adenoviruses, group 1	AGP, EIA	Yes	Slow
Avian encephalomyelitis virus	AGP, EIA	Yes	Rapid
Avian infectious bronchitis virus	HI, EIA	No	Rapid
Avian infectious laryngotracheitis virus	VN, EIA	No	Slow
Avian leucosis viruses	EIA for virus VN, EIA for antibody	Yes	Slow
Avian nephritis virus	IS	No	Slow
Avian orthoreoviruses	IS, EIA	Yes	Slow
Avian reticuloendotheliosis virus	AGP, IS, EIA	Yes	Slow
Chicken anaemia virus	IS, EIA, VN	Yes	Slow
Egg drop syndrome virus	HI, EIA	Yes	Slow
Infectious bursal disease virus	Serotype 1: AGP, EIA, VN serotype 2: VN	No	Rapid
Lnfluenza A virus	AGP, EIA, HI	No	Rapid
Marek's disease virus	AGP	No	Rapid
Newcastle disease virus	HI, EIA	No	Rapid
Turkey rhinotracheitis virus	EIA	No	Slow
Mycoplasma gallisepticum	Agg and HI to confirm a positive test EIA, HI	Yes	Slow
Mycoplasma synoviae	Agg and HI to confirm a positive test EIA, HI	Yes	Rapid
Salmonella pullorum	Agg	Yes	Slow

Agg agglutination; *AGP* agar gel precipitation, the technique is suitable when testing is carried out weekly; *HI* haemagglutination inhibition; *IS* immunostaining; *VN* virus neutralisation; *EIA* enzyme immunoassay

[a]Subject to agreement by the competent authority, other types of test may be used provided that they are at least as sensitive as those indicated and of appropriate specificity

8.8 Specific Pathogen Free (SPF) Avian Facility

A SPF unit needs to be able to operate to scale in order to meet the growing demands for chicken-based products (Chap. 17). An example of a SPF facility is illustrated in Fig. 8.2. Overall, the facility measures 2560 m^2 and is comprised of five avian housing rooms with a capacity of 800 birds per room. There is a requirement for showering facilities, a pass-through UV chamber and pass-through dunk tank,

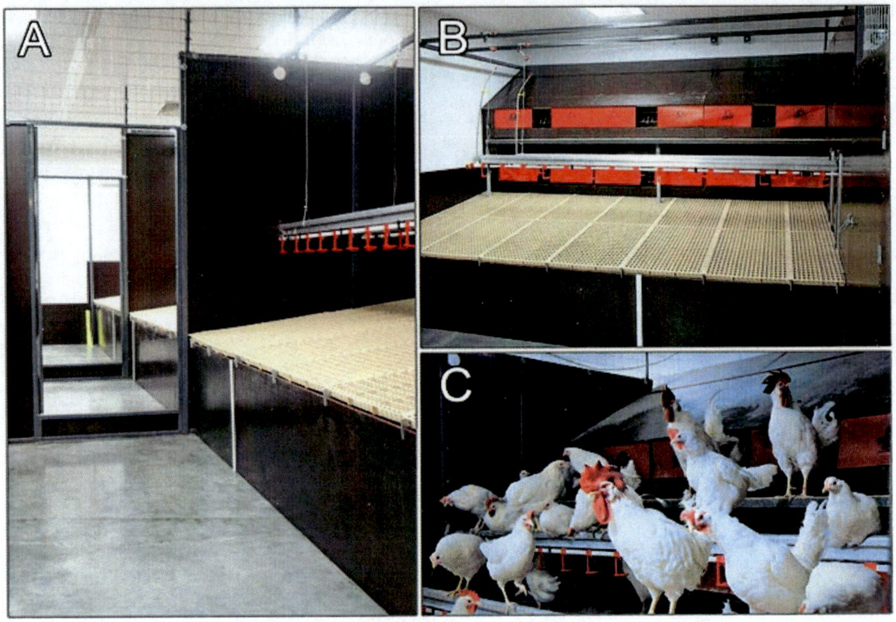

Fig. 8.2 State of the art SPF Avian Facility. A room subdivided into 3 sections supports the large-scale production of different IgY products in each section (**a**); access to this room is from a central controlled corridor. A single room providing accommodation suitable for 266 chickens, exceeding the space requirements of Directive 2010/63/EU, is shown in (**b**). White Leghorn chickens in the facility are shown in (**c**). Figure courtesy of Ovagen Group Ltd. (www.ovagen.ie)

controlled diet storage rooms, in such a facility; all are required to operate to SPF standards. The location of an avian facility is important so that there are no other chicken facilities or industrial premises nearby, thus ensuring excellent air quality. It is important to have access to a gamma-irradiation facility as all food and bedding needs to be sterilised.

Maintenance of SPF status relies on the use of filtered-air, positive-pressure (FAPP) poultry housing, irradiated food, sterile water and critical biosecurity measures. Heating is controlled by an air-conditioning unit using triple HEPA filtered air. Ammonia levels are kept to a minimum by frequent air exchanges, typically 20+ air changes per hour. Lights are electronically programmed and operated in strict adherence to recommendations and in keeping with the age of the chickens and laying cycle. For egg laying, a light cycle of 14 h light: 10 h dark is used at 20-30 lux with a two-hour dimming period prior to lights out. Nest boxes are fitted with a barrier which opens and closes and automatically expels the bird after laying resulting in clean eggs and prevents chickens from brooding. Eggs roll out the rear of the cage onto a conveyor belt which brings them to a collection bay external to the avian room. Moving chickens even temporarily from their cages, may negatively affect egg production or the chickens may undergo a molt, for example moving for cage cleaning (FASS 2010). Sterilised wood shavings are provided to the birds who are housed on a deep litter system. As required, additional sterilised shavings are provided to the birds. At the end of a production cycle, the building is thoroughly sanitised before re-stocking with new birds.

References

2010/63/EU (2010) Directive 2010/63/EU of the european parliament and of the council of 22 September 2010 on the protection of animals used for scientific purposes. OJL 276:33–79

98/58/EC (1998) Council directive concerning the protection of animals kept for farming purposes. OJL 221:23–27

99/74/EC (1999) Council directive laying down minimum standards for the protection of laying hens. OJL 203:53–57

Amro WA, Al-Qaisi W, Al-Razem F (2018) Production and purification of IgY antibodies from chicken egg yolk. J Genet Eng Biotechnol 16(1):99–103. https://doi.org/10.1016/j.jgeb.2017.10.003

Archer GS (2018) Effect of type of light source and location of light source on layer production, stress and fear during the start of lay. Int J Poult Sci. https://doi.org/10.3923/ijps.2018.92.99

Cherian G, Quezada N (2016) Egg quality, fatty acid composition and immunoglobulin Y content in eggs from laying hens fed full fat camelina or flax seed. J Anim Sci Biotechnol 7:15. https://doi.org/10.1186/s40104-016-0075-y

Cook ME, Trott DL (2010) IgY–immune component of eggs as a source of passive immunity for animals and humans. World's Poultry Sci J. Taylor & Francis 66:215–225. https://doi.org/10.1017/S0043933910000279

EFSA (2005) Opinion of the scientific panel on animal health and welfare on a request from the commission related to the welfare aspects of various systems of keeping hens. EFSA J 197:1–23

Erhard MH et al (2000) The humoral immune response and the productivity of laying hens kept on the ground or in cages. ATLA Alternatives to Laboratory Animals. https://doi.org/10.1177/026119290002800504

European Pharmacopoeia 7.0 (n.d.) EUROPEAN PHARMACOPOEIA 7.0 5.2.2. SPF chicken flocks for vaccines. Available at: https://www.valobiomedia.com/content/ep7.0_5.2.2_spf_flocks.pdf

FASS (2010) Guide for the care and use of agricultural animals in research and teaching'. In: Federation of Animal Sciences Societies. Champaign, IL., USA, pp 105–127

FAWC (1997) 'Report on the welfare of laying hens'. Farm Animal Welfare Council. Available at: http://www.fawc.org.uk/reports.htm

Framework (2014) 'Caring for animals aiming for better science–Publications Office of the EU'. Available at: https://op.europa.eu/en/publication-detail/-/publication/fca9ae7f-2554-11e9-8d04-01aa75ed71a1. Accessed: 17 November 2020

Hager-Theodorides AL et al (2014) Effects of dietary supplementation with quercetin on broiler immunological characteristics. Anim Feed Sci Technol. https://doi.org/10.1016/j.anifeedsci.2014.09.021

Hassan MR et al (2013) Effect of monochromatic and combined light colour on performance, blood parameters, ovarian morphology and reproductive hormones in laying hens. Ital J Anim Sci. Taylor & Francis 12(3):e56. https://doi.org/10.4081/ijas.2013.e56

Hawkins P et al. (2001) 'Laboratory birds: refinements in husbandry and procedures', 5th report of the BVAAWF/FRAME/RSPCA/UFAW Joint Working Group on Refinement Laboratory Animals, 35(Supplement 1), pp. S1–163

He JX et al (2014) Chronobiological studies of chicken IgY: monitoring of infradian, circadian and ultradian rhythms of IgY in blood and yolk of chickens. Vet Immunol Immunopathol 160 (3-4):266–272. https://doi.org/10.1016/j.vetimm.2014.05.016

Hofmann T et al (2020) Impact of housing environment on the immune system in chickens: a review. Animals *10*(7):1–26. https://doi.org/10.3390/ani10071138

Jeroch H (2011) Recommendations for energy and nutrients of layers: a critical review*. Lohmann Information 46(2):61–72

Kjærup RB et al (2017) Comparison of growth performance and immune parameters of three commercial chicken lines used in organic production. Vet Immunol Immunopathol 187:69–79. https://doi.org/10.1016/j.vetimm.2017.04.007

LASA (2016) 'Guiding principles for supervision and assessment of competence as required under EU and UK legislation', in Jennings, M Berdoy, M. (ed.). Available at: www.lasa.co.uk/publications. Accessed: 17 November 2020

LayWel (2006) Welfare implications of changes in production systems for laying hens. Available at: http://www.laywel.eu/web/xmlappservlet9a30.html. Accessed: 13 November 2019

Liu K et al (2018) Perch-shape preference and perching behaviors of young laying hens. Appl Anim Behav Sci 203:34–41. https://doi.org/10.1016/j.applanim.2018.02.009

Malomane DK et al (2019) The SYNBREED chicken diversity panel: a global resource to assess chicken diversity at high genomic resolution. BMC Genomics. BioMed Central Ltd. 20(1):345. https://doi.org/10.1186/s12864-019-5727-9

Marcq C et al (2013) Refinement of the production of antigen-specific hen egg yolk antibodies (IgY) intended for passive dietary immunization in animals. A review, Biotechnol Agron Soc Environ 13:295–308. https://popups.uliege.be:443/1780-4507/index.php?id=10243. Accessed: 19 November 2020

Matur E et al (2015) The effect of furnished cages on the immune response of laying hens under social stress. Poult Sci. Oxford University Press 94(12):2853–2862. https://doi.org/10.3382/ps/pev297

Moroki Y, Tanaka T (2016) A pecking device as an environmental enrichment for caged laying hens. Anim Sci J 87(8):1055–1062. https://doi.org/10.1111/asj.12525

Narat M (2003) Production of antibodies in chickens. Food Technol Biotechnol 41:259–267

Pauly D et al (2009) Monitoring of laying capacity, immunoglobulin Y concentration, and antibody titer development in chickens immunized with ricin and botulinum toxins over a two-year period. Poult Sci. https://doi.org/10.3382/ps.2008-00323

Ross M et al (2020) Hens with benefits: can environmental enrichment make chickens more resilient to stress? Physiol Behav. Elsevier Inc. 226:113077. https://doi.org/10.1016/j.physbeh.2020.113077

Rubin CJ et al (2010) Whole-genome resequencing reveals loci under selection during chicken domestication. Nature 464(7288):587–591. https://doi.org/10.1038/nature08832

Russell WMS, Burch RL (1959) The principles of humane experimental technique. Methuen, London

Schade R, Schwarzkopf C, Erhard M (2001) Pros and cons of obtaining antibodies from chickens' eggs rather than from mammals' serum. In: Chicken Egg Yolk Antibodies, Production and Application. Springer, Berlin Heidelberg, pp 211–217. https://doi.org/10.1007/978-3-662-04488-9_6

Schade R et al. (2005) 'Chicken egg yolk antibodies (IgY-technology): a review of progress in production and use in research and human and veterinary medicine', ATLA Alternatives to Laboratory Animals. FRAME, pp. 129–154. doi: https://doi.org/10.1177/026119290503300208

Schwarzkopf C, Staak C, Behn I, Erhard M (2001) Immunisaton. In: Chicken egg yolk antibodies, production and application: IgY technology. Springer Verlag, Berlin; Heidelberg, Germany, pp 25–64

Tixier-Boichard M et al (2012) A century of poultry genetics. World's Poultry Sci J. Cambridge University Press on behalf of World's Poultry Science Association 68:307–321. https://doi.org/10.1017/S0043933912000360

Zhang L et al (2015) The identification of loci for immune traits in chickens using a genome-wide association study. PLoS One. https://doi.org/10.1371/journal.pone.0117269

9 Other Avian Species: Ostrich, Quail, Turkey, Duck and Goose

Ricardo S. Vieira-Pires, Patricia M. Morgan, Tiago Ôchoa-Pires, and Marguerita Rosa

Abstract

The chicken has been the best studied of the avian species with regard to IgY production. Other agriculturally-important avian species present interesting alternatives. As more in-depth information became available on the immune systems and avian genomes, the biology and evolutionary relationships of these species have become clearer. Each of the species have three classes of immunoglobulin, IgM, IgA and IgY. Ducks and geese also have a truncated IgY, IgY (ΔFc). Antigen-specific IgY following immunization has been recovered from each of the species. Extraction and purification of IgY from the eggs follows similar protocols to that of the chicken. The amount of IgY recovered generally reflects the size of the egg, with the highest levels obtained from the ostrich which has the largest egg and the lowest from the quail. However, the smaller size of quail, shorter generation times, annual rate of eggs produced and the modular housing which has been developed make quail an ideal species for antibody discovery studies. Antigen specific IgY from each of the species is described as well as its current applications.

R. S. Vieira-Pires (✉) · T. Ôchoa-Pires
CNC – Center for Neuroscience and Cell Biology, University of Coimbra, Coimbra, Portugal
e-mail: ricardo.pires@biocant.pt

P. M. Morgan
Emerita Lecturer, School of Natural Sciences, National University of Ireland Galway, Galway, Ireland
e-mail: pat.morgan@nuigalway.ie

M. Rosa
CICECO–Aveiro Institute of Materials, Department of Chemistry, University of Aveiro, Aveiro, Portugal
e-mail: marguerita.rosa@ua.pt

X.-Y. Zhang et al. (eds.), *IgY-Technology: Production and Application of Egg Yolk Antibodies*, https://doi.org/10.1007/978-3-030-72688-1_9

Keywords

IgY · Avian species

9.1 Introduction

There are about 10,000 living species of birds which makes it the second most diverse group of living vertebrates with a history that spans over 160 million years (Mayr 2016). Of these large number of species, information on the biology and structure of IgY is available on just a few agriculturally relevant species: those of the landfowl of the order Galliformes and the family Phasianidae which includes chicken, turkey, and quail; and the order Struthioniformes and the family Struthionidae which include ostrich (Tully 2009); and those of waterfowl of the order Anseriformes and family Anatidae which includes ducks and geese (Routh and Sanderson 2009). The need for the development of effective vaccines for the poultry industry is a major driver for research in many avian species.

The evolutionary distance is greatest between the chicken and ostrich which is the most primitive bird species. The last common ancestor of chicken and ostrich was approximately 100 million years ago (Mya) (Brusatte et al. 2015). Ducks diverged from the chicken and turkey approximately 90 Mya (Huang et al. 2013). The last common ancestor of chickens and turkeys was between 28 Mya (Dimcheff et al. 2002) and 40 Mya (Dalloul et al. 2010); there appears to be no large-scale chromosomal rearrangements since their divergence; however, clear differences in genes of the immune system may reflect adaptations to different ecological niches (Dalloul et al. 2010). Ducks and geese had a common ancestor approximately 20.8 Mya (Huang et al. 2016). The reader is referred to a comprehensive comparison of the immune system of chickens with those of ducks, geese, turkeys and quail with more limited information on ostriches (Schultz and Magor 2013).

The availability of genomic information will allow more in-depth studies of avian species, their evolution and biology. The first avian genome to be sequenced was the chicken *Gallus gallus* (Hillier et al. 2004); genome sequences are now available for the turkey *Meleagris gallopavo* (Dalloul et al. 2010), Japanese quail *Coturnix japonica* (Kawahara-Miki et al. 2013), duck *Anas platyrhynchos* (Huang et al. 2013), goose *Anser anser* (Lu et al. 2015) and ostrich *Struthio camelus* (O'Connor et al. 2018). Of all the avian species the chicken has been most intensively studied. A recent report of use of experimental animals in Canada showed that the use of birds (26%) was only second to mice (31%); with chickens accounting for 93% of the avian studies and the combination of turkey, duck, quail and geese accounting for just 0.2% (Council on Animal Care 2019).

The generation of antibody diversity by gene conversion is a mechanism common to the majority of birds from the primitive ostrich (Huang et al. 2012) to ducks, geese, turkey and quail (Ratcliffe and Härtle 2014). All avian species have three classes of antibody, IgM, IgA and IgY (Härtle et al. 2014); IgY is selectively transferred from the serum to the egg yolk. The methodology for the extraction and purification of IgY follows that described for the chicken (Chap. 11). Descriptions of avian antibodies, with the exception of the chicken which has been extensively characterized and described (Chaps. 4 and 5) are rather limited. Ducks being a natural reservoir of influenza type A viruses and the more recent highly pathogenic H5N1 virus have been subjected to more intensive research (Sturm-Ramirez et al. 2005). The presence of the variant truncated IgY, IgY(ΔFc) in ducks and geese has previously been described (Chap. 5). The conserved mechanism of generation and the structural homology of IgY(ΔFc) in ducks and geese infers that it arose prior to the divergence of these two species (T. Huang et al. 2016). Species-specific differences in the glycans attached to glycoproteins or glycolipids have been demonstrated in avian species (Suzuki et al. 2013). The unique Galβ1-4Gal glycan epitope on IgY was shown to be present in ostrich, turkey, quail and duck but absent in chickens; an advantage may be that chickens are able to produce antibodies against carbohydrate antigens on pathogens as they do not express this epitope (Suzuki et al. 2013). Research pertaining to the production of IgY in each of these individual species will now be presented.

9.2 Ostrich *Struthio camelus*

The ecology and physiological parameters of ostriches in the wild, which should inform the husbandry of captive birds, has been described (Cooper et al. 2010). The incubation and hatching period of 42 days (Shanawany and Dingle 1999) is much longer than for the other avian species (Fig. 9.3). The start of lay is between 104 to 130 weeks and the age of maximum egg production ranges from 364 to 572 weeks (Kokoszyński 2017; Shanawany and Dingle 1999). Young female ostriches produce 10 to 20 eggs in the first year and a range of 40 to 130 eggs annually, but most often it is 40–60 eggs (Kokoszyński 2017). An analysis of the immunological transcripts in the ostrich showed the presence of IgM, IgA and IgY which are similar to those in other avian species (Huang et al. 2012). However, an earlier study showed that there is no cross-reactivity of chicken polyclonal or monoclonal antibodies with ostrich IgY suggesting some structural differences (Cadman et al. 1994). The molecular weight of ostrich IgY has been reported to be around 200 kDa (heavy chain, 64 kDa; light chain, 28 kDa) (Adachi et al. 2008) and thus heavier than chicken IgY (180 kDa) (Chap. 5). One of the main advantages of IgY production in ostrich is the size of the egg. Each egg weighs about 1.5 kg which is 30-fold larger than chicken eggs and can yield 2–4 g of IgY per egg; with an annual production of about 100 eggs this would yield about 200–400 g of IgY from one ostrich in a year (Adachi et al. 2008, 2011). In addition to the amount of IgY, another important factor is that it is from a single egg and therefore avoids any potential batch to batch variation when multiple eggs need to be collected to give sufficient material for purification. To date there have been few studies following immunization in the ostrich. An ostrich was shown to produce cross-reactive neutralization antibodies against the pandemic influenza virus A/H1NI following immunization with a seasonal flu vaccine; air-conditioning filters impregnated with these antibodies were proposed as a potential mechanism to slow the spread of the virus (Adachi et al. 2011). In another study, the in vitro inhibitory activity of ostrich IgY following immunization with enterotoxins on the growth of *S. aureus* and *E. coli* was demonstrated (Tobias et al. 2012). While the husbandry of ostriches may be challenging, the very high levels of IgY production in the ostrich may make it amenable for future industrial applications.

9.3 Japanese Quail *Coturnix japonica*

Japanese quail are used as a laboratory animal model for several areas of biological investigation, including immunology. The reader is referred to a detailed description of the biology, husbandry and breeding of quail which will not be discussed here (Baer et al. 2015). The advantages in using quails include their small size (body weight 80–300 g) (Kawahara-Miki et al. 2013); start of lay at about 39 days and 50% egg production at 45 days with an average of 253 eggs laid per year (Narinc et al. 2013). Egg weight averages about 10 g (Tserveni-Goussi and Fortomaris 2011) and eggs take 16 to 18 days to hatch (Baer et al. 2015). The positive attributes of quail include a minimum requirement of housing space; recently a modular housing system for quails, which adheres to the European Directive 2010/63 (2010/63/EU 2010) designed for antibody discovery studies was described (Ôchoa-Pires et al. 2020) (Fig. 9.1).

Quails can be immunized at weeks 6–8 of age (Ôchoa-Pires et al. 2020) and have been reported to respond to lower levels of antigen (15–60 μg of purified virus preparation) compared to that required to immunize chickens (Somowiyarjo et al. 1990). Formalin-inactivated *E. coli* were used to immunize quail via the thigh muscle, antibodies were isolated and purified by methods previously validated for chicken IgY (Chap. 11) and the molecular weight of the heavy and light chains was found to be 93 kDa, giving a molecular weight of 186 kDa for the intact IgY, which is slightly larger than for chicken IgY(Padmani et al. 2007). In another study, the production of specific IgY against *Salmonella typhimurium* and *Salmonella enteritidis* was reported (Esmailnejad et al. 2019a, b) and the antibodies could be stored at 4 °C for a period of 2 months with no loss of activity (Esmailnejad et al. 2019a, b). A recent study showed that aqueous two-phase extraction and purification of immunoglobulins by chromatography yielded ultra-pure IgY (Balaraman and Rathnasamy 2020). Overall, the yield of IgY in quail is lower than that from chicken. This is to be expected given that the egg yolk weight in quail is about 3 g (Murai et al. 2016), compared to 15 g in chicken (Singh et al. 2009); the reported levels of IgY in quail were in the range of 1–2.5 mg/g yolk as determined by ELISA which would give a total of 3–7.5 mg/egg (Murai et al. 2016). Consistently, results by Francisco et al. using size-exclusion chromatography analysis of chicken and quail egg yolk fractions also revealed –seven-fold lower concentration of IgY content in quails relative to chickens (Francisco et al. 2015)(Fig. 9.2).

Fig. 9.1 Modular housing system for egg-laying quails. Caging system for housing egg-laying quails. (**a**) Overall view of a 4-level housing rack, with up to two cages per level; (**b**) Frontal, (**c**) inside, (**d**) lateral and (**e**) back views of a rack level, showing details of the wired-floor drawer and removable polycarbonate panels; (**f**) Frontal egg-collecting section for easy monitoring and collection of eggs; (**g**) Egg cataloguing box; (**h**) LED light system for photoperiod control and (**i**) water system with automatic refilling with 4 water dispensers per level; both systems are independent from the main rack, resting attached to the facility wall. Figure adapted from original (Pinto and Vieira-Pires 2018) with permission

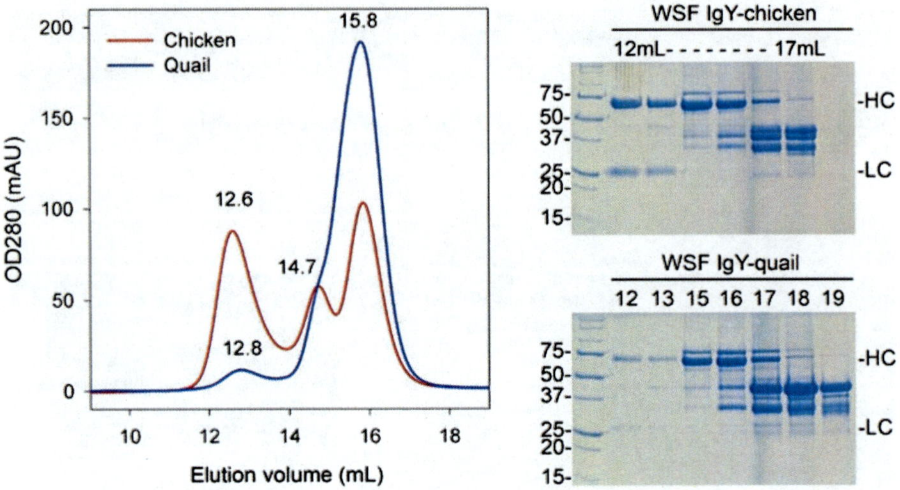

Fig. 9.2 Analysis of water-soluble protein fractions (WSF) obtained from chicken and quail egg yolks. Size exclusion chromatography reveals distinct elution profiles. Chicken WSF presents three main elution peaks (12.6, 14.7 and 15.8 mL) with the first one enriched in chicken IgY antibodies and the other two corresponding to the remaining protein impurities, as shown by SDS-PAGE. Quail WSF chromatographic profile is rather different, with a strikingly reduced quail IgY peak at 12.8 mL and a major convoluted peak for the remaining impurities (15.8 mL). HC, IgY heavy chain; LC, IgY light chain. Figure adapted from original (Francisco et al. 2015) with permission

9.4 Turkey *Meleagris gallopavo*

The performance of turkey hens was shown to be influenced by the strain (Siopes 1995). The turkey's reproductive cycle is shorter than other commercial avian species and their egg production is quite low (Tserveni-Goussi and Fortomaris 2011). Photo-stimulation between 29 and 31 weeks of age has been used to initiate egg production (Applegate 2002) with the first oviposition occurring within 2–3 weeks. The onset of lay is about the 32nd week of life, peaks about week 38th and ends about the 56th week of life (Kowalczyk et al. 2019). The mean egg weight is dependent on the strain of turkey; an average egg weight of 76 g for Beltsville Small Whites and 83 g in Broad Breasted Bronze turkey hens (Kosin and Abplanalp 1951); the mean yolk weight was recorded as 30.5 ± 2.89 g (Nisianakis et al. 2009). In an early study, the mean serum level of turkey IgY was reported to be 8.92 mg/mL and the IgY levels in yolk were reported to be 5.1 mg/mL (Goudswaard et al. 1978). A more recent in-depth study determined the levels of serum and egg yolk IgY after immunization against selected pathogens in breeder turkeys during different stages of life from start to end of lay (Kowalczyk et al. 2019). The average serum IgY concentration throughout the period of study was 22.04 ± 10.29 mg/mL which is higher than the average in chicken of 6.9 ± 1.68 mg/mL (Chap. 5). The mean IgY transfer from serum to egg yolk in the entire production cycle accounted for 59.5% which is about 13 mg/mL of egg yolk; this varied during the period of lay, being lowest during the period of highest laying intensity.

9.5 Duck *Anas platyrhynchos*

Between day 22 and 27 duck hatchlings penetrate their bill into the air-space and pips or cracks the shell and hatches (Oppenheim 1970). Mean egg weight has been reported to be 50.21 ± 2.74 g (Shah et al. 2008). The nutritional requirements of the main egg-type breed of ducks (Longyan laying ducks) has been reviewed: this breed matures early (110 days) and has an annual egg production of >280 eggs (Fouad et al. 2018). Other breeds of duck mature much later, and the annual production of eggs is lower (Biesiada-Drzazga et al. 2014). Ducks and geese have the unique capacity to produce both a full length IgY and a truncated isoform of IgY(Fc) from the same gene (Tong et al. 2011a, b) as described in Chap. 5. While both IgY and IgY(ΔFc) can be incorporated into the egg yolk, the full-length form (IgY) is preferentially incorporated (Munhoz et al. 2014). The IgY(ΔFc) retains the ability to connect to antigens but is incapable of antigen internalization and complement activation (Huang et al. 2016). It has been reported that the loss of the specific Fc domain could significantly reduce the severity and frequency of hypersensitivity reactions and enhance the safety of product, so IgY((ΔFc) might have some advantages over IgG or IgY in immunodiagnostics and immunotherapy. It has been demonstrated that immunization of ducks with as little as 100 μg of duck hepatitis B virus L protein plasmid generated 60 to 100 mg of highly neutralizing and purified IgY from each egg (Rollier et al. 2000); these antibodies protected offspring from a high titre duck hepatitis B viral challenges.

9.6 Goose *Anser anser*

Geese lay about 30–50 eggs each year, have a life span of about 20 years but their reproductive potential is at an acceptable level for about five seasons with egg number suppressed after the eight laying season (Tserveni-Goussi and Fortomaris 2011) and references therein. The average egg weight in Wulong breeder geese is 130 g (Zhang et al. 2020); in the same study it was shown that addition of 60 mg/kg Fe and 5.0×10^9 CFU/kg *B. subtilis* to the diet improved egg production. The age of maximum egg production is between 40–45 weeks and for specific breeds, such as those that originate from *Anser cycnoides* (e.g., the Kuban goose), egg production rates can be much higher (Tserveni-Goussi and Fortomaris 2011). It has been shown that geese can generate full length IgY as well as IgY(ΔFc) by a similar mechanism as in ducks; this combined with the sequence homology suggests that IgY(ΔFc) emerged before the divergence of ducks and geese approximately 20.8 Mya (Huang et al. 2016). Hydrophobic charge-induction chromatography (HCIC; Chap. 11) using new ligand, 2-mercapto-1-methyl-imidazole (MMI) was used to purify IgY (ΔFc) with a final purity of 98.6% and a yield of 85.0% from plasma (Tong et al. 2011a, b). Dengue virus (DENV), transmitted by an insect bite, is an increasingly global health problem. DENV specific goose IgY was generated in an inbreed hybrid (*Anser domesticus*) a cross between the German Embden, the Royal Chinese, and the Royal English goose breeds: vaccination with DEVN2 antigen induced a strong humoral response in the geese with antibody titres reaching as high as 1: 6,400,000; therapeutic efficacy was observed with 1–2 mg anti-DEVN2 administered to mice and protected against a lethal DENV2 challenge without antibody dependent enhancement (Fink et al. 2017). Recently, a goose antibody to DENV non-structural protein 1 was shown to neutralize DENV infection and protect against a lethal dengue virus challenge (O'Donnell et al. 2020). The same research team has demonstrated that goose zika virus-specific IgY protects against a lethal zika virus challenge without inducing antibody-dependent enhancement (O'Donnell et al. 2019). Goose, like chicken, are a good source of antibodies to snake venoms (Chap. 16) and the development of a point of care assay for the identification of the exact species of snake is desirable. It has been demonstrated that up to 2.77 mg of purified antibodies specific for Russell's vipers snakebite can be obtained from a single goose egg and that 10–15 goose eggs would provide enough antibody to make 10, 000 immunochromatographic diagnostic kits (Lin et al. 2020).

9.7 Summary

The potential to use other avian species, rather than the chicken, has been exploited in recent years. Further studies on the structure and biology of IgY from the avian species discussed in this chapter is warranted. The major parameters with regard to hatching times and start of lay as already discussed are presented pictorially in Fig. 9.3. Chicken breeds have been selected to lay an egg nearly every day and to reach maturity earlier. Some of the other species produce larger eggs but the number of eggs laid per year are is much lower (Fig. 9.3). The selection of a particular avian species for the production of IgY is dependent on whether the short generation time of the quail, the annual productivity of the chicken or size of the egg are the most important parameters to answer the research question or provide sufficient IgY for commercial purposes.

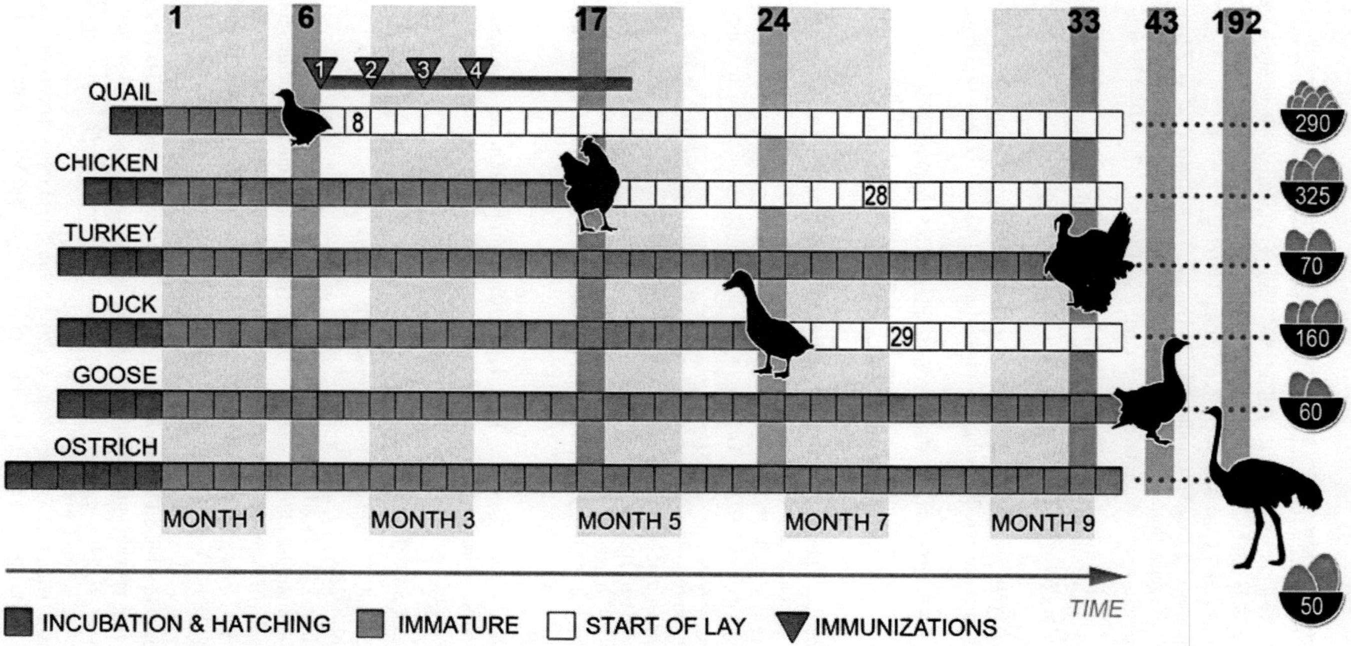

Fig. 9.3 Experimental timeline of avian species used for egg yolk IgY production. One (1) week periods are represented by each square in the timelines and one (1) month periods depicted by vertical background columns labelled (Month 1 to 9). The timeline periods for each species are indicated by the following: egg incubation and hatching periods (dark grey), period of immaturity until start of lay (light grey), egg laying periods (white). The average weeks of age until the beginning of regular egg-laying activity of females is depicted by vertical background columns top-labelled with the corresponding average week of age (6, 17, 24, 33, 43 and 192 weeks). The numbers indicate the start period of maximum egg production. Above the quail timeline an indicative 3-month immunization schedule (grey bar) in shown with four immunization events (inverted triangles, labelled 1 to 4) with 2-week intervals; in this case, quail egg collection may proceed for 2 to 3 months after the first immunization

References

2010/63/EU (2010) Directive 2010/63/EU of the European Parliament and of the Council of 22 September 2010 On the Protection of Animals used for Scientific Purposes. OJ L 276:33–79

Adachi K, Handharyani E, Sari DK, Takama K, Fukuda K, Endo I, Yamamoto R et al (2008) Development of neutralization antibodies against highly pathogenic H5N1 avian influenza virus using ostrich (Struthio Camelus) yolk. Mol Med Rep 1(2):203–209

Adachi K, Takama K, Tsukamoto M, Inai M, Handharyani E, Hiroi S, Tsukamoto Y (2011) Ostrich produce cross-reactive neutralization antibodies against pandemic influenza virus a/H1N1 following immunization with a seasonal influenza vaccine. Exp Ther Med 2(1):41–45. https://doi.org/10.3892/etm.2010.180

Applegate TJ (2002) Reproductive maturity of Turkey hens: egg composition, embryonic growth and hatchling transition. Avian Poult Biol Rev 13(1):31–41. https://doi.org/10.3184/147020602783698458

Baer J, Lansford R, Cheng K (2015) Chapter 22–Japanese quail as a laboratory animal model. In: Laboratory animal medicine, 3rd edn. Elsevier Inc. https://doi.org/10.1016/B978-0-12-409527-4.00022-5

Balaraman HB, Rathnasamy SK (2020) High selective purification of IgY from quail egg: process design and quantification of deep eutectic solvent based ultrasound assisted liquid phase microextraction coupled with preparative chromatography. Int J Biol Macromol 146 (March):253–262. https://doi.org/10.1016/j.ijbiomac.2019.12.242

Biesiada-Drzazga B, Charuta A, Banaszewska D (2014) Evaluation of particular traits of Pekin duck breed star 53 of french origin eggs during egg laying. Veterinarija Ir Zootechnika 67 (89):3–9

Brusatte SL, O'Connor JK, Jarvis ED (2015) The origin and diversification of birds. In: Current biology. Cell Press. https://doi.org/10.1016/j.cub.2015.08.003

Cadman HF, Kelly PJ, Dikanifura M, Carter SD, Azwai SM, Wright EP (1994) Isolation and characterization of serum immunoglobulin classes of the ostrich (Struthio Camelus). Avian Dis 38(3):616–620. https://doi.org/10.2307/1592087

Cooper RG, Horbańczuk JO, Villegas-Vizcaíno R, Sebei SK, Faki Mohammed AE, Mahrose KMA (2010) Wild ostrich (Struthio Camelus) ecology and physiology. Trop Anim Health Prod 42 (3):363–373. https://doi.org/10.1007/s11250-009-9428-2

Council on Animal Care, Canadian (2019) CCAC animal data report

Dalloul RA, Long JA, Zimin AV, Aslam L, Beal K (2010) Multi-platform next-generation sequencing of the domestic Turkey (Meleagris Gallopavo): genome assembly and analysis. PLoS Biology | www.plosbiology.org 16(9):1000475. https://doi.org/10.1371/journal.pbio.1000475

Dimcheff DE, Drovetski SV, Mindell DP (2002) Phylogeny of tetraoninae and other galliform birds using mitochondrial 12S and ND2 genes. Mol Phylogenet Evol 24(2):203–215. https://doi.org/10.1016/S1055-7903(02)00230-0

Esmailnejad A, Abdi-Hachesoo B, Nasab EH, Shakoori M (2019a) Production, purification, and evaluation of quail immunoglobulin Y against salmonella typhimurium and salmonella enteritidis. Mol Immunol 107(March):79–83. https://doi.org/10.1016/j.molimm.2019.01.012

Esmailnejad A, Hachesoo BA, Nasab EH, Shakoori M (2019b) Storage stability of anti-salmonella typhimurium immunoglobulin Y in immunized quail eggs stored at 4°C. Ind J Anim Sci 89 (12):1318–1321

Fink AL, Williams KL, Harris E, Alvine TD, Henderson T, Schiltz J, Nilles ML, Bradley DS (2017) Dengue virus specific IgY provides protection following lethal dengue virus challenge and is neutralizing in the absence of inducing antibody dependent enhancement. PLoS Negl Trop Dis 11(7):1–17. https://doi.org/10.1371/journal.pntd.0005721

Fouad AM, Dong R, Wang S, Chen W, Xia W, Zheng C (2018) Nutritional requirements of meat-type and egg-type ducks: what do we know? J Animal Sci Biotechnol. Bio Med Central Ltd 9:1. https://doi.org/10.1186/s40104-017-0217-x

Francisco R, Tavares APM, Bernardo SC, Silva D, Laranjeira J, Curto P, Simões I, Ribeiro T, Fidalgo-Carvalho I, Freire MG, Vieira-Pires RS (2015) IgY-technology revisited. In: PEGS Europe 2015 summit, 2nd-6th November 2015. Lisbon, Portugal

Goudswaard J, Noordzij A, Van Dam RH, Van Der Donk JA (1978) Quantification of Turkey immunoglobulins IgA, IgM and IgG in serum and secretions. Zeitschrift Fur Immunitatsforschung 154(3):248–255. https://doi.org/10.1016/s0340-904x(78)80019-0

Härtle S, Magor KE, Gobel TW, Davison F, Kaspers B (2014) Structure and evolution of avian immunoglobulins. In: Pete K, Schat KA, Kaspers B (eds) Avian immunology: second edition, 2nd edn. Elsevier Inc, pp 103–114

Hillier LW, Miller W, Birney E, Warren W, Hardison RC, Ponting CP, Bork P et al (2004) Sequence and comparative analysis of the chicken genome provide unique perspectives on vertebrate evolution. Nature 432(7018):695–716. https://doi.org/10.1038/nature03154

Huang Y, Li Y, Burt DW, Chen H, Zhang Y, Qian W, Kim H et al (2013) The duck genome and transcriptome provide insight into an avian influenza virus reservoir species. Nat Genet. https://doi.org/10.1038/ng.2657

Huang T, Wu K, Yuan X, Shao S, Wang WY, Wei S, Cao G (2016) Molecular analysis of the immunoglobulin genes in goose. Dev Comp Immunol. https://doi.org/10.1016/j.dci.2016.02.024

Huang T, Zhang M, Wei Z, Wang P, Sun Y, Hu X, Ren L et al (2012) Analysis of immunoglobulin transcripts in the ostrich Struthio Camelus, a primitive avian species. PLoS One 7(3):1–10. https://doi.org/10.1371/journal.pone.0034346

Kawahara-Miki R, Sano S, Nunome M, Shimmura T, Kuwayama T, Takahashi S, Kawashima T, Matsuda Y, Yoshimura T, Kono T (2013) Next-generation sequencing reveals genomic features in the Japanese quail. Genomics 101(6):345–353. https://doi.org/10.1016/j.ygeno.2013.03.006

Kokoszyński D (2017) Chapter 4–Guinea fowl, goose, Turkey, ostrich, and emu eggs. In: Patricia Y B T–Egg innovations and strategies for improvements hester (ed) Egg innovations and strategies for improvements. Academic, San Diego, pp 33–43. https://doi.org/10.1016/B978-0-12-800879-9.00004-4

Kosin IL, Abplanalp H (1951) The pattern of egg laying in turkeys. Poult Sci 30(2):168–179. https://doi.org/10.3382/ps.0300168

Kowalczyk J, Śmiałek M, Tykałowski B, Dziewulska D, Stenzel T, Koncicki A (2019) Field evaluation of maternal antibody transfer from breeder Turkey hens to egg yolks, egg whites, and poults. Poult Sci 98(8):3150–3157. https://doi.org/10.3382/ps/pez126

Lin J-H, Lo C-M, Chuang S-H, Chiang C-H, Wang S-D, Lin T-Y, Liao J-W, Hung D-Z (2020) Collocation of avian and mammal antibodies to develop a rapid and sensitive diagnostic tool for russell's vipers snakebite. Edited by Jean-Philippe Chippaux. PLoS Negl Trop Dis 14(9): e0008701. https://doi.org/10.1371/journal.pntd.0008701

Lu L, Chen Y, Wang Z, Li X, Chen W, Tao Z, Shen J et al (2015) The goose genome sequence leads to insights into the evolution of waterfowl and susceptibility to fatty liver. Genome Biol 16(1). https://doi.org/10.1186/s13059-015-0652-y

Mayr G. (2016) An introduction to birds, the geological settings of their evolution, and the Avian Skeleton. 1–17. https://doi.org/10.1002/9781119020677.ch1

Munhoz LS, Vargas GD'Á, Fischer G, De Lima M, Esteves PA, Hübner SDO (2014) Avian IgY antibodies: characteristics and applications in immunodiagnostic. Ciência Rural 44(1):153–160. https://doi.org/10.1590/S0103-84782014000100025

Murai A, Kakiuchi M, Hamano T, Kobayashi M, Tsudzuki M, Nakano M, Matsuda Y, Horio F (2016) An ELISA for quantifying quail IgY and characterizing maternal IgY transfer to egg yolk in several quail strains. Vet Immunol Immunopathol 175:16–23. https://doi.org/10.1016/j.vetimm.2016.04.013

Narinc D, Karaman E, Aksoy T, Firat MZ (2013) Investigation of nonlinear models to describe long-term egg production in Japanese quail. Poult Sci 92(6):1676–1682. https://doi.org/10.3382/ps.2012-02511

Nisianakis P, Giannenas I, Gavriil A, Kontopidis G, Kyriazakis I (2009) Variation in trace element contents among chicken, Turkey, duck, goose, and pigeon eggs analyzed by inductively coupled plasma mass spectrometry (ICP-MS). Biol Trace Elem Res 128(1):62–71. https://doi.org/10.1007/s12011-008-8249-x

O'Connor RE, Farré M, Joseph S, Damas J, Kiazim L, Jennings R, Bennett S et al (2018) Chromosome-level assembly reveals extensive rearrangement in saker falcon and budgerigar, but not ostrich, genomes. Genome Biol 19(1):171. https://doi.org/10.1186/s13059-018-1550-x

O'Donnell KL, Espinosa DA, Puerta-Guardo H, Biering SB, Warnes CM, Schiltz J, Nilles ML, Li J, Harris E, Bradley DS (2020) Avian anti-NS1 IgY antibodies neutralize dengue virus infection and protect against lethal dengue virus challenge. Antivir Res 183(November):104923. https://doi.org/10.1016/j.antiviral.2020.104923

O'Donnell K, Meberg B, Schiltz J, Nilles M, Bradley D (2019) Zika virus-specific IgY results are therapeutic following a lethal zika virus challenge without inducing antibody-dependent enhancement. Viruses 11(3):301. https://doi.org/10.3390/v11030301

Ôchoa-Pires T, Rosa M, Laranjeira J, Francisco R, Silva D, Vieira-Pires RS (2020) Antibody discovery using Japanese quail: towards new anti-infective strategies. Ecronicon Microbiology 16.6:29–36. https://www.ecronicon.com/ecmi/pdf/ECMI-16-00936.pdf

Oppenheim RW (1970) Some aspects of embryonic behaviour in the duck (Anas Platyrhynchos). Anim Behav 18(PART 2):335–352. https://doi.org/10.1016/S0003-3472(70)80046-X

Padmani G, Anbu Gomez L, Vani C, Michael A. (2007) Isolation and characterization of antibodies challenging Escherichia Coli from Japanese quail. International Journal of Innovative Research in Science, Engineering and Technology (An ISO 3297). https://doi.org/10.15680/IJIRSET.2016.0508005

Pinto C, Vieira-Pires RS (2018) Avian biotech research: a new housing system for egg laying quails. Press Office, University of Coimbra, Portugal. https://www.igytechnology.com/avian-biotech-research-a-new-housing-system-for-egg-laying-quails/

Ratcliffe MJH, Härtle S (2014) B cells, the bursa of fabricius and the generation of antibody repertoires. In: Schat KA, Kaspers B, Kaiser P (eds) Avian immunology: second edition. Elsevier Inc, pp 65–89. https://doi.org/10.1016/B978-0-12-396965-1.00004-2

Rollier C, Charollois C, Jamard C, Trepo C, Cova L (2000) Maternally transferred antibodies from DNA-immunized avians protect offspring against hepadnavirus infection. J Virol 74(10):4908–4911. https://doi.org/10.1128/jvi.74.10.4908-4911.2000

Routh A, Sanderson S (2009) Waterfowl. In: Handbook of avian medicine. Elsevier Ltd, pp 275–308. https://doi.org/10.1016/B978-0-7020-2874-8.00012-2

Schultz U, Magor KE (2013) Comparative immunology of agricultural birds. In: Avian immunology, 2nd edn. https://doi.org/10.1016/B978-0-12-396965-1.00021-2

Shah GM, Jan U, Ahanger FA, Bhat BA, Fazili MF (2008) Egg laying, egg parameters and clutch size in mallard Anas Platyrhynchos. Indian Birds 4(3):106–108

Shanawany MM, Dingle J (1999) Ostrich production systems. In FAO animal production and health paper, 15–54, 126–28. Food and Agriculture Organisation of the United Nations

Singh R, Cheng KM, Silversides FG (2009) Production performance and egg quality of four strains of laying hens kept in conventional cages and floor pens. Poult Sci 88(2):256–264. https://doi.org/10.3382/ps.2008-00237

Siopes TD (1995) Turkey breeder hen performance by strain during consecutive lay periods. Poult Sci 74(8):1269–1279. https://doi.org/10.3382/ps.0741269

Somowiyarjo S, Sako N, Nonaka F (1990) Production of avian antibodies to three potyviruses in coturnix quail. J Virol Methods 28(2):125–132. https://doi.org/10.1016/0166-0934(90)90026-C

Sturm-Ramirez KM, Hulse-Post DJ, Govorkova EA, Humberd J, Seiler P, Puthavathana P, Buranathai C et al (2005) Are ducks contributing to the endemicity of highly pathogenic h5n1 influenza virus in Asia? J Virol 79(17):11269–11279. https://doi.org/10.1128/jvi.79.17.11269-11279.2005

Suzuki N, Nawa D, Tseng Hsiung S, Lin CW, Khoo KH, Yamamoto K (2013) Distribution of the Galβ1-4Gal epitope among birds: species-specific loss of the glycan structure in chicken and its relatives. PLoS One 8(3). https://doi.org/10.1371/journal.pone.0059291

Tobias FL, Garcia LNN, Kanashiro MM, Medina-Acosta E, Brom-de-Luna JG, de Almeida CMC, Junior RRA, Lemos M, Vieira-da-Motta O (2012) Growth inhibition of staphylococcus aureus and Escherichia Coli strains by neutralizing IgY antibodies from ostrich egg yolk. Braz J Microbiol 43(2):544–551. https://doi.org/10.1590/S1517-83822012000200015

Tong HF, Lin D, Pan Y, Yao SJ (2011a) A new purification process for goose immunoglobulin IgY (fc) with hydrophobic charge-induction chromatography. Biochem Eng J 56:205–211. https://doi.org/10.1016/j.bej.2011.06.010

Tong HF, Lin DQ, Pan Y, Yao SJ (2011b) A new purification process for goose immunoglobulin IgY(ΔFc) with hydrophobic charge-induction chromatography. Biochem Eng J 56(3):205–211. https://doi.org/10.1016/j.bej.2011.06.010

Tserveni-Goussi A, Fortomaris P (2011) Production and quality of quail, pheasant, goose and Turkey eggs for uses other than human consumption. In Improving the safety and quality of eggs and egg products: egg chemistry, Production and consumption. Woodhead Publishing Limited

Tully TN (2009) Ratites. In: Handbook of avian medicine. Elsevier Ltd, pp 258–274. https://doi.org/10.1016/B978-0-7020-2874-8.00011-0

Zhang B, Sui F, Wang B, Yang W, Li W (2020) Dietary combined supplementation of iron and bacillus subtilis enhances reproductive performance, eggshell quality, nutrient digestibility, antioxidant capacity, and hematopoietic function in breeder geese. Poult Sci 99 (11):6119–6127. https://doi.org/10.1016/j.psj.2020.06.077

Immunization of Hens

10

Álvaro Ferreira Júnior, Shikun Ge, Rao Wu, and Xiaoying Zhang

Abstract

Immunization of hens is the critical step in generating biologically active antibodies of high specificity and avidity in large amounts. A number of aspects each require careful consideration during the design of a study: the breed of hens and rearing conditions; the type and source of antigens; hapten designs for antigens of small molecules; methods of antigen inactivation; the amount of antigen applied; the type of adjuvants; the choice of antigen delivery routes; and immunization intervals. Although intramuscular injection has been the most commonly used antigen administration route in hens, other methods such as gene gun immunization and group immunization have been increasingly applied. The antibody titre needs to be carefully monitored during the immunization period, and the immunization program should be adjusted according to the antibody response observed.

Á. Ferreira Júnior (✉)
Escola de Veterinária e Zootecnia, Universidade Federal de Goiás, Goiânia, Brazil
e-mail: alvaro.ferreira@ufg.br

S. Ge
CBMA, University of Minho, Braga, Portugal

College of Biological Science and Engineering, Shaanxi University of Technology, Hanzhong, China

R. Wu · X. Zhang
College of Biological Science and Engineering, Shaanxi University of Technology, Hanzhong, China
e-mail: zhxying@snut.edu.cn; xzhang67@uoguelph.ca; zhang@bio.uminho.pt

X.-Y. Zhang et al. (eds.), *IgY-Technology: Production and Application of Egg Yolk Antibodies*, https://doi.org/10.1007/978-3-030-72688-1_10

Keywords

Immunization program · Antigen · Adjuvant · Antigen delivery · Antibody titre

The successful outcome of immunization of hens is dependent on a number of different criteria in order to obtain highly specific and high titre IgY antibodies. Depending on the desired outcome, such as long-term harvest of a large amount of specific IgY the protocols may be different and complex in comparison to simple chicken vaccine immunization.

10.1 Maintenance of Hens

The anatomy and physiology of hens and the legislation governing the rearing of hens for the production of IgY are outlined in Chaps. 2, 8, respectively. Hens to be selected for immunization should be producing six to seven eggs per week which are uniform and have strong eggshells without fractures (Chap. 8). The chickens should be monitored to ensure they are alert, drinking fresh water and maintained on an appropriate diet under the guidance of veterinarian. Healthy hens have bright eyes, bright rosy red combs and good plumage; pale combs and bristly feathers are indicators of ill health (Mack and Bell 1990). According to welfare procedures, immunized hens should be monitored daily for at least 7-days post-immunization. The first side effect of immunization could be the sudden interruption of laying eggs which can persist for one week (Schade et al. 1994). Inflammatory reactions at inoculation site(s) may cause both local and systemic deleterious effects. As a consequence, energy supplied for egg development is switched to repairing the inflamed hen tissues.

The conditions in which hens are raised could also influence the response to immunization, particularly when certain antigens/pathogens are present in the environment and the hens have been constantly challenged before immunization; in this case it will be difficult to generate a satisfactory specific antibody response to those pathogens (Zimmermann and Curtis 2019). Alternatively, specific-pathogen free (SPF) chickens could be used to avoid possible experimental failure. SPF chickens are those in which certain microorganisms (bacteria, viruses, fungi) and corresponding antibodies are undetectable (Chap. 8) but may contain other microorganisms. In particular, the use of SPF chickens should be considered in antibody studies targeting the antigens of common pathogens or their analogues in birds. The immunization procedure of SPF hens is the same as that of ordinary laying hens, but the response to antigen and antibody response will be different (Schade et al. 1994). The biosafety conditions and the immunization protocol should also ensure that the SPF status of the hens is maintained.

Genetic differences exist between different breeds of hens that may play a role in both egg production (Chap. 2) and antibody response; these differences result in different immunological adaptations as determined in whole blood, timing of egg laying and egg laying capacity (Bílková et al. 2017).

Hens are subject to various biorhythms (Chap. 7) which may also affect IgY production.

10.2 Source of Antigens

Rational antigen design and high antigen quality are prerequisites in obtaining a high titre of specific IgY antibody. In the scope this book, antigens are non-self-chicken substances delivered to the chicken immune system and capable of inducing the production of IgY antibodies. Immunogenicity (the ability to induce cellular and humoral immune responses) and antigenicity (the ability to be specifically recognized by the antibodies generated as a result of the immune response) are two desirable antigen features for the development of higher performance antibodies.

10.2.1 Whole Virus or Bacteria

The entire composition of a whole virus may comprise dozens of antigens capable of causing a response by the host immune system (Chap. 3). Bacteria have outer cell walls and membranes (some also have pili and capsule), and contain many proteins and nucleic acids, which together form thousands of antigens. An inactivated antigen (usually refers to whole virus or bacteria) (Fig. 10.1) preserves antigenicity and immunogenicity, but has no capacity to replicate in the blood, tissues, or cells of an inoculated host, nor retain the infectivity, virulence and pathogenicity of the original pathogen (Herrera-Rodriguez et al. 2019). In some circumstances such as for a zoonotic pathogen, inactivation is important for preservation of the immunized chickens' welfare and to protect the laboratory team. Before inactivating an antigen for IgY production, it is recommended to check whether some antigenic regions remain after the inactivation protocol (Toinon et al. 2019).

Other methods used are treatment with Beta-propiolactone (BPL), formaldehyde (FA) and physical methods. BPL is commonly used for whole virus chemical inactivation (Fu et al. 2006; Nie et al. 2019); however, it drastically reduces the immunization performance (Toinon et al. 2019). There have been no studies of IgY antibody production by antigens inactivated using BPL.

Although the whole virus chemical inactivation by FA has lower efficiency than BPL, for whole bacterial inactivation the aldehydes are the most commonly used treatment (Hochel et al. 2004). Frequently, the aldehydes cause structural protein modifications and consequently reduce specific antibody production by the immunized host (Fan et al. 2015). In experiments using aldehyde-inactivated

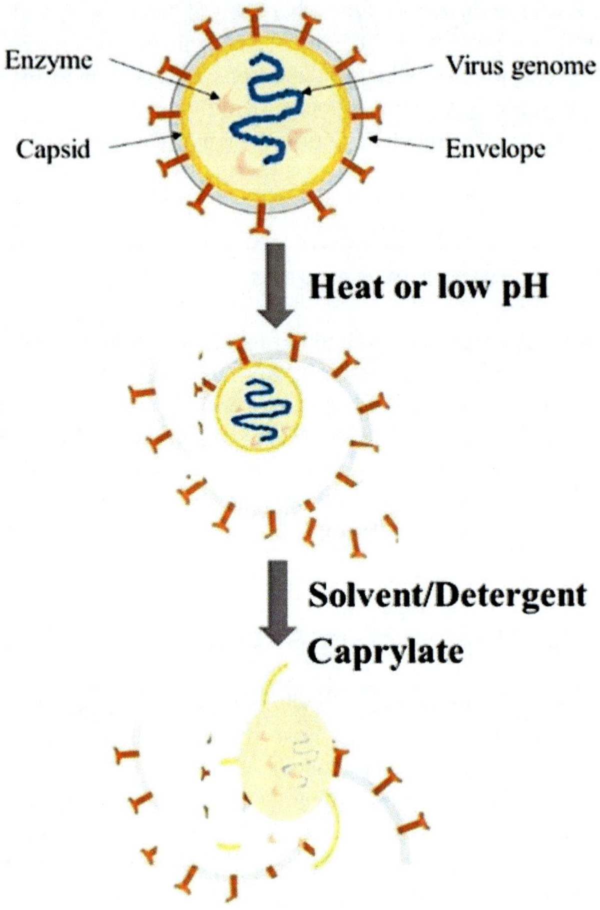

Fig. 10.1 Commonly used methods for inactivation of viruses Heat: Disruption of envelope/capsid. Low pH: ionic disruption of envelope and destructive conformational change of structural proteins. Solvent/Detergent: S/D-mediated disruption of envelope. Caprylate: partitioning of non-ionized molecule into envelope and disruption of envelope. (Source: XYZ-IgY lab)

bacteria it is necessary to compare them to the native proteins to confirm the remaining antigenicity. Water-soluble fractions (WSF) of protein containing IgY were isolated from the egg yolk of hens initially immunized with FA inactivated bacterium V. *parahaemolyticus* (Dong et al. 2008). IgY showed high affinity for their corresponding antigens with high titres from day 28 onwards. Protein contents and total IgY concentrations remained stable throughout the immunization period, whereas specific IgY concentrations increased steadily and reached a plateau at day 49 (Kassim et al. 2012).

Inactivation by physical methods are the most common methodologies for recovering whole parasite antigens (protozoa or helminths). Antigens can be extracted by freezing-thawing cycles (Kobayashi et al. 2004; Ferreira Júnior et al. 2012), sonication cycles (Manhani et al. 2011) and extraction by detergent treatment (Lopes et al. 2019; Carrara et al. 2020). Normally, antigens that are obtained by these processes do not suffer a radical inactivation; however, enormous quantities of antigen proteins are required for a satisfactory extraction and there are inevitable antigen losses (de Faria et al. 2019, Souza et al. 2020).

10.2.2 Recombinant Proteins

Recombinant protein systems have opened new avenues through the recombination of a selected fragment of pathogenic DNA into the genome of the expression host. Recombinant proteins are prepared by selecting and cloning the desired DNA fragment through PCR and the amplicon is inserted into an expression vector which is transformed in *E. coli* strains (prokaryotic system) or mammalian cells (eukaryotic system) that is induced to express a recombinant protein (Seto 2010; Doxey et al. 2018). Plasmids have a set of prokaryotic regions (for example, replication, antibiotic resistance) which are critical for amplification in a host; it also has an eukaryotic expression cassette (promoter, polyadenylation sequence) which is necessary for expression in eukaryotic cells (Prazeres and Monteiro 2014).

FanC (K99) gene of *Enterotoxigenic Escherichia coli* (ETEC) was cloned into pET32a (+) vectors, and recombinant vector was transferred into *E. coli* BL21 CodonPlus (DE3) to express protein. The specific IgY was obtained by immunizing the hens with the recombinant FanC protein and showed binding specifically to the FanC protein of ETEC (Nasiri et al. 2016). In another recent study, polymerase chain reaction was carried out to amplify InvG gene with InvG one pair of specific primer of *Salmonella typhimurium*. Recombinant pET32a-InvG plasmid was transferred to *E. coli* BL21 (DE3) to express recombinant InvG protein. The recombinant InvG protein was injected to laying hens, and obtained the anti-InvG IgY corroborated to bind specifically with the InvG protein of *Salmonella typhimurium* (Jahandar et al. 2019).

10.2.3 Haptens

Haptens are organic or inorganic small molecules (<10 kDa) with low/no immunogenicity (Gandhi et al. 2009), which can be converted to antigens by linking the hapten to a carrier protein (Li et al. 2013). Three carrier proteins have been routinely used: hemocyanin, ovalbumin (OVA) and bovine serum albumin (BSA) (Li et al. 2013, 2017 Liu et al. 2014).

Generally, the selection of the method for coupling hapten derivatives with carrier proteins is dependent on the types of active groups on the hapten derivative (Song et al. 2010):

1. The hapten derivative contains a carboxylic group (-COOH). The hapten derivative can be coupled with a carrier protein either by the carbodiimide, mixed anhydride or active ester methods;
2. The hapten derivative contains an amino group (-NH_2). The hapten derivative can be coupled with a carrier protein by either the glutaraldehyde, di-isocyanate,

halogenated nitrobenzene, imide ester, carbonyl chloride (phosgene) or diazotization methods;
3. The hapten derivative contains a hydroxyl group (-OH). The hapten derivative can be coupled with carrier protein either by succinic anhydride, azo benzoic acid or chlorine sodium acetate methods;
4. The hapten derivative contains a sulfhydryl group (-SH). The hapten derivative can be coupled with a carrier protein via a disulphide bond by using SAMSA (S-acid-based mercapto succinic anhydride) reaction;
5. The hapten derivative contains aldehyde (-CHO) or ketone groups (-CO-). The hapten derivative is first reacted with o-(carboxymethyl) hydroxylamine or p-hydrazine to form an intermediate containing a carboxylic group and then coupled based on that carboxylic group as outlined above.

Haptens and the corresponding hapten-carrier conjugates have been essential to the development of sensitive quantitative and qualitative immunoassays (Lemus and Karol 2008). However, due to the structural instability of the conjugate, the immunogenicity of artificial antigens may be quite different, and some artificial antigens cannot fully meet the requirements of immunoassay.

Haptens have been used in IgY generation. A specific IgY has been developed to detect florfenicol amine (FFA) residues (Liu et al. 2014). FFA was conjugated to BSA by glutaraldehyde (GA) and the conjugate was used to immunize hens. The titre of anti-FFA IgY antibody reached a peak of 1:128,000 after three booster injections. An indirect competitive enzyme-linked immunosorbent assay (icELISA) using specific anti-FFA IgY showed that the IC_{50} value was 12.30 ng/mL and the regression curve equation was $y = -13.71x + 64.95$ ($R^2 = 0.945$) (Liu et al. 2014).

10.2.4 Virus-Like Particles

Virus-like particles (VLPs) are highly effective types of subunit immunogens that mimic the overall structure of viral particles but without the infectious genetic material (Boschetti et al. 2005; Frietze et al. 2016). Predicted antigenic determinants are synthesized using genetic engineering vector techniques and heterologous protein expression systems. Cloning technology is used to construct the necessary structural genes and protein expression systems are used to produce VLPs; these are made up of monomers and are able to multimerize VLPs, and display the antigenic determinants of target pathogens on their surface (Noad and Roy 2003; Syomin and Ilyin 2019). VLPs have been recognized as safe and effective to induce specific antibody responses effectively with high titre and lymphocyte proliferation (Syomin and Ilyin 2019); VLPs have been increasingly used for vaccine design (Noad and Roy 2003). However, the production of VLPs still faces challenges, such as difficulties in forming the correct assembly, low recovery and purity during the preparation process. The insertion of epitopes in the protein loops on the surface of VLPs produces structural instability, and the whole protein cannot be tolerated during VLP assemblage, resulting in disruption of the capsid and/or the protein antigen (Brune and Howarth 2018).

VLPs have been applied in the generation of IgY. Hens were immunized with VLPs of stomatitis vesicular virus expressing recombinant antigens of human Norovirus to prevent and treat enteric diseases (Zhu et al. 2019). The VLPs were diluted in DMEM medium, delivered by an intramuscular route and boosted once at a 2-week interval. No side effects were observed in the hens with regard to food intake and egg production, and the specific IgY percentage increased from 0.7% to 5% (Zhu et al. 2019). In another recent study, hens were immunized with canine parvovirus virus-like particles (CPV-VLP). The IgY single chain fragment variables (IgY-scFv) were generated by a T7 phage display system and expressed in *E. coli*. The titre of the primary scFv library reached 1.5×10^6 pfu/mL, and 95% of the phage contained the target fragments (Ge et al. 2020).

10.2.5 Bacterial Ghosts

Bacterial ghosts (BGs) are the empty Gram-negative envelopes with the rigid peptidoglycan layer (Langemann et al. 2010). BGs are an appropriate immunogen for both oral or intranasal administration, and it functions as an adjuvant as it contains intrinsic properties such as lipopolysaccharide (LPS), monophosphoryl lipid A (MPL), peptidoglycan or flagella (Langemann et al. 2010; Lin et al. 2015; Si et al. 2017).

BGs have been produced by controlled expression of the cloned lysis gene *E*. The growth phase in a sample fermentation with *E. coli* harbouring plasmids for temperature-inducible E-lysis is conducted at 35 °C, pH 7.2 and aeration parameters sufficient for exponential growth. E-lysis of the culture is induced by temperature up-shift from 35 to 42 °C. The BGs product is harvested from the fermenter via tangential flow filtration (TFF) in a 0.2 μm hollow fibre module at a temperature of 15 °C. The TFF procedure as described above is an alternative to harvesting and washing BGs via centrifugation. The advantage of the implementation of TFF for harvesting and washing of BGs keeps all processes in a closed system and reduces the risk of cross-contamination during the handling procedure (Langemann et al. 2010).

Chickens were immunized subcutaneously with Newcastle disease virus (NDV) LPS-BGs adjuvanted vaccine and challenged with NDV through the oculo-nasal route on day 14 after the first vaccination. The titres of specific IgY and serum levels of IL-2 and IFN-gamma were significantly increased (Si et al. 2017).

10.2.6 Virosomes

Virosomes, used as an immunomodulatory nanocarriers to deliver various substances, such as vaccines, therapeutic drugs, or even genetic material to particular cells through receptor-mediated specific endocytosis, are a promising and successful method of application (Hofer et al. 2009; Blom et al. 2017). The size of the virosome impacts on the efficiency of immunization: mononuclear phagocytic systems efficiently engulf particles from 2 to 3 μm diameter by phagocytosis; $<$ 0.5 μm by pinocytosis; whereas $>$10 μm particles are not efficiently captured (Li et al. 2018).

While there is no reported use of virosomes to produce IgY antibodies, theoretically, virosome particles decorated with chicken phagocytosis stimulating macrophage molecules have the potential to induce high IgY titres (Dong et al. 2016). Chicken IgY is naturally recognized by receptors on macrophages and stimulates phagocytosis, and the generation of a chimeric molecule consisting of the *Bordetella avium* outer membrane protein and IgY-Fc fragment upgraded phagocytosis and the humoral and cellular immune response in immunized chicken (Dong et al. 2016).

10.2.7 Nucleic Acid Vaccines as Immunogens

Nucleic acid vaccines were introduced decades ago and have already been widely used to treat infectious and malignant diseases. Nucleic acid vaccination is achieved by injection with genetically engineered DNA (as a plasmid) or RNA (as mRNA). In contrast to recombinant bacteria or virus vaccines, nucleic acid vaccines consist only of DNA or RNA, which is taken up by cells and transformed into protein (Karimkhanilouyi and Ghorbian 2019).

An important strategy to achieve an immune response is to use DNA plasmids having antigens encoded on them. This antigen-encoding DNA plasmid can induce humoral and cellular immune response against parasites, bacteria and disease-producing viruses (Khan 2013).

Genes of interest coupled with a suitable promoter are injected directly into muscle or coated into gold micro particles and "shot" into the skin by pressurized gas using a gene gun. This can induce cellular and humoral immunity in experimental animals for a longer period of time. The mechanism appears to be through uptake and expression of the DNA in antigen presenting cells (APCs) (Khan 2013).

A number of features have to be considered in the design of DNA immunogens. The selection of the antigen sequence, the vector, the delivery route, the dose, adjuvants, and boosting agents will all affect the outcome of immunization. The selection of a suitable target antigen from the pathogen should be given first priority while designing a DNA vaccine; whether the gene is mutated or wild type, intracellular or membrane-bound or secreted. After the selection of the desired gene, modifications can be made to achieve the desired immunogenicity of the DNA vaccine (Khan 2013).

The use of mRNA has many beneficial features over subunit vaccines, killed or attenuated vaccines, as well as DNA-based vaccines. Exogenous mRNA is inherently immunostimulatory, as it is recognized by a variety of cell surface, endosomal and cytosolic innate immune receptors (Pardi et al. 2018). It is potentially advantageous for vaccination because in some cases it may provide adjuvant activity to drive dendritic cell (DC) maturation and thus elicit robust T and B cell immune responses. Although the paradoxical effects of innate immune sensing on different formats of mRNA vaccines are incompletely understood, some progress has been made in recent years in elucidating these phenomena (Pardi et al. 2018).

10.2.8 Vaccine Products as Antigens

Commercial vaccine products which include a variety of vaccine preparation techniques can be used as antigens for IgY production. There are factors that influence humoral and cellular vaccine responses in animals and there is substantial variation between individuals in the immune response to vaccination.

A vaccine product used as antigen may reduce the cost and difficulty of antigen preparation and is able to provide large amount of immunogen easily. By using the vaccine product, it may also offer the convenience to generate antibody targeting to multiple antigens at the same time. However, the vaccine product may fail to induce high antibody titre in the host as the purpose of vaccine design is initially to generate systemic protective effects, instead of single antibody titre development. Furthermore, it may also limit the refined design of immunogen of need, and the composition of polyclonal antibodies to multiple targets can also complicate the evaluation of specific antibodies as well as therapeutic effects.

A dual IgY against both Rotavirus (RV) and norovirus (NoV) was developed through immunization of hens with a divalent vaccine comprising neutralizing antigens of both RV and NoV (Dai et al. 2013). Approximately 45 mg of IgY was readily obtained from each yolk with high titres of anti-P particle and anti-VP8 antibodies. Reduction of RV replication were observed with viruses treated with the IgY before and after inoculation into cells, suggesting an application of the IgY as both a prophylactic and a therapeutic treatment (Dai et al. 2013).

10.3 Routes of Administration

Due to the biological, anatomical and physiological characteristics of avian species, there are various immunization routes in chickens for example: intramuscular injection (i.m.) (Schade et al. 2005), subcutaneous injection (s.c.), intravenous injection (i.v.) (Cai et al. 2010), nasal drip, drinking water (Tokarzewski 2002) and eye drop and aerosol (Noormohammadi and Whithear 2019). The best route of administration should be chosen according to the characteristics of the antigen and the nature of the adjuvant to ensure an immune response.

The most common multipoint injection route for immunization of hens is i.m. injection (Schade et al. 2005). The specific IgY antibody titres were compared after injections of the same antigen via an i.m., s.c. or i.v. route (Schwarzkopf and Thiele 1994; Schade et al. 2005). It was shown that the s.c. antigen injection provoked a higher titre than injection via the i.m. route. The i.v. should only be used without adjuvants. A combination of i.m. antigen injections with a final i.v. injection, frequently significantly increased the IgY titre. However, i.v. injection (without adjuvant) must be performed very slowly (approximately 500 μL over 15 minutes) to avoid an anaphylactic reaction.

Group immunization generally includes oral immunization, nasal drip immunization, drinking water immunization and aerosol immunization. Oral immunization is possible (Hedlund and Hau 2001; Schade et al. 2005). At days 1, 14 and 33 of an immunization protocol, two hens per group were orally given 100 μg of human IgG, together with various adjuvants (Cholera Toxin B-subunit, dimethyl dioctadecyl ammonium bromide or Softigen®) carried on poly (D, L-lactide-co-glycolide) microspheres. Significant IgY titres were provoked by using Softigen® and Cholera Toxin B-subunit, but these titres were considerably lower than those in i.m. inoculated hens (Schade et al. 2005). Nasal drip immunization is commonly used in chicks to stimulate the body to produce antibodies locally or systemically through mucosal absorption, and is mostly applicable to live, weakened vaccines. Drinking water immunization is mostly suitable for attenuated vaccines, which are easy to use and safe, and the general dose should be 2–3 times the amount injected. Aerosol immunization is suitable for large intensive farms, where the method is simple, quick and particularly effective with vaccines that are respiratory-friendly. SPF chickens were vaccinated with *Mycoplasma gallisepticum* (MG) vaccine strains 6/85 by aerosol inoculation, systemic antibody response following aerosol vaccination was initially weak but increased gradually over time (Noormohammadi and Whithear 2019).

A gene gun immunization involves the delivery of plasmid-coated nonporous inert microparticles (0.1 to 5 μm) by a helium-propelled particle bombardment methodology which is able to delivery DNA-vaccines (–1 to 10 μg DNA) to antigen presenting cells (Dendritic cells) that home to the top layers of skin (Gary and Weiner 2020). The basic principle of the gene gun is the foreign DNA coated on the surface of tiny gold particles and tungsten particles, and the particles are injected into the recipient cells or tissues at high speed under high pressure. The skin of the chest and thigh sites of hens are suitable for gene gun-based DNA immunization. Feathers at the inoculation site should be removed, then the stratum corneum should be scraped off with a blade, and finally disinfected with alcohol. All processes should be carried out in a dry and clean environment. The operator should wear disposable gloves, carefully put the "bullet" into the magazine, connect the gene gun to the helium cylinder, adjust the air pressure to 300 psi, and pull the trigger within 15 seconds after pressing the safety valve (Fig. 10.2) (Sanford et al. 1987).

Gene gun technology has the advantages of high efficiency and convenience and has attracted increasing attention. As a purely physical gene transfer method, the gene gun technology has the following advantages: it is simple and can transfer genes quickly and easily; it has almost no requirements for target cells; it requires a small number of therapeutic genes (Sanford et al. 1987). The level of antibodies in the hen's blood after gene gun immunization is 1.5 times that of intramuscular injection, while the amount of DNA used is only 3% of that for intramuscular injections (Witkowski et al. 2009).

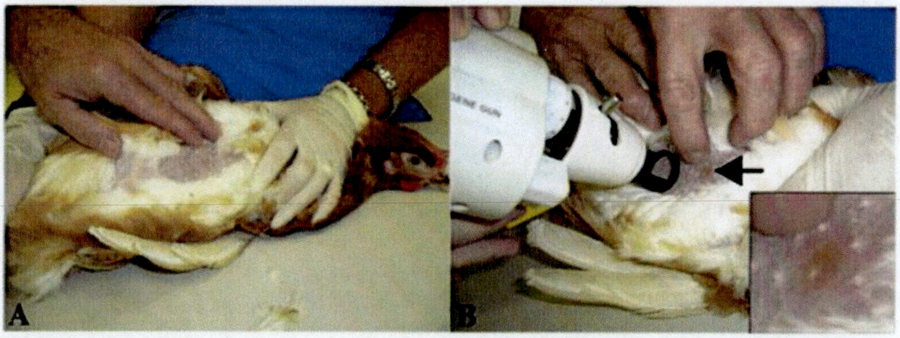

Fig. 10.2 Gene gun immunisation of a chicken (**a**) For DNA immunisation the chicken was fixed horizontally to access an Arteriae (feather-less region). (**b**) The feather-free area should be sufficiently large to allow two non-overlapping applications per breast side. The arrow shows gold ammunition on the subject's skin—a control for vaccine delivery. For greater detail a scale-up of the area can be seen in the lower right corner of panel B. Figure from (Witkowski et al. 2009) with publisher's permission

10.4 Amount of Antigen

The immune response, antibody concentration and antibody titre are significantly affected by the antigen dose. The optimal immune dose (antigen concentration) varies with the immunogen, and within a certain dosage range, the immune effect will be enhanced with an increase in antigen concentration. The optimal immune dose for different antigens should be obtained through specific experiments. In general, small antigens may be required in large doses or coupled to the carrier (amino acid).

A number of studies of the immune response of hens to varying doses of antigen have been performed. The results of a study by Behn et al. showed that a 0.1 mg dose of mouse IgG was more effective than a 1 mg dose in inducing an antibody response in hens (Behn et al. 1996). It is generally accepted from studies of Erhard et al. that the immune effect is best at an antigen dose of 10–1000 μg (Erhard et al. 2000). The results of a Schade et al. study found that immune injection of 5.0 mg or 2.5 mg human albumin significantly inhibited antibody titre expression compared with 1.0 mg or 0.5 mg human albumin (Schade et al. 2005).

10.5 Adjuvant

Adjuvant is a kind of non-specific immune enhancer that can enhance or alter the immune response of the host; it can be delivered to the host in advance or together with the antigens to promote a stronger immune response. Although there is no direct correlation between this stimulus and the antigen, the selection of different adjuvants will also result in different immunological effects. In the selection of an adjuvant one

Table 10.1 Adjuvants used in the immunization of chickens

Adjuvant	Chemistry composition	Major activity	Case	Reference
FCA/FIA	Heat-inactivated *Mycobacterium tuberculosis*, paraffin oil and lanolin	Activity from sustained release of immunogens in oil droplets	*Cryptosporidium parvum* infection in scid mice	Kobayashi et al. (2004)
AHBA	Aluminum hydroxyl phosphate complex	AHBA delays the consumption of antigen and prolongs the duration of stimulation of the immune system	Influenza (H9N2) and Newcastle inactive vaccines	Jafari et al. (2017)
Emulsigen	Dimethyldioctadecyl ammonium bromide (DDA)	Induces the enhancement of immune responses by increasing the surface area of antigens in oil-in-water emulsions	*Staphylococcus aureus* Efb and map proteins	Grzywa et al. (2015)
Chitosan	Deacetylation product of chitin	Enhances Th1 immunity	Inactivated Newcastle disease vaccine	Yang et al. (2020)
Montanide	Mineral oil and surfactant from mannide monooleate family	The immunostimulatory activity works by depot formation at the site of injection resulting in a slower release of antigens	EF-1α co-administered with chicken IL-7 DNA vaccine	Panebra and Lillehoj (2019)
Alginate	(1–4)-linked β-D-mannuronic acid and α-L-guluronic acid	Enhances the immunogenicity of vaccines	Diphtheria toxoid	Sarei et al. (2013)
Solid lipid nanoparticles	Fatty acids or mono-, di-, or triglycerides	High physical stability and high encapsulation efficiency	*Mycoplasma bovis* PG 45 R9 antigen	Olbrich et al. (2002)

Notes: *FCA* Freund Compete Adjuvant; *FIA* Freund Incomplete Adjuvant does not contain *Mycobacterium tuberculosis antigens* needs to consider the possible side effects to the inoculated animal, including local or systemic cytotoxicity and inflammatory reactions (Apostólico et al. 2016). Adjuvants currently used in chickens to enhance the efficacy of immunization are summarised in Table 10.1.

Lymphocytes comprising white blood cells, are divided into T cells and B cells, and both of which can be stimulated by adjuvants. T cells and secreted cytokines are associated with adaptive or cell-mediated immune responses, whereas B cells and antibodies are key factors in the humoral immune response (Noh et al. 2019, Maqbool et al. 2019, Chap. 3). Adjuvants which act in several non-mutually exclusive ways to augment the adaptive immune response and to generate effective more specific immunological memory B cells, can affect the migration, maturation, antigen presentation and expression of costimulatory molecules by dendritic cells (DCs), and in turn improve the responses to antigen of B cells (McKee et al. 2007). Furthermore, some adjuvants enhance cross presentation by Dendritic cells of MHC I-restricted antigens to $CD8^+$ T cells, act directly on B cells, improving their proliferation and/or conversion into memory B cells that are essential for the successful immunization, and stimulate the rapid and sustained production of high titres of antibodies with high avidity (McKee et al. 2007; Coffman et al. 2010).

The development of novel adjuvants that retain or improve immunization efficacy while avoiding potentially serious side effects is a worthwhile undertaking (Moore and Loxton 2019). It is an interesting point to discuss that the use of adjuvant for antibody generation may be somehow different to the development of a vaccine as antibody generation focuses more on B cell activation and immunoglobulin related pathways, while vaccine design requires more systematic effects of adjuvant.

10.6 Immunization Intervals

There is little consensus on the selection of the optimal interval between multiple immunizations. The general rule is that the interval between the first and second antigen injections should be at least four weeks to allow enough time to generate "immune memory". The study of Messerschmid & Sarkar supports this principle by immunizing chickens at week 0 (day 1), week 10, and week 15 to obtain high concentrations antibody of 1: 160,000 (Messerschmid and Sarkar 1993). Similar results were obtained by Hilnak and Schwarzkopf who recommended long interval immunizations (Hlinak et al. 1996; Schwarzkopf et al. 2000). If the immune interval is too short, it may cause "immune suppression".

Longer term intervals between immunizations (>2 months) may be advisable when hens are planned to use for life-long period of time to minimize the experimental costs and maximize the IgY harvest. When monitored over a 2-year period, stable antibody titres of 1: 100,000 to 1: 1,000,000 were obtained following up to 10 booster injections of botulinum toxin with the strategy of longer booster intervals (11 injections of 10 to 20 μg of immobilized native toxin) (Pauly et al. 2009).

10.7 Monitoring the IgY Titre

In mammals, the initial and secondary immune responses occur sequentially after initial and secondary immunizations. A similar process is observed in the immune response of chickens: IgY begins to appear in the serum approximately 4 days after the first antigen injection; reaches a maximum concentration in 6–8 days and then decreases; there is a significant increase in the amount of IgY with a booster injection of the antigen (Woolley and Landon 1995). The concentration of IgY in the yolk is essentially constant (10–20 mg/mL) and is 1.23 times the serum concentration (Woolley and Landon 1995). A delay of 3 to 4 days is observed for the appearance of specific IgY in yolk after first appearance of the specific IgY in the serum of a hen (Woolley and Landon 1995). There is a significant circaseptan rhythm in yolk IgY and circaquattran rhythm in serum IgY. The serum IgY concentration reaches a peak in the morning, decreases to a minimum during the daytime and increases again at night revealing a significant circadian rhythm was superimposed by an ultradian rhythm (He et al. 2014) and Chap. 7.

10.8 Immunization for the Generation of IgY-scFv

For the purpose of developing IgY-scFv (Chap. 13), the immunization scheme is in principle the same as for the generation of polyclonal IgY. After the last immunization (usually the third or fourth booster immunization) when a satisfactory polyclonal IgY titre is observed, the hen spleen is harvested for RNA extraction.

10.9 Discussion

The effect of immunization is affected by multiple parameters as discussed above. Determining the optimum protocol for immunization will rely on the consideration of each of the parameters discussed above and their impact on each other. Each laboratory may explore the most suitable immunization program according to their conditions, skills and research needs. There are scientific needs to further optimize the immunization program, and to further develop novel immunogens, adjuvants and delivery systems, particularly as the impact of avian characteristics on immunization is taken into consideration.

References

Apostólico JS, Lunardelli VA, Coirada FC, Boscardin SB, Rosa DS (2016) Adjuvants: classification, modus operandi, and licensing. J Immunol Res 2016:1459394

Behn II, Hommel U, Oertel M, Hauschildt S (1996) Kinetics of IgY formation after immunisation of hens with different protein antigens. ALTEX 13(5):18–21

Bílková B, Swiderska Z, Janda J, Zita L, Vinkler M (2017) Different breeds, different blood: cytometric analysis of whole blood cellular composition in chicken breeds. Vet Immunol Immunopathol 188:71–77

Blom RAM, Amacker M, Moser C, Van Dijk RM, Bonetti R, Seydoux E, Hall SRR, Garnier CV, Blank F (2017) Virosome-bound antigen enhances DC-dependent specific CD4+ T cell stimulation, inducing a Th1 and Treg profile in vitro. Nanomed Nanotechnol Biol Med 13 (5):1725–1737

Boschetti N, Stucki M, Späth PJ, Kempf C (2005) Virus safety of intravenous immunoglobulin: future challenges. Clin Rev Allergy Immunol 29(3):333–344

Brune KD, Howarth M (2018) New routes and opportunities for modular construction of particulate vaccines: stick, click, and glue. Front Immunol 9:1432

Cai YC, Chen JX, Guo J, Chen SH, Tong XM, Tian LG (2010) Dynamic observation of chicken egg yolk antibodies against soluble egg antigen of Schistosoma japonicum obtained by two immunization routes. Chinese J Parasitol Parasitic Dis 28(2):121–124

Carrara GMP, Silva GB, Faria LS, Nunes DS, Ribeiro VS, Lopes CA, Gonçalves-Pires M, Santos MM, Borges IP, Ferreira-Junior Á, Ávila VMR, Costa-Cruz JM (2020) IgY antibody and human neurocysticercosis: a novel approach on immunodiagnosis using Taenia crassiceps hydrophobic antigens. Parasitology 147(2):240–247

Coffman RL, Sher A, Seder RA (2010) Vaccine adjuvants: putting innate immunity to work. Immunity 33(4):492–503

Dai YC, Zhang XF, Tan M, Huang P, Lei W, Fang H, Zhong W, Jiang X (2013) A dual chicken IgY against rotavirus and norovirus. Antivir Res 97(3):293–300

Dong D, Liu H, Xiao Q, Li R (2008) Affinity purification of egg yolk immunoglobulins (igy) with a stable synthetic ligand. J Chromatogr 870(1):51–54

Dong W, Zhang H, Huang H, Zhou J, Hu L, Lian A, Zhu L, Ma N, Yang P, Wei K, Zhu R (2016) Chicken IgY fc linked to Bordetella avium ompA and Taishan Pinus massoniana pollen polysaccharide adjuvant enhances macrophage function and specific immune responses. Front Microbiol 7:1708

Doxey AC, Mansfield MJ, Montecucco C (2018) Discovery of novel bacterial toxins by genomics and computational biology. Toxicon 147:2–12

Erhard MH, Schmidt P, Zinsmeister P, Hofmann A, Münster U, Kaspers B, Wiesmüller KH, Bessler WG, Stangassinger M (2000) Adjuvant effects of various lipopeptides and interferon-gamma on the humoral immune response of chickens. Poult Sci 79(9):1264–1270

Fan YC, Chiu HC, Chen LK, Chang GJ, Chiou SS (2015) Formalin inactivation of Japanese encephalitis virus vaccine alters the antigenicity and immunogenicity of a neutralization epitope in envelope protein domain III. PLoS Negl Trop Dis 9(10):e0004167

Faria LS, de Souza DLN, Ribeiro RP, de Sousa JEN, Borges IP, Ávila VMR, Ferreira-Júnior Á, Goulart LR, Costa-Cruz JM (2019) Highly specific and sensitive anti-Strongyloides venezuelensis IgY antibodies applied to the human strongyloidiasis immunodiagnosis. Parasitol Int 72:101933

Ferreira Júnior Á, Santiago FM, Silva MV, Ferreira FB, Macêdo Júnior AG, Mota CM, Faria MS, Filho HHS, Silva DA, Cunha-Júnior JP, Mineo JR, Mineo TW (2012) Production, characterization and applications for toxoplasma gondii-specific polyclonal chicken egg yolk immunoglobulins. PLoS One 7(7):e40391

Frietze KM, Peabody DS, Chackerian B (2016) Engineering virus-like particles as vaccine platforms. Curr Opin Virol 18:44–49

Fu CY, Huang H, Wang XM, Liu YG, Wang ZG, Cui SJ, Gao HL, Li Z, Li JP, Kong XG (2006) Preparation and evaluation of anti-SARS coronavirus IgY from yolks of immunized SPF chickens. J Virol Methods 133(1):112–115

Gandhi S, Caplash N, Sharma P, Suri CR (2009) Strip-based immunochromatographic assay using specific egg yolk antibodies for rapid detection of morphine in urine samples. Biosens Bioelectron 25(2):502–505

Gary, Weiner (2020) DNA vaccine: prime time is now. Curr Opin Immunol 65:21–27. https://doi.org/10.1016/j.coi.2020.01.006

Ge SK, Xu L, Li B, Zhong FG, Liu X, Zhang XY (2020) Canine parvovirus is diagnosed and neutralized by chicken IgY-scFv generated against the virus capsid protein. Vet Res 51(1):110

Grzywa R, Walczak M, Łupicka-Słowik A, Bobrek K, Boivin S, Brown EL, Gaweł A, Stefaniak T, Oleksyszyn J, Sieńczyk M (2015) Adjuvant-dependent immunogenicity of Staphylococcus aureus Efb and map proteins in chickens. Vet Immunol Immunopathol 166(1–2):50–56

He JX, Thirumalai D, Schade R, Zhang XY (2014) Chronobiological studies of chicken IgY: monitoring of infradian, circadian and ultradian rhythms of IgY in blood and yolk of chickens. Vet Immunol Immunopathol 160(3–4):266–272

Hedlund GP, Hau J (2001) Oral immunisation of chickens using cholera toxin B subunit and Softigen as adjuvants results in high antibody titre in the egg yolk. In Vivo (Atnens, Greece) 15 (5):381–384

Herrera-Rodriguez J, Signorazzi A, Holtrop M, de Vries-Idema J, Huckriede A (2019) Inactivated or damaged? Comparing the effect of inactivation methods on influenza virions to optimize vaccine production. Vaccine 37(12):1630–1637

Hlinak A, Schrödl W, Witt S, Schade R (1996) Krüger M. production of egg yolk antibodies against human cell-associated antigens. ALTEX 13(5):76–77

Hochel I, Viochna D, Škvor J, Musil M (2004) Development of an indirect competitive elisa for detection of campylobacter jejuni subsp.jejuni o:23 in foods. Folia Microbiol 49(5):579

Hofer U, Lehmann AD, Waelti E, Amacker M, Gehr P, Rothen-Rutishauser B (2009) Virosomes can enter cells by non-phagocytic mechanisms. J Liposome Res 19(4):301–309

Jafari M, Pour MM, Taghizadeh M, Masoudi S, Bayat Z (2017) Comparative assessment of humoral immune responses of aluminum hydroxide and oil-emulsion adjuvants in influenza (H9N2) and Newcastle inactive vaccines to chickens. Artificial Cells, Nanomed Biotechnol 45 (1):84–89

Jahandar MH, Nassiri M, Nasiri K, Haghparast A (2019) Production and purification of specific igy against invg protein of salmonella typhimurium. Int J Infect 6(2):e87683

Karimkhanilouyi S, Ghorbian S (2019) Nucleic acid vaccines for hepatitis B and C virus. Infect Genet Evol 75:103968

Kassim N, Mtenga AB, Shim WB, Chung DH (2012) The in vitro and in vivo efficacy of hen IgY against Vibrio parahaemolyticus and Vibrio vulnificus. J Microbiol Biotechnol 22 (10):1423–1431

Khan KH (2013) DNA vaccines: roles against diseases. Germs 3(1):26–35

Kobayashi C, Yokoyama H, Nguyen SV, Kodama Y, Kimata T, Izeki M (2004) Effect of egg yolk antibody on experimental Cryptosporidium parvum infection in scid mice. Vaccine 23 (2):232–235

Langemann T, Koller VJ, Muhammad A, Kudela P, Mayr UB, Lubitz W (2010) The bacterial ghost platform system: production and applications. Bioengineered Bugs 1(5):326–336

Lemus R, Karol MH (2008) Conjugation of haptens. Methods Mol Med 138:167–182

Li T, Liu H, Cai K, Tian M, Wang Q, Shi J, Gao X, Wang H (2013) Hypersensitive detection and quantitation of BoNT/a by IgY antibody against substrate linear-peptide. PLoS One 8(3):e58908

Li H, Tatematsu K, Somiya M, Iijima M, Kuroda S (2018) Development of a macrophage-targeting and phagocytosis-inducing bio-nanocapsule-based nanocarrier for drug delivery. Acta Biomater 73:412–423

Li C, Zhang Y, Eremin SA, Yakup O, Yao G, Zhang X (2017) Detection of kanamycin and gentamicin residues in animal-derived food using IgY antibody based ic-ELISA and FPIA. Food Chem 227:48–54

Lin A, Merkley ED, Clowers BH, Hutchison JR, Kreuzer HW (2015) Effects of bacterial inactivation methods on downstream proteomic analysis. J Microbiol Methods 112:3–10

Liu XS, Yang WJ, He JX, Zhao JZ, Justo GC, Zhang XY (2014) Preparation of artificial antigen and egg yolk-derived immunoglobulin (IgY) of florfenicol amine for ELISA assay. J Environ Sci Health B 49(2):109–115

Lopes CA, de Faria LS, de Sousa JEN, Borges IP, Ribeiro RP, Bueno LL, Rodrigues Ávila VM, Ferreira Júnior Á, Costa-Cruz JM (2019) Anti-Ascaris suum immunoglobulin Y as a novel biotechnological tool for the diagnosis of human ascariasis. J Helminthol 94:e71

Mack NO, Bell D (1990) Commercial chicken production manual, 4th edn. AVI Publishing Company, USA

Manhani MN, Ribeiro VS, Cardoso R, Ueira-Vieira C, Goulart LR, Costa-Cruz JM (2011) Specific phage-displayed peptides discriminate different forms of neurocysticercosis by antibody detection in the serum samples. Parasite Immunol 33(6):322–329

Maqbool B, Wang Y, Cui X, He S, Guan R, Wang S, Wang Y, Hu S (2019) Ginseng stem-leaf saponins in combination with selenium enhance immune responses to an attenuated pseudorabies virus vaccine. Microbiol Immunol 63(7):269–279

McKee AS, Munks MW, Marrack P (2007) How do adjuvants work? Important considerations for new generation adjuvants. Immunity 27(5):687–690

Messerschmid S, Sarkar S (1993) Raising polyclonal antibodies against tmv in a hen utilizing the immunological memory. Anzger für Schdlingskunde Pflanzenschutz Umweltschutz 66 (8):145–147

Moore DK, Loxton AG (2019) Regulatory B lymphocytes: development and modulation of the host immune response during disease. Immunotherapy 11(8):691–704

Nasiri K, Zibaee S, Nassiri M, Tahmoorespur M, Haghparast A (2016) Production of specific IgY antibody to the recombinant FanC protein produced in Escherichia coli. Iran J Basic Med Sci 19 (8):883–889

Nie W, Zhao C, Guo X, Sun L, Meng T, Liu Y, Song X, Xu K, Wang J, Li J (2019) Preparation and identification of chicken egg yolk immunoglobulins against human enterovirus 71 for diagnosis of hand-foot-and-mouth disease. Anal Biochem 573:44–50

Noad R, Roy P (2003) Virus-like particles as immunogens. Trends Microbiol 11(9):438–444

Noh EM, Kim JM, Lee HY, Song HK, Joung SO, Yang HJ, Kim MJ, Kim KS, Lee YR (2019) Immuno-enhancement effects of Platycodon grandiflorum extracts in splenocytes and a cyclophosphamide-induced immunosuppressed rat model. BMC Complement Altern Med 19 (1):322

Noormohammadi AH, Whithear KG (2019) Comparison of the short-term and long-term efficacies of the mycoplasma gallisepticum vaccines ts-11 and 6/85. Avian Pathol 48(3):238–244

Olbrich C, Müller RH, Tabatt K, Kayser O, Schulze C, Schade R (2002) Stable biocompatible adjuvants–a new type of adjuvant based on solid lipid nanoparticles: a study on cytotoxicity, compatibility and efficacy in chicken. Altern Lab Anim 30(4):443–458

Panebra A, Lillehoj HS (2019) Eimeria tenella elongation factor-1α (EF-1α) Coadministered with chicken IL-7 (chIL-7) DNA vaccine emulsified in Montanide gel 01 adjuvant enhanced the immune response to E. acervulina infection in broiler chickens. Avian Dis 63(2):342–350

Pardi N, Hogan MJ, Porter FW, Weissman D (2018) mRNA vaccines–a new era in vaccinology. Nature reviews. Drug Discovery 17(4):261–279

Pauly D, Dorner M, Zhang X, Hlinak A, Dorner B (2009) Monitoring of laying capacity, immunoglobulin Y concentration, and antibody titer development in chickens immunized with ricin and botulinum toxins over a two-year period. Poult Sci 88(2):281–290

Prazeres DMF, Monteiro GA (2014) Plasmid biopharmaceuticals. Microbiol Spect 2(6):1–18

Sanford JC, Klein TM, Wolf ED (1987) Delivery of substances into cells and tissues using a particle bombardment process. J Particulate Sci Technol 5:27–37

Sarei F, Dounighi NM, Zolfagharian H, Khaki P, Bidhendi SM (2013) Alginate nanoparticles as a promising adjuvant and vaccine delivery system. Indian J Pharm Sci 75(4):442–449

Schade R, Bürger W, Schöneberg T, Schniering A, Schwarzkopf C, Hlinak A, Kobilke H (1994) Avian egg yolk antibodies. The egg laying capacity of hens following immunisation with antigens of different kind and origin and the efficiency of egg yolk antibodies in comparison to mammalian antibodies. ALTEX 11(2):75–84

Schade R, Calzado EG, Sarmiento R, Chacana PA, Porankiewicz-Asplund J, Terzolo HR (2005) Chicken egg yolk antibodies (IgY-technology): a review of progress in production and use in research and human and veterinary medicine. Altern Lab Anim 33(2):129–154

Schwarzkopf C, Staak C, Behn I (2000) Immunisation. In: Chicken egg yolk antibodies, production and application. IgY-technology. Springer Lab Manuals, Berlin, Heidelberg, New York, pp 25–64

Schwarzkopf C, Thiele B (1994) Effectivity of different methods for the extraction and purification of IgY. ALTEX 13(5):35–39

Seto D (2010) Viral genomics and bioinformatics. Viruses 2(12):2587–2593

Si W, Yu S, Liu H, Wang C, Chen L, Wang G, Wu J, Liu S (2017) A bacterial ghost improves the immunological efficacy of a Newcastle disease virus inactivated vaccine. Vet Microbiol 203:189–195

Song J, Wang RM, Wang YQ, Tang YR, Deng AP (2010) Hapten design, modification and preparation of artificial antigens. Chin J Anal Chem 38(8):1211–1218

Souza DC, de Faria LS, Sousa JEN, Lopes CA, Ribeiro VDS, da Silva VJ, Ribeiro RP, Rabelo ÉML, Ávila VMR, Ferreira Júnior Á, Costa-Cruz JM (2020) Use of polyclonal IgY antibodies to detect serum immune complexes in patients with active hookworm infection. Parasitology 147(6):715–720

Syomin BV, Ilyin YV (2019) Virus-like particles as an instrument of vaccine production. J Mol Biol 53(3):323–334

Toinon A, Moreno N, Chausse H, Mas E, Nicolai MC, Guinchard F, Jaudinaud I, Guinet-Morlot F, Riou P, Manin C (2019) Potency test to discriminate between differentially over-inactivated rabies vaccines: agreement between the NIH assay and a G-protein based ELISA. Biologicals 60:49–54

Tokarzewski S (2002) Influence of enrofloxacin and chloramphenicol on the level of IgY in serum and egg yolk after immunostimulation of hens with salmonella enteritidis antigens. Pol J Vet Sci 5(3):151–158

Witkowski PT, Bourquain DR, Hohn O, Schade R, Nitsche A (2009) Gene gun-supported DNA immunisation of chicken for straightforward production of poxvirus-specific IgY antibodies. J Immunol Methods 341(1–2):146–153

Woolley JA, Landon J (1995) Comparison of antibody production to human interleukin-6 (IL-6) by sheep and chickens. J Immunol Methods 178(2):253–265

Yang Y, Xing R, Liu S, Qin Y, Li K, Yu H, Li P (2020) Chitosan, hydroxypropyltrimethyl ammonium chloride chitosan and sulfated chitosan nanoparticles as adjuvants for inactivated Newcastle disease vaccine. Carbohydr Polym 229:115423

Zhu Y, Ma Y, Lu M, Zhang Y, Li A, Liang X, Li JR (2019) Efficient production of human norovirus-specific IgY in egg yolks by vaccination of hens with a recombinant vesicular stomatitis virus expressing vp1 protein. Viruses 11(5):444

Zimmermann P, Curtis N (2019) Factors that influence the immune response to vaccination. Clin Microbiol Rev 32(2):e00084–e00018

Extraction and Purification of IgY 11

Patricia M. Morgan, Mara G. Freire, Ana Paula M. Tavares, Antonysamy Michael, and Xiaoying Zhang

Abstract

Immunoglobulin Y (IgY) is an antibody present in the yolk of the egg and to obtain it at high purity levels, while retaining its biological activity, multi-step processes are required. The initial step is removal of the lipid fraction from the yolk. The delipidation methods routinely used are water dilution or precipitation by either polyethylene glycol (PEG), anionic polysaccharides or organic solvents. The second step is the extraction of the IgY from the remaining proteins in the mixture, which can be achieved by precipitation with a variety of salts, or additional PEG steps or by aqueous biphasic systems. The final step in IgY purification involves chromatographic methods based on different separation techniques such as cation exchange, hydrophobic charge-induction and various affinity systems giving IgY with >90% purity. In this chapter, comparisons

P. M. Morgan (✉)
Emerita Lecturer, School of Natural Sciences, National University of Ireland Galway, Galway, Ireland
e-mail: pat.morgan@nuigalway.ie

M. G. Freire · A. P. M. Tavares
CICECO-Aveiro Institute of Materials, Department of Chemistry, University of Aveiro, Aveiro, Portugal
e-mail: maragfreire@ua.pt; aptavares@ua.pt

A. Michael
Department of Microbiology, PSG College of Arts and Science, Coimbatore, Tamil Nadu, India

X. Zhang
College of Biological Science and Engineering, Shaanxi University of Technology, Hanzhong, China
e-mail: zhxying@snut.edu.cn; xzhang67@uoguelph.ca; zhang@bio.uminho.pt

X.-Y. Zhang et al. (eds.), *IgY-Technology: Production and Application of Egg Yolk Antibodies*, https://doi.org/10.1007/978-3-030-72688-1_11

among different methods of IgY extraction applied to a single batch of eggs and the commercial extraction kits currently available are discussed. Finally, the techniques for the characterization of the IgY with regard to purity and activity are outlined as well as the appropriate storage conditions.

Keywords

Extraction · Purification · Chromatography · Characterization · Storage

11.1 Introduction

In this chapter, various methods for IgY extraction and purification from hen egg yolk are discussed. The knowledge of the composition of the original biological matrix and properties of the solutes are fundamental to rationalise the best extraction and purification conditions aiming at attaining antibodies of high purity and biological activity (Walsh 2013). The formation of a chicken egg is a complex process, starting from the growth of a single follicle and culminating in the passage of the egg through the oviduct and finally egg laying (Chap. 2). The egg must contain all the nutrients (proteins, lipids, minerals and vitamins) necessary for the development of the chick and antibodies, namely IgY, for its protection until it develops its own immune response. These nutrients are derived from the hen plasma and transported to the ovary and then into the egg (Schjeide et al. 1963). The yolk is surrounded by a two-layer vitelline membrane (Sunwoo and Gujral 2015). The composition of the yolk is approximately 48% water, 33% lipids and 17% proteins (Burley 1991). The yolk, after aqueous dilution, can be separated into two main fractions by low speed centrifugation (Fig. 11.1): (i) the major fraction is an aqueous

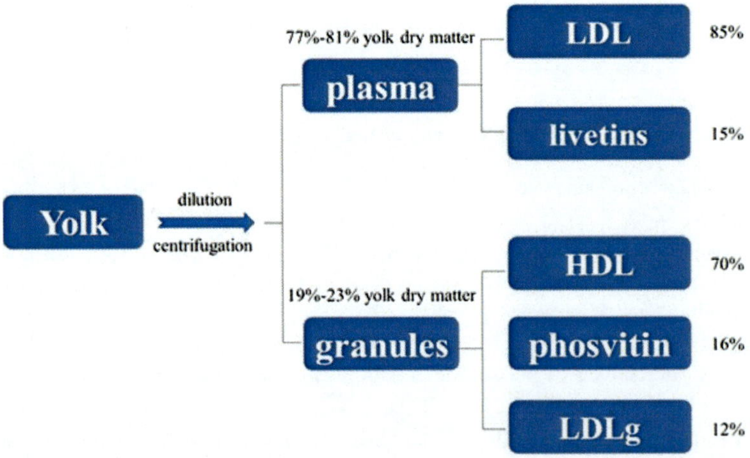

Fig. 11.1 Constituents of plasma and granules from hen egg yolk (LDL, low-density lipoprotein; HDL, high-density lipoprotein). Adapted from Anton (2013)

plasma consisting of 85% low density lipoproteins and the remainder α-livetin (serum albumin), β-livetin (α2-glycoprotein), and γ-livetin (IgY); (ii) the second fraction contains granules of varying sizes composed of 70% high-density lipoproteins (α- and β-lipovitellins), 16% phosvitin (glycophosphoprotein), and 12% low-density lipoproteins (LDL) (Anton 2013). All of the livetins are water-soluble and correspond to serum proteins (Williams 1962).

Low density lipoproteins are protein-lipid complexes which are spherical nanoparticles (17–60 nm) with a lipid core composed of 71% triacylglycerol, and 4% cholesterol esters surrounded by a monofilm of phospholipids and proteins; due to their low density (0.982) LDLs are soluble in aqueous solutions (Anton et al. 2003).

11.1.1 Properties of IgY

Numerous methods of extraction and fractionation of proteins are based on their physicochemical and structural characteristics such as solubility, hydrophobicity, molecular weight, and isoelectric point (pI). The structure and biology of IgY is reviewed in Chap. 5. However, the properties of particular interest for the extraction of IgY are summarized in Table 11.1. Generally, protein solubility is affected by the pH of the aqueous medium and is a result of the electrostatic and hydrophobic interactions between the protein molecules. Solubility increases if the electrostatic repulsion is greater than the hydrophobic interactions. At the isoelectric point proteins have a net zero charge and molecules tend to associate resulting in insolubility and precipitation out of solution (Scopes 1993). Above or below the pI, the net charge is negative or positive, respectively and solubility is enhanced. Values for the pI of IgY have been reported to be in the range of 6.7 ± 0.9 (Dávalos-Pantoja et al. 2000) and 5.5–7.0 (Chalamaiah et al. 2017) and 5.8 (Polson et al. 1980). It would be worthwhile to determine the exact value experimentally as this would provide a better theoretical basis for the extraction.

Table 11.1 Properties of IgY important for extraction

Property	Value
Molecular weight [a]	180 kD
Isoelectric point [b]	5.5–7.6
Molar extinction coefficient [c]	1.094, 1.33, 1.35
pH stability [d]	4–11
Solubility [e]	> 86% at pH 2–5 and 8–12
Thermostability [f]	4 °C > 6 months
	65 °C > 2 months
	100 °C > 6 months
Pepsin stability [g]	pH 4.0, 91% for 1 h; 63% after 10 h

Notes: Data according to [a] Amro et al. 2018; [b] Dávalos-Pantoja et al. 2000; [c] Stålberg and Larsson 2001, Pauly et al. 2011, Hodek et al. 2013 respectively; [d] Schade et al. 2005; [e] Chalamaiah et al. 2017 [f] Ren et al. 2016; Ge et al. 2020; [g] Gadde et al. 2015

11.1.2 General Considerations for the Extraction of IgY

A variety of processes may be used, alone or in combination, to extract IgY from egg yolk. The choice is influenced by a number of factors such as: the preservation of the activity of the immune specific immunoglobulins and the degree of purity; the scale of extraction either laboratory or industrial; the costs and the equipment required; and the potential use of other yolk products after the extraction of the IgY (Huang and Ahn 2019). There is a balance to be struck between the purity and the amount of IgY as every purification step results in some loss of the protein (Schade et al. 2001). Two earlier review articles about the methods for the isolation and purification of IgY are published (Meulenaer and Huyghebaert 2001; Kovacs-Nolan and Mine 2004). A comparison of the different methods of extraction and purification is made difficult as the yield and purity of IgY are not always specified as the emphasis in many of the publications is on the biological activity of the antibody against its target antigen. The most valuable information on the choice of a method for the isolation of IgY is obtained when researchers have compared the different methods on the extraction of a single batch of egg yolks. These comparative studies are presented in Sect. 11.6. In general, care must be taken in reviewing the yields as although the data are usually expressed as mg of IgY/mL of egg yolk or mg IgY/mg egg yolk, it may also be expressed as mg IgY mg/mL water in yolk (Kitaguchi et al. 2008) or mg IgY/mL of the water soluble fraction (Kassim et al. 2012); these latter two reporting methods will give yields which appear to be much higher.

The protein content of the various fractions obtained during the extraction are also estimated by a variety of methods. To determine the protein concentration based on the UV absorbance at 280 nm (Table 11.1) different percent extinction coefficients (now referred to as molar absorptivity) of 1.35 for a 1% solution of IgY (Stålberg and Larsson 2001) or an extinction coefficient of 1.33 for IgY (Pauly et al. 2011) or 1.094 (Hodek et al. 2013) were used; this probably reflects the solvent in which the molar absorptivity is determined as there is a solvent effect even in water (Bohman and Arnold 2016) so care needs to be taken to ensure the correct value is used. Protein concentrations are also determined using a number of assays; the Lowry assay using BSA as a standard protein (Deignan et al. 2000); the Biuret method (Akita and Nakai 1993) or the Bradford assay (Chang et al. 2000). Different methods of protein estimation, which interact with different specific amino acids of the protein (Bocian et al. 2020), are likely to give different results and may explain some of the variation in IgY levels recorded. The standard curves for the protein estimation should ideally be performed with a purified IgY rather than another protein such as BSA which is often used.

11.2 Egg Collection and Separation of Yolk from Egg White

Eggs are collected daily from the hens after immunization (Chap. 10). The egg shells are numbered, recorded and stored at 4 °C for two to three weeks, or at room temperature for days until used. This avoids fouling of the material (Fishman and

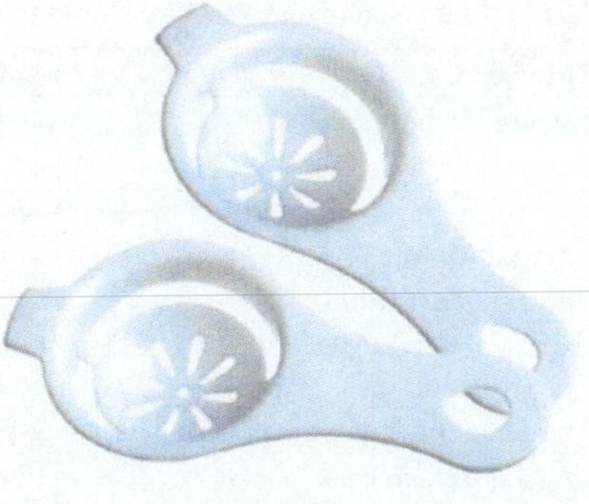

Fig. 11.2 An egg separator used to remove the egg white from the yolk

Berg 2018). Before removing the egg from the shell, the egg is washed with clean water. Alternatively, for aseptic collection of the egg yolk, the egg can be washed with clean water, soaked in a solution of dodecyl dimethyl benzyl ammonium bromide for 1–3 min, wiped dry with a light cotton fabric (Etamine), and finally cleaned with 75% ethanol to allow for natural air drying. For industrial scale processes, a commercial egg washing machine and sheller is used. The first step in the extraction of IgY from yolk, is the separation of the yolk from the egg white; this is an important step as egg white is mainly composed of proteins. The major proteins in egg white are ovalbumin (54%), ovotransferrin (12%), ovomucoid (11%), G2 and G3 globulin (4% each), ovomucin (3.5%) and lysozyme (3.4%) (Ji et al. 2020). The shell is carefully cracked, and the egg white is initially removed using an egg separator (Fig. 11.2).

To remove any adhering egg white, the yolk may be rinsed in flowing water or a buffer or rolled on filter paper (Schade et al. 2001). The outer layer of the vitelline membrane surrounding the egg yolk is composed of ovomucin, lysozyme and vitelline membrane outer proteins I and II while the inner layer is principally composed of glycoproteins (Sunwoo and Gujral 2015). The yolk contents are released by puncturing the vitelline membrane, which is retained on filter paper, and the volume of egg yolk is measured.

11.3 Delipidation of Egg Yolk

One of the main challenges in the extraction of IgY is the removal of the lipids present in the egg yolk. The general strategy for IgY extraction from yolk involves a preliminary step where the granules are removed by one of a number of different methods and the IgY remains in the plasma (Fig. 11.1; Table 11.2).

11.3.1 Water Dilution

Depending on the potential use of the IgY it may be preferable to extract using biocompatible methods. For example, the water dilution method is based on the

Table 11.2 Summary of methods used to remove lipids from egg yolk

Method	Variable conditions
Water dilution method	Degree of dilution
	+/− pH adjustment
	+/− freeze/thaw cycle
3.5% Polyethylene glycol precipitation (PEG)	+/− Chloroform
Anionic polysaccharide precipitation	Dextran, Na-alginate, λ-carrageenan, Na-carboxymethyl cellulose, Pectin
Organic solvent extraction	2-propanol and acetone
Specific chemicals	Phosphotungstic acid and magnesium chloride Caprylic acid

Note: +/− indicates with/without

aggregation of yolk lipoproteins at low ionic strengths resulting in their being insoluble in water (Sunwoo and Gujral 2015). The water-soluble fraction (WSF) is then separated from the lipid components by centrifugation or filtration. Two main factors have been assessed for delipidation using water dilution, the extent of dilution and the pH of the solution. The following conditions have been used: addition of deionized water to give a ten-fold dilution (Jensenius et al. 1981; Nilsson et al. 2008); addition of tap water (eight-fold dilution) followed by freezing, and filtration (Hodek et al. 2013); six-fold water dilution at pH between 5.0–5.2 for 6 hours at 4 °C (Akita and Nakai 1992). Acidic conditions have been shown to effect changes in the yolk granules which increases their ability to bind lipids and thus at the lower pH of 5, the recovery of IgY increases and the amount of LDL in the WSF decreases (Sunwoo and Gujral 2015). Freezing of the diluted yolk results in structural changes in the lipoprotein particles, due to less availability of water, and leads to the formation of lipid aggregates which are large enough to be removed by low-speed centrifugation (Zivkovic et al. 2009). Water dilution methods have yielded 9.8 mg IgY/mL of egg yolk with 93%–96% recovery (Akita and Nakai 1992) and 13.1 mg of IgY per mL of egg yolk at a purity of 71% (Deignan et al. 2000). The water dilution method has several advantages for the purification of IgY for oral administration (Nilsson et al. 2008); these include no toxic products are used in the purification; it is rapid, efficient and suitable for large-scale production. Removal of the aggregated yolk lipids is most commonly achieved by the following methods: centrifugation (Jensenius et al. 1981; Akita and Nakai 1992) or filtration (Sect. 11.4.3).

11.3.2 Polyethylene Glycol Precipitation

Polyethylene glycol (PEG), a polymer of ethylene oxide with the general formula of $H(OCH_2CH_2)_nOH$, can be synthesised to yield straight chain or branched polymers and is one of the most useful protein precipitants (Arakawa and Timasheff 1985).

PEG is non-toxic and non-immunogenic and as such has been widely used in the biotechnology industry. Polson were the first to develop this method where the addition of 3.5% PEG was used to extract IgY from egg yolk (Polson et al. 1980). At this concentration of PEG, the lipid is displaced from the solution and removed by centrifugation. This method was superseded by the addition of chloroform to a dilute suspension of egg yolk which resulted in nearly a three-fold increase in the IgY recovered (Polson 1990) without denaturing the protein. A modification of the chloroform PEG method in which PEG was replaced by the addition of 25% trichloroacetic acid was reported to yield twice as much IgY (200–250 mg IgY/egg yolk) compared to the chloroform PEG method (Asemota et al. 2013).

11.3.3 Anionic Polysaccharides

The use of the anionic polysaccharides, such as dextran sulphate, which interacts with the egg yolk lipoproteins to form a precipitate has also been used; the excess dextran sulphate is then precipitated by the addition of calcium chloride (Jensenius et al. 1981). IgY recovery was compared to the addition of the anionic polysaccharides Na-alginate, λ-carrageenan, Na-carboxymethyl cellulose, and pectin to six-fold diluted yolk at various pH levels (Chang et al. 2000). They reported that the interactions of the polysaccharides and lipoproteins were determined by ionic bonding, hydrophobic interactions and hydrogen bonding. The combination of pectin at 0.15% level at pH 5.0 gave an IgY recovery of 74%. Polysaccharides, as safe and nontoxic substance, can be applied for the isolation of the IgY in biomedical fields.

11.3.4 Organic Solvents

Organic solvents which are miscible with water, such as ethanol (C_2H_5OH), acetone ($(CH_3)_2CO$) and 2-propanol (C_3H_8O) are classical protein precipitants. However, care must be exercised to avoid denaturation of the protein by the solvent which involves working at −20 °C and finally careful removal of the solvent is necessary. The results of Deignan et al. showed that a yield of 7.11 mg (range 6.20–7.80) per mL of egg yolk with a purity of 57%IgY was obtained after precipitation using 2-propanol followed by acetone (Deignan et al. 2000). In general, the requirement to work at very low temperatures and the use of organic solvents which are deemed undesirable in the food, pharmaceutical and nutraceutical environment restricts the scale of use of this method.

11.3.5 Specific Chemicals

Phosphotungstic acid and magnesium chloride have been used to precipitate lipoproteins (Vieira et al. 1984). Peak IgY yields of 15.1 mg of IgY per mL of egg yolk, with a purity above 60% purity, were obtained after lipid removal using phosphotungstic acid and magnesium chloride (Deignan et al. 2000). Some acids such as caprylic acid were also successfully used to eliminate lipids (Mclaren et al. 1994).

11.4 Extraction of IgY Following Delipidation

The concentrated fraction containing IgY obtained after the delipidation step, contains other water-soluble proteins and some minor lipids or lipoproteins (Fig. 11.1). An additional step to remove remaining lipoproteins is achieved by addition of 0.1% charcoal at pH 4.0 (Ko and Ahn 2007). The major protein components of the plasma are IgY, α-livetin (chicken serum albumin), and β-livetin (α-2-glycoprotein) and the relative amounts in the yolk are 3:5:2 (Chalamaiah et al. 2017). A more detailed proteomic analysis of egg yolk (Mann and Mann 2008) identified 119 proteins of which 100 were in the plasma fraction (Fig. 11.1). The plasma proteins in highest abundance were as follows: serum albumin, vitellogenin-cleavage products, apovitellenins, IgY, ovalbumin and a 12 kDa serum protein with cross-reactivity to β2-microglobulin. A number of methods are commonly used to extract IgY from the delipidated plasma such as precipitation, filtration, column chromatography and aqueous biphasic systems (Table 11.3).

11.4.1 Salt Precipitation

The second step to extract IgY after delipidation is carried out by salt precipitation using saturated salt solutions. Selective precipitation may be achieved by adjusting the pH of the solution to the isoelectric point of IgY (pI = 5.7) where it is least soluble (Chalamaiah et al. 2017). Proteins are usually soluble in solutions with low concentrations of ions which ensures the protein remains folded and stable (Scopes

Table 11.3 Summary of IgY extraction methods

Method	Variable conditions
Salt precipitation	Ammonium sulphate Lithium sulphate Sodium sulphate Sodium chloride
PEG precipitation	8.5% and 12%
Filtration methods	Funnel Column gel filtration Ultrafiltration
Aqueous biphasic systems	Good's buffer and polypropylene glycol 400 Phosphate salt and Triton X 100 Polyethylene glycol (PEG) 1500 and potassium phosphate

1993). At the lower salt concentrations, the anions and cations neutralize the charge on the protein, and it is soluble; at higher salt concentrations the surface will become charged again and aggregate which is the salting out effect. The salting out ability of anions and cations usually follows the Hofmeister series shown below:

$$PO_4^{3-} > SO_4^{2-} > CH_3COO^- > Cl^- > Br^- > ClO_4^- > I^- > SCN^-NH^{4+}$$
$$> Rb^+ > K^+ > Na^+ > Li^+ > Mg^{2+} > Ca^{2+} > Ba^{2+}$$

Diverse salt solutions have been used to purify IgY such as ammonium sulphate (Ko and Ahn 2007; Xu et al. 2013), lithium sulphate (Bizhanov et al. 2004), sodium sulphate (Jensenius et al. 1981; Deignan et al. 2000) and sodium chloride (Hodek et al. 2013). Comparisons are made among these salts based on the yield and purity of the IgY obtained. Ammonium sulphate, $(NH_4)_2SO_4$, is often the first choice for salting out because it is soluble at high ionic strength and the cost is comparatively low (Duong-Ly and Gabelli 2014). Ammonium sulphate precipitation with two steps (50% and 30%) yielded 9.2 mg IgY/mL egg yolk and a purity of 98% (Xu et al. 2013); two steps (55% and 32%) yielded 4.7–9.2 mg/mL egg yolk however, the purity was not determined (Dai et al. 2013); two steps (both at 40% at pH 9.0) yielded 70–80% of IgY present in the WSF with higher purity (units not given) (Ko and Ahn 2007). Lithium sulphate precipitation in two steps (34% each step) yielded 7.7 mg IgY/mL egg yolk and a purity of 94% (Bizhanov et al. 2004). Sodium sulphate yielded 10–15 mg IgY/mL egg yolk representing a recovery of 70–80% of the total yolk IgY (Jensenius et al. 1981). Sodium chloride at a final concentration of 8.8%. yielded 8.9 mg of IgY per mL of yolk and a purity of 97% purity (Hodek et al. 2013).

Sodium sulphate precipitation performed after various delipidation methods showed yields of IgY from 4.9 to 7.5 mg of IgY/mL of yolk with purities ranging from 87% to 94% (Akita and Nakai 1993). Removal of the salt at the end may be achieved by dialysis, often for prolonged periods of time, depending on the salt utilized, or by the quicker method of gel filtration chromatography.

11.4.2 PEG Precipitation

This method involves the extraction of IgY from the delipidated solution by means of two further PEG precipitation steps (Polson et al. 1980; Pauly et al. 2011). In the initial step 3.5% PEG-6000 is used to remove lipids. At higher concentrations of PEG, the protein molecules are excluded from the solvent region occupied by the PEG molecules and as result the protein becomes concentrated and precipitates out of solution when its solubility limit is exceeded (Kumar et al. 2009). In this method there are two further precipitation steps: the addition of 8.5% PEG-6000 yields a pellet following centrifugation for 20 min at 10,000 g; in the final step, 12% PEG-6000 is added to the recovered pellet, the solution is centrifuged and the final pellet is recovered and dissolved in buffer and stored at −20 °C. The IgY purity of the extract is around 80%, which means, depending on the age of the laying hen (Chap. 2), 40–80 mg per egg. Other reports on the use of this method showed a IgY yield of 4.9 mg/mL per egg yolk, at a purity of 89% (Akita and Nakai 1993) and a range of 8.39–8.83 mg of IgY per mL of egg yolk. Moreover, SDS-PAGE analysis shows no contaminating bands (Deignan et al. 2000). IgY PEG precipitation at 3.5% and twice at 12% followed by cryo-ethanol yielded 4.9 mg/mL egg yolk with a purity of 89% (Akita and Nakai 1993). The addition of the cryo-ethanol served to remove the PEG from the pellet containing the IgY and any contaminating proteins which have a higher solubility than IgY in the alcohol (Polson et al. 1985). This is a mild method and can stabilize the three-dimensional confirmation of the protein which is beneficial in the extraction of bioactive proteins. PEG is a relatively inexpensive raw material and thus can be effectively used in scaled up processes. Resolubilization of the final pellet after centrifugation steps, rather than filtration, is sometimes more time consuming (Gagnon et al. 1996). Contamination of the pellet with PEG may alter the retention behaviour in ion exchange chromatography (Gagnon et al. 1996).

11.4.3 Filtration

Filtration is a separation process based on a barrier that retains some species based on their size and shape while allowing other species in the solution to pass through it (Saxena et al. 2009). The common filtration methods for IgY purification are funnel filtration, gel filtration on a column and ultrafiltration through membranes with different pore sizes (Meulenaer and Huyghebaert 2001). A filtration system was proposed by Kim and Nakai 1998 to separate IgY from egg yolk. A WSF was delipidated using a cellulose powder column in conjunction with an octadecyl column (Kim and Nakai 1998); IgY was precipitated with 1.5 M NaCl pH 9.0 and purified by ultrafiltration through different membranes with cut-offs of 300 kDa–30 kDa with the following results: Amicon, 74%–99% purity, 80%–85% recovery; Harp, 81%–84% purity, 72%–74% recovery; A/G, 89% purity, 75% recovery (Kim and Nakai 1998).

IgY was purified from the WSF by ultrafiltration using polyethersulfone and modified polyethersulfone membranes (Hernández-Campos et al. 2010). The authors concluded that both pH and ionic strength affect the purification and recovery of IgY. When NaCl was added (150 and 1500 mM) the purity of IgY decreased, mainly at pH values close to or higher than the isoelectric point of IgY. The best results were obtained in the absence of salt at pH values of 5.7 (purification factor > 3.5 and IgY recovery of 83%) for polyethersulfone and pH 6.7 (purification factor > 4 and IgY recovery of 94%) for and modified polyethersulfone membranes (Hernández-Campos et al. 2010). Filtration methods are considered feasible for IgY purification for large-scale and industrial applications and have the added advantage of concentrating the IgY solution (Akita and Nakai 1992; Meulenaer and Huyghebaert 2001).

11.4.4 Aqueous Biphasic Systems (ABS)

Liquid-liquid extraction applying aqueous biphasic systems (ABS) represents a viable alternative for the extraction of biomolecules (Freire et al. 2010; Pereira et al. 2015, 2016). Due to their biocompatible water-rich environment, ABS are a promising alternative to classical liquid-liquid extraction methodologies where volatile organic solvents are often used (Albertsson 1970). In addition, ABS allow the combination of extraction, purification and concentration in a single step. Typical ABS consist of two immiscible aqueous-rich phases based on polymer/polymer, polymer/salt or salt/salt combinations (Freire et al. 2012). The use of ABS for the purification of IgG antibodies was largely investigated using conventional polymer-polymer- or polymer-salt-based ABS (Azevedo et al. 2009) and as there is a restricted polarity difference between the coexisting phases this has limited their success in the purification of IgG antibodies (Ayyar et al. 2012). To overcome this limitation, the addition of specific ligands to the polymer used in the ABS creation or the use of hybrid processes have been described (Ruiz-Ruiz et al. 2012; Ferreira et al. 2016). More recently, ionic-liquid-(IL)-based ABS have been proposed as more efficient separation processes than polymer-based ABS, particularly due to the possibility of tailoring the phases polarities and affinities by a proper manipulation of the IL cation/anion combinations (Freire et al. 2012). ILs are organic salts, most of them water-soluble, with circa millions of different cation-anion combinations (Shi and Wang 2016). The effective use of ABS for antibody purification requires knowledge of the molecular-level mechanisms behind the biomolecules partitioning between the two aqueous phases (Freire et al. 2012). Usually, the biomolecule partition is relatively complex due to solute-solvent interactions, such as van der Waals, hydrogen-bonding, and electrostatic effects (Freire et al. 2010). Although less investigated, some reports can be found in the literature on the use of ABS to extract IgY. In this work, novel ABS constituted by cholinium (Ch)-based Good's buffer ILs (GB-ILs) ([Ch] [Tricine], [Ch][HEPES], [Ch][MES], and [Ch][TES]) and polypropylene glycol 400 (PPG 400) were used for the recovery of IgY from the WSF of egg yolk. In all the ABS evaluated, IgY revealed a preferential partitioning to the IL-rich phase. Computational tools were used to confirm that hydrogen bonding and van der Waals interactions between IgY and the IL were the main interactions responsible for the partition of IgY to the IL-rich phase. Extraction efficiencies of IgY in the range of 79%–94% were reported (Taha et al. 2015).

Micelle-based ABS composed of a phosphate salt, NaCl and Triton X-100 have been described for the extraction of IgY. Lipids were extracted into the surfactant-rich top phase, whereas IgY migrated to the phosphate-rich bottom phase. At optimal conditions, the yield of IgY was 97%, corresponding to 11.1–14.9 mg IgY/g egg yolk. The amount of lipids in the bottom phase was 22.9% of the total amount in the egg yolk (Stålberg and Larsson 2001). An integrated system for the recovery of IgY from egg yolk was developed by (Priyanka et al. 2014). In this work, IgY was extracted from immunized chicken egg yolks using aqueous two-phase and aqueous three-phase systems followed by a precipitation step with PEG 6000. Polyethylene glycol (PEG) 1500 and potassium phosphate were used as phase-forming

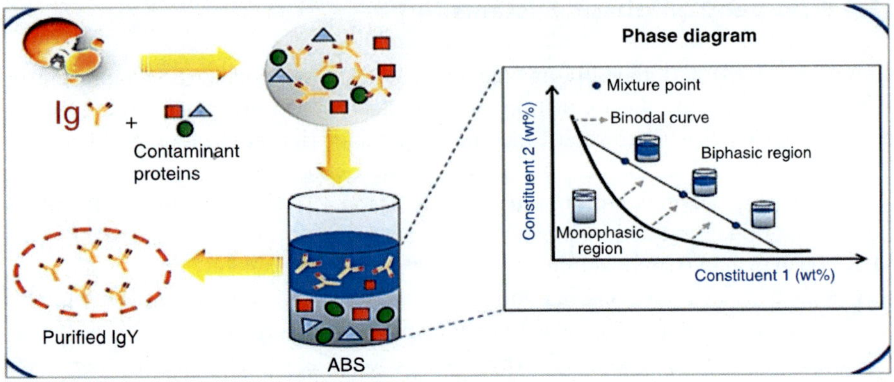

Fig. 11.3 Schematic diagram representing the extraction of IgY from egg yolk using aqueous biphasic systems. Note: The picture in the right side corresponds to an example of an ABS phase diagram shown in an orthogonal representation

constituents of ABS. Butanol and ammonium sulphate were added to the diluted yolk to obtain a three-phase system (two liquid phases and one solid intermediate phase) in two consecutive stages: (i) composed of butanol and 19% ammonium sulphate; and (ii) composed of butanol and 14% ammonium sulphate. In the two-phase system, IgY preferentially migrated to the PEG-rich top phase with a yield of 9.0 mg/mL of egg yolk, while for the three-phase system the yield was 6.0 mg/mL. The authors suggested that the lower yield obtained with the three-phase system was due to the multiple stages involved during the IgY extraction. It was demonstrated that in both approaches the IgY purity increased, as addressed by SDS-PAGE (Priyanka et al. 2014) (Fig. 11.3).

In summary, the results achieved to date with ABS reveal that these systems can be designed to act as effective extraction platforms for IgY. Improved ABS can be achieved by a tailoring of the phases' polarities and affinities by the proper selection of the ABS phase-forming constituents and mixture compositions.

11.5 Purification of IgY

There is ample background information available as the widespread manufacturing of antibodies as biopharmaceuticals has led to the use of chromatographic techniques in the downstream stage due to their high resolution and selectivity, efficiency and applicability in continuous processing (Liu et al. 2010; Cramer and Holstein 2011). To obtain a sample of IgY at the highest purity, additional polishing steps are required after its extraction. In general, chromatographic methods are usually used. The various chromatographic methods rely on the relative affinities of the charged protein for a stationary or mobile phase.

11.5.1 Cation Exchange Chromatography

Ion exchange chromatography is a separation method where the resin is either positively (anion) or negatively (cation) charged and the analyte to be separated has the opposite charge (Pratap 2017). Ko and Ahn evaluated the performance of cation exchange chromatography for IgY purification from diluted egg yolk against the ammonium sulphate precipitation method (Ko and Ahn 2007). The column was packed with pre-swollen carboxymethyl cellulose. The purity of IgY obtained from the ammonium sulphate method was 70%–80%, and much higher than the 30%–40% obtained by cation exchange chromatography. For IgY purification ammonium sulphate precipitation can be scaled up whereas cation exchange chromatography is limited by the volumes that can be separated (Ko and Ahn 2007).

11.5.2 Hydrophobic Charge-Induction Chromatography

Hydrophobic charge-induction chromatography (HCIC) is characterized by the adsorption of proteins over a hydrophobic surface in the presence of high concentrations of lyotropic salts (e.g. ammonium sulphate) (Burton and Harding 1998). In this type of chromatography, ligands may establish hydrophobic, thiophilic and electrostatic interactions with target proteins (Tong et al. 2011). HCIC chromatography was used to purify IgY antibodies from egg yolk, specifically applying macroporous cellulose-tungsten carbide composite beads activated by allyl bromide or divinyl sulfone, and then coupled with 4-mercaptoethyl-pyridine hydrochloride, 2-mercapto-l-methyl-imidazole, and 2-mercapto-benzimidazole as the HCIC ligands (Feng et al. 2008). High adsorption capacities of IgY between 100.6 and 137.6 mgIgY/mL adsorbent were reported for four adsorbents evaluated (Feng et al. 2008). Packed bed and expanded bed HCIC with a macroporous cellulose-tungsten carbide composite activated with allyl bromide and 4-mercapto-ethyl-pyridine hydrochloride were additionally investigated to improve the separation efficiency of IgY (Xia et al. 2012). For the packed bed mode, IgY was extracted with a purity of 92% and 58.4% yield, whilst for the expanded bed mode, a purity of 90% and a yield of 59.5% was obtained. The authors concluded that the expanded bed mode is a more appropriate purification platform for IgY (Xia et al. 2012).

11.5.3 Affinity Chromatography

Among the several chromatographic approaches, affinity chromatography is one of the most selective and efficient processes for purification of antibodies (Ayyar et al. 2012; Coskun 2016). This chromatographic method exploits the interactions of a specific ligand to an antibody; if a highly selective ligand is identified it will retain the antibody while contaminants flow through the column; this leads to high levels of purity in a single step (Liu et al. 2010; Fitzgerald et al. 2011; Ayyar et al. 2012). Native immunoglobulin binding proteins, such as protein A and protein G, effective

Table 11.4 Affinity chromatography methods developed for IgY purification

Ligand	pH of elution buffer	Purity (%)	Recovery (%)	Reference
Elastin-like polypeptide (ELP)-fused C2 domain of streptococcal protein G	2.5–9.5	96.3	64.0	(Xia et al. 2017)
A synthetic ligand (TG19318)	3.0	90.0	90.0–95.0	(Verdoliva et al. 2000)
A stable synthetic ligand[a]	7.0–7.2	92.1	78.2	(Dong et al. 2008)
Protein M[b]	2.8–3.5	98.7		(Jiang et al. 2016)
IgY-binding peptides [c]	3.0	93.0	70.0	(Khan et al. 2017)

[a] Detailed ligand information is not provided; [b] A transmembrane protein from human mycoplasma; [c] IgY-specific peptides identified by T7 phage display technology, with amino acid sequences: GVKCTWSSIVDWVCVDM, GTRCDWSAAYGWLCYDY, RSVCVWTAVTGWDCRND

in the purification of IgG, are not suitable for IgY purification because of the differences between the specific amino acid sequence of IgY-Fc and IgG-Fc (Jiang et al. 2016). A number of ligands have been developed specifically for the affinity purification of IgY (Table. 11.4).

The use of an elastin-like polypeptide-tagged immunoglobulin-binding domain of streptococcal protein G to purify IgY was shown to give high levels of purity (96.3%) and a recovery which was significantly higher than that by ammonium sulphate precipitation or ethanol fractionation; the purification could be completed within 3 hours which is an additional advantage (Xia et al. 2017). In another study, a synthetic ligand (TG19318) was used to purify IgY, after extraction by the water dilution method, and gave a purity of greater than 90% and a yield of IgY of 10.2 mg/mL egg yolk (Verdoliva et al. 2000). A wide screening of 700 synthetic ligands, synthesized by epichlorohydrin and cyanuric chloride methods was evaluated for IgY purification from the chloroform extract of egg yolk (Dong et al. 2008). A highly efficient ligand for IgY was identified, leading to a purity level of 92.1% with a recovery yield of 78.2%. This ligand has an adsorption capacity of 74.8 mg IgY/mL (Dong et al. 2008). A specific IgY binding peptide (without cross-reactivity to human or mouse/rat IgG) was developed (Khan et al. 2017). The peptide-conjugated column was prepared by immobilization of the biotinylated Y4–4 peptide in a HiTrap Streptavidin HP column. The column was used to purify IgY which had been extracted by the water dilution method followed by ammonium sulphate precipitation (35%) and gave a purity 93% and recovery of 70% (Table 11.4). Purified protein M coupled to NHS-activated Sepharose 4FF was shown to bind to the IgY variable region and IgY from the WSF was recovered with 98.7% purity from egg yolk (Jiang et al. 2016). An approximate 125 times increase in the effective IgY in the eluent was obtained by this method. The method is also applicable to the purification of monoclonal and engineered antibodies. The capacity for IgY adsorption (2 mg IgY/mL agarose) was low when compared to HiTrap Protein G HP for IgG adsorption (25 mg IgG/mL agarose) and could be further optimised (Jiang et al. 2016). Barroso achieved a recovery of 96% of IgY by using an affinity chromatographic column 4B Sepharose gel activated with divinyl sulfone (DVS), with an adsorption capacity of 26 mg of IgY/mL of gel (Barroso et al. 2005). Metal affinity chromatography for IgY purification from chicken egg yolk was also used in an IDA-Cu^{2+}-cryogel system. The maximum adsorption capacity of IDA-Cu^{2+} cryogel was 27 mg of IgY/g of resin (Junior et al. 2015), similar to the values provided by Barroso (Barroso et al. 2005). Chen attempted the purification of IgY from immunized chicken egg white lysozyme (LS) by an immunoaffinity column LS-Sepharose 4. The capacity of the column for specific IgY against LS was 0.68 mg of IgY/mL of wet gel (0.54 mg of IgY/mg of LS). IgY was effectively isolated with a purification factor of 3380 (Chen et al. 2002). The advantages of affinity chromatography with regard to high specificity and purity of the IgY, often in a single step, are likely to promote continuing research in this domain. Finally, it is important to highlight the use of a different chromatographic approach, namely a three-zone simulated moving-bed (SMB) platform for the separation of IgY from egg yolk. The three-zone SMB equipment was set up by connecting three columns. The highest purity of IgY achieved was 98% with a productivity ca. 0.0067 g IgY/(g adsorbent h) (Song and Kim 2013).

11.6 Comparison of Methods Used for Delipidation and Extraction of IgY

As discussed earlier (Sect. 11.1.2) there are two criteria related to extraction of IgY, which are the variety of methods used for protein determination which are likely to yield different results and how the concentration of IgY is reported, that make comparison of results of different methods difficult. However, a number of researchers have undertaken comparative studies of extraction methods on a single batch of egg yolk and these studies will now be discussed (Akita and Nakai 1993; Deignan et al. 2000; Bizhanov and Vyshniauskis 2000; Ren et al. 2016). Comparisons are made within each study but there is also some difficulty in comparing across studies due to the various methods for presentation of results. The results of the first comparative study in 1993 showed that the water dilution method was the most efficient for the extraction of IgY in terms of yield and functionally active protein and can be readily scaled up (Table 11.5). The procedure

Table 11.5 Summary of results of methods compared

Method	Water dilution	PEG	Dextran sulphate	Xanthan gum
Yield (mg IgY/mL egg yolk)	9.8	4.9	7.5	7.3
Purity	94%	89%	87%	89%

Data according to (Akita and Nakai 1993)

Table 11.6 Summary of results of methods compared

Method	Freeze & Thaw, pH 7.0	3.5% PEG	Dextran sulphate & calcium chloride	Phosphotungstic acid & magnesium chloride	2-propanol & Acetone
Yield mg IgY/mL egg yolk	13.1	11.0	15.6	15.1	4.0
Purity	71%	57.1%	64.3%	69.8%	57%
12% PEG	+		+	+	+
Yield mg IgY/mL egg yolk	7.49		8.8	8.6	2.0

+ Indicates addition of PEG. Data according to (Deignan et al. 2000)

Table 11.7 Summary of results of methods compared

Method	PEG	Chloroform	Dextran blue & calcium chloride
Yield mg IgY/mL egg yolk	4.7	9.0	4.1
Purity (IgY/protein%)	94%	52%	49%
Major bands by SDS-PAGE (kDa)	66	46, 75	35, 41, 66
Minor bands by SDS PAGE (kDa)	6 in total. Ranges 41–57 and 75–80	46, 50, 75	45

Data according to (Bizhanov and Vyshniauskis 2000)

also allows for the use of the rest of the egg yolk as a food product or the extraction of other biologically active molecules.

A comparison of five methods for delipidation, followed by three methods of IgY precipitation was reported by Deignan (Deignan et al. 2000); Table 11.6). Based on yield and purity of IgY, the two best methods of delipidation were precipitation with dextran sulphate and calcium chloride and phosphotungstic acid and magnesium chloride. For Ig precipitation 12% PEG gave the highest yield of protein compared to precipitation with sodium sulphate or ammonium sulphate (results not shown). However, freezing and thawing at pH 7.0 followed by 12% PEG, while giving slightly lower yields of similar purity, represents a very cost-effective method.

A comparison of three IgY extraction methods by Bizhanov (Bizhanov and Vyshniauskis 2000) showed a higher IgY yield using chloroform but with a lower level of purity (Table 11.7). Similar levels of specific activity of the IgY preparations were obtained for all three methods suggesting the integrity of the antigen binding sites was maintained during the extraction. Each of the methods resulted in multiple bands by SDS-PAGE (Sect. 11.8.1) indicating the presence of protein contaminants.

Six principal extraction methods were compared and the highest yield was obtained by the water dilution or carrageenan methods by Ren (Ren et al. 2016) Table 11.8). The organic solvents (chloroform and phenol) gave the lowest lipid

Table 11.8 Summary of results of methods compared

Method	Water Dilution	PEG	Caprylic acid	Chloroform	Phenol	Carrageenan
Yield (units not given)	++					++
SDS-PAGE purity		++				++
% Lipid residue	10.8	10.0	9.1	7.3	7.8	11.2
Filtration rate	++++	+++	+++	+++++	+++	++
Protein concentration mg/mL	+++	++	+++	++++	++	+

Data according to (Ren et al. 2016)

residues but the antibody titre was particularly affected by storage at −20 °C for one month. The water dilution method and caprylic acid yielded protein concentrations of 20 mg/mL while the carrageenan extraction gave only 10 mg/mL. The IgY preparation extracted by using the water dilution method had the highest titre compared to the other five methods.

The final choice of a method to extract IgY depends on the use of the product, and the water dilution and PEG methods are the most commonly used. For some uses, extraction may not be necessary. For example, IgY used for passive immunization of animals (Chap. 15), may be delivered either as liquid yolk or egg yolk powder (Ge et al. 2020).

11.7 Commercially Available IgY Extraction Kits

A number of companies supply kits containing all the reagents necessary to extract and purify IgY. The components of the kits are not disclosed, due to commercial sensitivity. However, a review of the protocols presented in Table 11.9 shows that two kits involve a delipidation step, followed by IgY precipitation with relatively high yields and purity. The other methods rely on thiophilic adsorption chromatography (Chen et al. 2002). The term "thiophilic" refers to an affinity for sulfone groups that lie in close proximity to thioether groups. Thiophilic adsorption chromatography is essentially a resin-based alternative to ammonium sulphate precipitation and yields a concentrated, essentially salt-free, highly purified IgY. The gentle binding and elution conditions ensure a high protein recovery with excellent preservation of antibody activity. The binding capacity is in the range of 20–30 mg IgY/mL of settled beads depending on the supplier and the columns can be regenerated and reused a number of times.

Table 11.9 Commercially available IgY extraction kits

Company	Steps	Amount of IgY	Yield	Purity
Thermo Scientific™ Pierce™ ChickenIgY Purification Kit	Delipidation Precipitation	80 to 120 mg IgY per egg yolk	80%	85% ~ 95%
Exalpha Biologicals, Inc. IgY EggsPress Purification Kit	Delipidation Precipitation	4–7 mg IgY per mL egg yolk. 60–105 mg/ egg yolk		90%
HiTrap IgY Purification HP	Delipidation step* Column separation based on thiophilic adsorption		78%	>70%
Affiland egg yolk purification kit	Column separation of egg yolk	50–100 mg IgY/egg yolk		98%
BioVision's Hi-Bind™ T-Gel Agarose	Thiophilic Adsorption Chromatography			
G-Biosciences' Thiophilic Resin	Thiophilic Adsorption Chromatography			
Pierce™ Thiophilic Adsorption Kit	Thiophilic Adsorption Chromatography			76% by HPLC

aReagents not supplied with kit

11.8 Methods Used to Confirm Purity and Activity of IgY

Antibodies may be characterized by different methods to determine their degree of purity and their biological activity. These parameters are determined independently and confirmation of the result by a single analytical method is usually acceptable for polyclonal antibodies. For antibodies used to treat human diseases the degree of characterization required is much greater with regard to structure, charge state and microheterogeneity and each property must be determined by at least two analytical methods (Staub et al. 2011).

11.8.1 Molecular Weight and Structure

The molecular weight of IgY is accepted to be180 kDa (Table 11.1) and after extraction and purification of IgY it is usual to confirm the molecular weight experimentally. The sequence of amino acids (primary structure) in a protein determines how it folds into secondary and tertiary structures; IgY also has a quaternary structure composed of two heavy and two light chains and is glycosylated. To confirm the molecular weight of IgY, it is necessary to disrupt the secondary and tertiary structures, and this is achieved by the addition of sodium dodecyl sulphate (SDS) and then by electrophoresis which separates charged macromolecules under an electric field. SDS-polyacrylamide gel electrophoresis (SDS-PAGE) separation is then directly related to the molecular weight of the intact protein. For example, SDS-PAGE of IgY recovered by affinity purification exhibited a single band near 180 kDa (Edupuganti et al. 2013). SDS-PAGE may also be performed under reducing conditions, which disrupts the di-sulphide bonds between the heavy and light chains (Chap. 5) which are then separated by electrophoresis giving bands at 65–67 kDa (Heavy chain) and 21–23 kDa (Light chain) (Sheng et al. 2018). The appearance of a number of additional bands after SDS-PAGE indicates that other protein contaminants are present (Table 11.6). The molecular weight can also be determined using mass spectrometry and the average molecular weight of IgY was determined in an early study as 167 kDa, with heavy chains of 65 kDa and the light chains of 19 kDa (Sun et al. 2001); however, it is generally accepted that the molecular weight of IgY is approximately 180 kDa with heavy chains of 67 kDa and light chains of 25 kDa (Sheng et al. 2018). A more detailed analysis of the purity of an IgY preparation delipidated using the water dilution method was performed: the protocol involved two-dimensional gel electrophoresis and nanoflow liquid chromatography coupled offline to matrix-assisted laser desorption/ionization time-of-flight tandem mass spectrometry; 25 individual proteins were identified in that preparation as well as IgY (Nilsson et al. 2008). Capillary electrophoresis, operates on the same principal as gel electrophoresis, but has the advantages of a smaller sample size, improved resolution, decreased separation time and real-time detection (Staub et al. 2011). It has been used to investigate IgY separated by Protein M affinity chromatography (Jiang et al. 2016). Circular dichroism (CD) is a technique used to evaluate the secondary structure of purified IgY. The work of Liu et al. showed that, after two-stage ultrafiltration steps (Liu et al. 2010). IgY as evaluated by CD displayed a typical β-sheet curve (~45% of the sheet content), suggesting that the IgY folded with a reasonable secondary structure (Liu et al. 2010). The effects of enzymatic deglycosylation on the conformation and stability of IgY were also evaluated by CD (Sheng et al. 2017). In this work, IgY was recovered from egg yolk by precipitation with PEG 6000, followed by further purification by gel permeation chromatography. The CD spectra of IgY and deglycosylated IgY suggested a β-sheet content for native IgY of 42%, and an α-helix structure in IgY (Sheng et al. 2017). However, conformational changes after enzymatic treatment were observed, with a decrease in the α-helix and β-sheet contents of 23.2% and 23.3%, respectively (Sheng et al. 2017). On the other hand, the random coil increased from 5% to 28.5%, indicating that the IgY structure became more flexible and disordered after the removal of *N*-glycan, meaning that glycosylation is fundamental to maintain the structure and stability of IgY (Sheng et al. 2017).

11.8.2 Biological Activity of IgY

After extraction and purification it is necessary to show that the IgY binds to the specific antigen used in the immunization protocol (Chap. 10). The most common bioassay used is an *enzyme-linked immunosorbent assay* (ELISA) where binding of the IgY to the antigen of interest is determined quantitatively. It is also important to ensure that the IgY does not bind other possible contaminants present in the sample to be measured. *Western blot analysis* may also be used to qualitatively characterize the purified IgY from egg yolk. In this methodology, the protein bands from the SDS-PAGE gel are eluted into a nitrocellulose filter, allowing the identification of specific proteins in a complex matrix by antibody labelling (Yang and Mahmood 2012). For example Western blot analysis of IgY purified by the PEG 6000 precipitation method, specifically recognized the purified canine IgG used but did not recognize IgG from other animal species, such as cat IgG, guinea pig IgG, rabbit IgG, goat IgG, sheep IgG and horse IgG (Santos et al. 2014). The activity of the IgY may also be determined by *immunodiffusion assays*. These methods, for the measurement of antibody and antigen binding, are confirmed by the presence of a precipitation arc following diffusion through a gel. The two related methods used are Ouchterlony's double diffusion method (Ouchterlony 1949) or radial immunodiffusion (Hudson and Hay 1980). Both methods have been applied to evaluate IgY after extraction and purification (Chen et al. 2002). The real-time analysis of the interaction of an antibody with an antigen is possible using *surface plasmon resonance analysis* (SPR). IgY is immobilized on a prepared gold sensor surface and binding of the antigen results in a change in polarized light as the molecules bind and dissociate resulting in a real-time sensogram (Edupuganti et al. 2013; Khan et al. 2017). The specific conditions for the immobilization of IgY at a flow rate of 5 μL/min, pH 4.0 at concentration of 100 μg/mL gave a linear range over 50–500 ng/mL and a detection limit of 40 ng/mL (Chun-Yu et al. 2012).

11.9 Purification and Characterization of Monoclonal IgY

The methods used for the production of monoclonal IgY and antibody fragments are outlined in Chap. 14. The first step is to harvest the antibody from the cell culture by removal of the cells and cell debris, either by centrifugation, depth filtration and sterile filtration to yield a clarified fluid suitable for chromatography (Liu et al. 2010).

While IgY does not bind to protein A or G, protein M has been shown to very effective in the purification of scFv IgY (Jiang et al. 2016) as well as specific antigen affinity chromatography (Khan et al. 2017). In common with polyclonal IgY, monoclonal IgY antibodies are routinely assayed by ELISA, Western blotting and SPR. There are a range of other techniques which are applied to the analysis of monoclonal antibodies which due to space limitations we cannot describe in detail. Spectroscopy is commonly used to assess the secondary and tertiary structure of antibodies. Numerous techniques are available, such as X-ray, nuclear magnetic resonance, absorption, fluorescence, Circular Dichroism (CD), dynamic light scattering, and infrared spectroscopy (Baudys and Kim 2000). CD spectroscopy is the technique of choice for studying chirality, particularly for monitoring and characterizing protein interactions in solution (Bertucci et al. 2010). IgY aggregates can also be analysed by CD (Joshi et al. 2014). Capillary isoelectric focussing may be used to determine the pI of proteins, characterize impurities, and monitor protein charge heterogeneity of monoclonal antibodies. It can distinguish between two proteins whose pI differs by as few as 0.005 pH units (Fonslow et al. 2010).

11.10 The Storage and Stability of IgY Product

It is important to maintain the stability and bioactivity of IgY in different storage conditions. IgY solutions can be stored for months to years at 4 °C, ideally with the addition of bacterial growth inhibitors such as sodium azide, thimerosal and gentamicin (Schade et al. 2005). Two important points should be considered when storing IgY. IgY preparations stored with sodium azide need to be dialyzed for use when labelling with horseradish peroxidase as it can inhibit its activity at dilutions of primary antibody of lower than 1:1000 (Schade et al. 2005). The repeated freeze-thawing of IgY antibody between −70 °C and room temperature results in a 50% loss of activity, which can be avoided by the addition of glycerol or by freezing in a small aliquots which once thawed are stored at 4 °C (Staak et al. 2000; Gadde et al. 2015). Longer term storage of large scale antibody preparations is usually by lyophilization (freeze-drying) (Fishman and Berg 2019). Consideration should be given to the concentration of the protein solution to be freeze dried and the inclusion of molecules which can help stabilize the protein, such as sucrose, trehalose or polysorbate 80 (Fishman and Berg 2019).

IgY solutions show a slight tendency toward self-aggregation, but this does not seem to influence the overall IgY antibody activity, and the activity of highly purified IgY decreases faster than that of partially purified IgY (Schade et al. 2005).

For oral delivery of IgY, it is notable that pepsin causes higher inactivation of IgY compared to trypsin (Gadde et al. 2015). Several methods have been developed to protect IgY from degradation by pH and pepsin present in the stomach (Chap. 12), allowing it to arrive unscathed at the small intestine, which is especially important for the use of IgY against enteric pathogens (Gadde et al. 2015). IgY in liquid yolk preparations retains its activity for up to 12 weeks at 4 °C but may become contaminated by bacterial growth (Ge et al. 2020). Yolk powder is prepared by dilution of the yolk with an equal volume of distilled water at 4 °C, freezing at −80 °C and lyophilization at −70 °C and can be stored free from bacterial contamination for up to 1 year (Ge et al. 2020). Further extraction of IgY from yolk, either by the water dilution or the PEG methods, reduced the antibody titre as compared to the liquid yolk or egg powder and suggests that for veterinary products IgY delivered as egg yolk powder may be optimal (Ge et al. 2020).

References

Akita EM, Nakai S (1992) Immunoglobulins from egg yolk: isolation and purification. J Food Sci 57(3):629–634

Akita EM, Nakai S (1993) Comparison of four purification methods for the production of immunoglobulins from eggs laid by hens immunized with an enterotoxigenic E. coli strain. J Immunol Methods 160(2):207–214
Albertsson PA (1970) Partition of cell particles and macromolecules in polymer two-phase systems. Adv Protein Chem 24(24):309–341
Amro WA, Al-Qaisi W, Al-Razem F (2018) Production and purification of IgY antibodies from chicken egg yolk. J Genet Eng Biotechnol 16(1):99–103
Anton M (2013) Egg yolk: structures, functionalities and processes. J Sci Food Agric 93 (12):2871–2880
Anton M, Martinet V, Dalgalarrondo M, Beaumal V, David-Briand E, Rabesona H (2003) Chemical and structural characterisation of low-density lipoproteins purified from hen egg yolk. Food Chem 83(2):175–183
Arakawa T, Timasheff SN (1985) Mechanism of poly(ethylene glycol) interaction with proteins. Biochemistry 24(24):6756–6762
Asemota H, Curtello S, Vaillant A, Mohammed W, Vuma S, Av C, Kurhade A, Kissoon S, Smikle M, Wisdom B, Kurhade G (2013) Purification of Avian IgY with Trichloroacetic Acid (TCA). J Chromatogr Sep Tech 4(9):1000205
Ayyar BV, Arora S, Murphy C, O'Kennedy R (2012) Affinity chromatography as a tool for antibody purification. Methods 56(2):116–129
Azevedo AM, Rosa PAJ, Ferreira IF, Pisco AMMO, Vries J, Korporaal R, Visser TJ, Aires-Barros M (2009) Affinity-enhanced purification of human antibodies by aqueous two-phase extraction. Sep Purif Technol 65(1):31–39
Barroso P, Murcia H, Vega N, Pérez G (2005) Purification of IgY against Salvia bogotensis lectin. Biomedica: Revista del Instituto Nacional de Salud 25(4):496–510
Baudys M, Kim SW (2000) Peptide and protein characterization. In: Frokjaer SL (ed) Hovgaard (eds) Pharmaceutical formulation: development of peptides and proteins. Taylor & Francis, London, pp 41–69
Bertucci C, Pistolozzi M, De Simone A (2010) Circular dichroism in drug discovery and development: an abridged review. Anal Bioanal Chem 398(1):155–166
Bizhanov G, Jonauskienė I, Hau J (2004) A novel method, based on lithium sulfate precipitation for purification of chicken egg yolk immunoglobulin Y, applied to immunospecific antibodies against Sendai virus. Scand J Lab Anim Sci 31:121–130
Bizhanov G, Vyshniauskis G (2000) A comparison of three methods for extracting IgY from the egg yolk of hens immunized with Sendai virus. Vet Res Commun 24(2):103–113
Bocian A, Sławek S, Jaromin M, Hus KK, Buczkowicz J, Łysiak D, Petrílla V, Petrillova M, Legáth J (2020) Comparison of methods for measuring protein concentration in venom samples. Animals 10(3):448
Bohman A, Arnold MA (2016) Molar absorptivity measurements in absorbing solvents: impact on solvent absorptivity values. Appl Spectrosc 71(3):446–455
Burley RW (1991) The avian egg. Chemistry and biology. J Agric Sci 116(1):169–170
Burton SC, Harding DRK (1998) Hydrophobic charge induction chromatography: salt independent protein adsorption and facile elution with aqueous buffers. J Chromatogr A 814(1–2):71
Chalamaiah M, Esparza Y, Temelli F, Wu J (2017) Physicochemical and functional properties of livetins fraction from hen egg yolk. Food Biosci 18:38–45
Chang HM, Lu TC, Chen CC, Tu YY, Hwang JY (2000) Isolation of immunoglobulin from egg yolk by anionic polysaccharides. J Agric Food Chem 48(4):995–999
Chen CC, Tu YY, Chen TL, Chang HM (2002) Isolation and characterization of immunoglobulin in yolk (IgY) specific against hen egg white lysozyme by immunoaffinity chromatography. J Agric Food Chem 50(19):5424–5428
Chun-Yu W, Feng F, Yun-Feng B, Ze-Zhong C, Mao-Zhong T, Jian-Ding Q, Yong-Nian N (2012) Determination of transferrin in human serum samples using chicken egg yolk antibodies (IgY) by surface plasmon resonance. Chem J Chinese Univ -Chinese Edition 33:1177–1181
Coskun O (2016) Separation techniques: chromatography. Northern Clin Istanbul 3(2):156–160

Cramer S, Holstein M (2011) Downstream bioprocessing: recent advances and future promise. Curr Opin Chem Eng 1:27–37

Dai YC, Zhang XF, Tan M, Huang P, Lei W, Fang H, Zhong W, Jiang X (2013) A dual chicken IgY against rotavirus and norovirus. Antivir Res 97(3):293–300

Dávalos-Pantoja L, Ortega-Vinuesa JL, Bastos-González D, Hidalgo-Alvarez R (2000) A comparative study between the adsorption of IgY and IgG on latex particles. Journal of biomaterials science. Polymer edition 11(6):657–673

Deignan T, Kelly J, Alwan A, O'Farrelly C (2000) Comparative analysis of methods of purification of egg yolk immunoglobulin. Food Agric Immunol 12:77–85

Dong D, Liu H, Xiao Q, Li R (2008) Affinity purification of egg yolk immunoglobulins (IgY) with a stable synthetic ligand. J Chromatogr B Analyt 870(1):51–54

Duong-Ly K, Gabelli S (2014) Salting out of proteins using ammonium sulfate precipitation. Methods Enzymol 541:85–94

Edupuganti SR, Edupuganti OP, O'Kennedy R, Defrancq E, Boullanger S (2013) Use of T-2 toxin-immobilized amine-activated beads as an efficient affinity purification matrix for the isolation of specific IgY. J Chromatogr B, Analyt Technol Biomed life Sci 923-924:98–101

Feng XH, Qiang LD, Ping WL, Jie CZ, Jing YS (2008) Preparation and evaluation of cellulose adsorbents for hydrophobic charge induction chromatography. Ind Eng Chem Res 47 (23):9566–9572

Ferreira AM, Faustino VFM, Mondal D, Coutinho JOAP, Freire MG (2016) Improving the extraction and purification of immunoglobulin G by the use of ionic liquids as adjuvants in aqueous biphasic systems. J Biotechnol 236:166–175

Fishman JB, Berg EA (2018) Isolation of IgY from chicken eggs. Cold Spring Harb Protoc 2018(6): pdb.prot099150

Fishman J, Berg E (2019) Antibody purification and storage. Cold Spring Harb Protoc 2019(5):pdb. top099101

Fitzgerald J, Leonard P, Darcy E, O'Kennedy R (2011) Immunoaffinity chromatography. Protein Purificat Protoc 244(1):167–177

Fonslow BR, Kang SA, Gestaut DR, Graczyk B, Davis TN, Sabatini DM, Yates JR 3rd (2010) Native capillary isoelectric focusing for the separation of protein complex isoforms and subcomplexes. Anal Chem 82(15):6643–6651

Freire MG, Cláudio AFM, Araújo JMM, Coutinho JAP, Rebelo LPN (2012) Aqueous biphasic systems: a boost brought about by using ionic liquids. ChemInform 41(14):4966–4995

Freire MG, Neves CMSS, Marrucho IM, Lopes JNC, Rebelo LPN, Coutinho JAP (2010) High-performance extraction of alkaloids using aqueous two-phase systems with ionic liquids. Green Chem 12(10):1715–1718

Gadde U, Rathinam T, Lillehoj HS (2015) Passive immunization with hyperimmune egg-yolk IgY as prophylaxis and therapy for poultry diseases--A review. Anim Health Res Rev 16 (2):163–176

Gagnon P, Godfrey B, Ladd D (1996) Method for obtaining unique selectivities in ion-exchange chromatography by addition of organic polymers to the mobile phase. J Chromatogr A 743:51–55

Ge S, Yang Y, Chelliappan B, Michael A, Zhong F, Zhang X (2020) Evaluation of different igy preparation methods and storage stability as potential animal feed supplement. Pak J Zool 52 (6):2305–2311

Hernández-Campos FJ, Brito-De E, Nchez BT-S (2010) Purification of egg yolk immunoglobulin (IgY) by ultrafiltration: effect of pH, ionic strength, and membrane properties. J Agric Food Chem 58(1):187–193

Hodek P, Trefil P, Šimůnek J, Hudecek J, Stiborova M (2013) Optimized protocol of chicken antibody (IgY) purification providing electrophoretically homogenous preparations. Int J Electrochem Sci 8:113–124

Huang X, Ahn DU (2019) How Can the Value and Use of Egg Yolk Be Increased? J Food Sci 84 (2):205–212

Hudson L, Hay F (1980) Practical Immunology, vol 222, 2nd edn. Blackwell Scientific Publications, Oxford

Jensenius JC, Andersen I, Hau J, Crone M, Koch C (1981) Eggs: conveniently packaged antibodies. Methods for purification of yolk IgG. J Immunol Methods 46(1):63–68

Ji S, Ahn DU, Zhao Y, Li K, Li S, Huang X (2020) An easy and rapid separation method for five major proteins from egg white: successive extraction and MALDI-TOF-MS identification. Food Chem 315:126207

Jiang X, Diraviyam T, Zhang X (2016) Affinity purification of egg yolk immunoglobulins (IgY) using a human mycoplasma protein. J Chromatogr B Analyt Technol Biomed Life Sci 1012-1013:37–41

Joshi V, Shivach T, Yadav N, Rathore AS (2014) Circular dichroism spectroscopy as a tool for monitoring aggregation in monoclonal antibody therapeutics. Anal Chem 86(23):11606–11613

Junior WF, Cano R, Totola AH, Carvalho LM, Cerri MO, Coimbra JSDR, Carvalho GGP, Carvalho BMA (2015) Adsorption of immunoglobulin Y in supermacroporous continuous cryogel with immobilized Cu2+ ions. J Chromatogr A 1395:16–22

Kassim N, Mtenga AB, Shim WB, Chung DH (2012) The in vitro and in vivo efficacy of hen IgY against Vibrio parahaemolyticus and Vibrio vulnificus. J Microbiol Biotechnol 22 (10):1423–1431

Khan KH, Himeno A, Kosugi S, Nakashima Y, Rafique A, Imamura A, Hatanaka T, Kato DI, Ito Y (2017) IgY-binding peptide screened from a random peptide library as a ligand for IgY purification. J Peptide Sci 23(10):790–797

Kim H, Nakai S (1998) Simple separation of immunoglobulin from egg yolk by ultrafiltration. J Food Ence 63(3):485–490

Kitaguchi K, Minoura M, Noritake M, Mizutani M, Kinoshita K, Horio F, Murai A (2008) Determination of immunoglobulin Y concentration in yolk extract prepared by water dilution method: comparisons among three strains of chickens. J Poultry Sci 45:82–87

Ko KY, Ahn DU (2007) Preparation of immunoglobulin Y from egg yolk using ammonium sulfate precipitation and ion exchange chromatography. Poult Sci 86:400–407

Kovacs-Nolan JY, Mine Y (2004) Avian egg antibodies: basic and potential applications. Avian Poult Biol Rev 15:25–46

Kumar V, Sharma VK, Kalonia DS (2009) Effect of polyols on polyethylene glycol (PEG)-induced precipitation of proteins: Impact on solubility, stability and conformation. Int J Pharm 366 (1–2):38–43

Liu HF, Ma J, Winter C, Bayer R (2010) Recovery and purification process development for monoclonal antibody production. MAbs 2(5):480–499

Mann K, Mann M (2008) The chicken egg yolk plasma and granule proteomes. Proteomics 8 (1):178–191

Mclaren RD, Prosser CG, Grieve RCJ, Borissenko M (1994) The use of caprylic acid for the extraction of the immunoglobulin fraction from egg yolk of chickens immunised with ovine alpha-lactalbumin. J Immunol Methods 177(1–2):175–184

Meulenaer B, Huyghebaert A (2001) Isolation and purification of chicken egg yolk immunoglobulins: a review. Food Agric Immunol 13:275–288

Nilsson E, Hanrieder J, Bergquist J, Larsson A (2008) Proteomic characterization of IgY preparations purified with a water dilution method. J Agric Food Chem 56(24):11638–11642

Ouchterlony O (1949) Antigen-antibody reactions in gels. Acta Pathol Microbiol Scand 26 (4):507–515

Pauly D, Chacana PA, Calzado EG, Brembs B, Schade R (2011) IgY technology: extraction of chicken antibodies from egg yolk by polyethylene glycol (PEG) precipitation. J Vis Exp 1 (51):3084

Pereira M, Cruz R, Almeida M, Lima Á, Coutinho J, Freire M (2016) Single-step purification of ovalbumin from egg white using aqueous biphasic systems. Process Biochem 51(6):781–791

Pereira MM, Pedro SN, Quental MV (2015) Enhanced extraction of bovine serum albumin with aqueous biphasic systems of phosphonium- and ammonium-based ionic liquids. J Biotechnol 206:17–25

Polson A (1990) Isolation of IgY from the yolks of eggs by a chloroform polyethylene glycol procedure. Immunol Investig 19(3):253–258

Polson A, Coetzer T, Kruger J, von Maltzahn E, van der Merwe KJ (1985) Improvements in the isolation of IgY from the yolks of eggs laid by immunized hens. Immunol Investig 14 (4):323–327

Polson A, Von Wechmar MB, van Regenmortel MHV (1980) Isolation of viral IgY antibodies from yolks of immunized hens. Immunol Commun 9(5):475–493

Pratap GJS (2017) HPLC for peptides and proteins: principles, methods and applications. Pharm Method 8(1):139–144

Priyanka BS, Abhijith KS, Rastogi NK, Raghavarao KSMS, Thakur MS (2014) Integrated approach for the extraction and purification of IgY from chicken egg yolk. Sep Ence Technol 49(4):562–568

Ren H, Yang W, Thirumalai D, Zhang X, Schade R (2016) A comparative evaluation of six principal IgY antibody extraction methods. Altern Lab Anim: ATLA 44(1):11–20

Ruiz-Ruiz F, Benavides J, Aguilar O, Rito-Palomares M (2012) Aqueous two-phase affinity partitioning systems: current applications and trends. J Chromatogr A 1244:1–13

Santos FN, Brum BC, Cruz PB, Molinaro CM, Silva VL, de Miranda Chaves SA (2014) Production and characterization of IgY against canine IgG: prospect of a new tool for the immunodiagnostic of canine diseases. Braz Arch Biol Technol 57(4):523–531

Saxena A, Tripathi BP, Kumar M, Shahi VK (2009) Membrane-based techniques for the separation and purification of proteins: an overview. Adv Colloid Interf Sci 145(1–2):1–22

Schade R, Behn I, Erhard M, Hlinak A, Staak C (2001) Isolation of IgY from yolk. In: Schade R, Behn I, Erhard M, Hlinak A, Staak C (eds) Chicken egg yolk antibodies, production and application. Springer Lab Manual, pp 65–107

Schade R, Calzado EG, Sarmiento R, Chacana PA, Porankiewicz-Asplund J, Terzolo HR (2005) Chicken egg yolk antibodies (IgY-technology): a review of progress in production and use in research and human and veterinary medicine. Altern Lab Anim: ATLA 33(2):129–154

Schjeide OA, Myrna W, Mccandless RG, Munn R, Marion P, Carlsen E (1963) Liver synthesis, plasma transport, and structural alterations accompanying passage of yolk proteins. Integr Comp Biol 3(2):167–184

Scopes RK (1993) Protein purification: principles and practice. Springer, New York (N.Y.)

Sheng L, He Z, Chen J, Liu Y, Ma M, Cai Z (2017) The impact of N-glycosylation on conformation and stability of immunoglobulin Y from egg yolk. Int J Biol Macromol 96:129–136

Sheng L, He Z, Liu Y, Ma M, Cai Z (2018) Mass spectrometry characterization for N-glycosylation of immunoglobulin Y from hen egg yolk. Int J Biol Macromol 108:277–283

Shi R, Wang Y (2016) Dual ionic and organic nature of ionic liquids. Sci Rep 6:19644

Song SM, Kim IH (2013) A three-zone simulated moving-bed for separation of immunoglobulin Y. Korean J Chem Eng 30(8):1527–1532

Staak C, Schwarzkopf C, Behn I, Hommel U, Hlinak A, Schade R, Erhard M (2000) Isolation of IgY from yolk. In: Schade R, Behn I, Erhard M, Hlinak A, Staak C (eds) Chicken egg yolk antibodies, production and application: IgY technology. Springer Lab Manuals, Berlin, Germany, Heidelberg, Germany, & New York, USA, pp 65–107

Stålberg J, Larsson A (2001) Extraction of IgY from egg yolk using a novel aqueous two-phase system and comparison with other extraction methods. Ups J Med Sci 106(2):99–110

Staub A, Guillarme D, Schappler J, Veuthey JL, Rudaz S (2011) Intact protein analysis in the biopharmaceutical field. J Pharm Biomed Anal 55(4):810–822

Sun S, Mo W, Ji Y, Liu S (2001) Preparation and mass spectrometric study of egg yolk antibody (IgY) against rabies virus. Rapid Commun Mass Spectrom 15(9):708–712

Sunwoo H, Gujral N (2015) Chemical composition of eggs and egg products. In: Cheung PCK, Mehta BM (eds) Handbook of food chemistry. Springer, Berlin, Heidelberg, pp 331–363

Taha M, Almeida MR, Silva FAE, Domingues P, Ventura SPM, Coutinho JOAP, Freire MG (2015) Novel biocompatible and self-buffering ionic liquids for biopharmaceutical applications. Chem A Euro J 21(12):4781–4788
Tong HF, Lin DQ, Pan Y, Yao SJ (2011) A new purification process for goose immunoglobulin IgY (ΔFc) with hydrophobic charge-induction chromatography. Biochem Eng J 56(3):205–211
Verdoliva A, Basile G, Fassina G (2000) Affinity purification of immunoglobulins from chicken egg yolk using a new synthetic ligand. J Chromatogr B, Biomed Sci Appl 749(2):233–242
Vieira JG, Oliveria MA, Russo EM, Maciel RM, Pereira AB (1984) Egg yolk as a source of antibodies for human parathyroid hormone (hPTH) radioimmunoassay. J Immunoass 5 (1–2):121–129
Walsh G (2013) Pharmaceutical biotechnology: concepts and applications. Wiley
Williams J (1962) Serum proteins and the livetins of hen's-egg yolk. Biochem J 83(2):346
Xia HF, Lin DQ, Chen ZM (2012) Purification of immunoglobulin of egg yolk with hydrophobic charge induction chromatography: comparison of operation modes with packed bed and expanded bed. Sep Ence Technol 47(16):2366–2372
Xia W, Lu H, Li Y, Cao J, Zhou X, Zhang X, Xia X, Sun H (2017) Purification of chicken IgY by binding capture using elastin-like polypeptide-tagged immunoglobulin-binding domain of streptococcal protein G. Vet Immunol Immunopathol 192:13–19
Xu J, Ren C, Wang S, Liu D, Cao L, Tao J (2013) Protection efficacy of multivalent egg yolk immunoglobulin against eimeria tenella infection in chickens. Iran J Parasitol 8(3):449–458
Yang PC, Mahmood T (2012) Western blot: Technique, theory, and trouble shooting. N Am J Med Sci 4(9):429–434
Zivkovic AM, Wiest MM, Nguyen UT, Davis R, Watkins SM, German JB (2009) Effects of sample handling and storage on quantitative lipid analysis in human serum. Metabolomics 5 (4):507–516

12 IgY Delivery and Dosage Form

Rao Wu and Xiaoying Zhang

Abstract

Egg Yolk antibodies (IgY) have many advantages for use as therapeutic and prophylactic drugs or health products for human and veterinary applications. However, antibodies are associated with loss of bioactivity when used in various forms. As proteins, gastrointestinal acidity and the action of digestive enzymes might cause structural alterations while they pass through the gastrointestinal tract impacting on their biological activity. Apart from some conventional dosage forms (powder, liquid, and injection), different gastrointestinal administrations and parenteral administrations have been developed to improve IgY delivery. Some novel drug delivery methods and materials have been applied in improve IgY product design in the past decade. It is reasonable to believe that IgY products will incorporate novel dosage form technologies for macromolecules in order to provide better therapeutic and health effects.

Keywords

IgY · Dosage form · Drug delivery system · Gastrointestinal tract · Gastrointestinal administrations · Parenteral administrations

12.1 Introduction

IgY antibodies have been developed as licensed drug and health products for human and animal applications (Chaps. 15, 16, and 17), however, successful IgY delivery and host therapeutic or prophylactic intervention could be challenging even when

R. Wu · X. Zhang (✉)
College of Biological Science and Engineering, Shaanxi University of Technology, Hanzhong, China
e-mail: zhxying@snut.edu.cn; xzhang67@uoguelph.ca; zhang@bio.uminho.pt

X.-Y. Zhang et al. (eds.), *IgY-Technology: Production and Application of Egg Yolk Antibodies*, https://doi.org/10.1007/978-3-030-72688-1_12

the specific IgY is well developed and validated. From this perspective, the design and application of different dosage forms of IgY for human and animal purposes remains a critical topic. The central concept of dosage form design, or, drug delivery systems, is to deliver the precise amount of drug or active component to the desired location of the host, in order to reach a designed optimal effect in the time period of need. Different dosage forms can be designed and adjuvants (i.e., excipient, solubilization, stability, preservation and emulsification) can be applied accordingly in order to preserve, protect and deliver the active compounds.

IgY antibodies are proteins with a three-dimensional structure and post-translational modifications such as glycosylation which are necessary for its bioactivity (Chap. 5). Different routine dosage forms have been applied in IgY delivery, including oral administrations (i.e., powder, liquid, injection), which are majorly targeting to the gastrointestinal tract, as well as intramuscular and intravenous injections in some veterinary practices (Pérez de la Lastra et al. 2020; Rahman et al. 2012).

12.2 Oral Administrations

IgY technology offers great opportunities for designing prophylactic and therapeutic strategies against infectious gastrointestinal diseases in animals and humans (Chaps. 15 and 16), such as bovine and human rotaviruses, bovine coronavirus, Yersinia ruckeri, enterotoxigenic Escherichia coli, *Salmonella spp*, *Edwardsiella tarda*, *Staphylococcus* and *Pseudomonas* (Mine and Kovacs-Nolan 2002). However, despite the numerous biological advantages, it is still difficult for IgY to deliver systematic therapeutic effects as well as full gastrointestinal effect when administered orally due to the rapid inactivation of gastric conditions including low pH value and digestive enzymes including pepsin, trypsin and chymotrypsin (Fig. 12.1), IgY may need to reach and target to low gastrointestinal tract retaining high bioactivity for some diseases such as colitis, or, IgY is expected a full gastrointestinal adsorption and bioreaction. The degraded IgY proteins may lose their bioactivity before they reach their targets; however, there is also study which indicates that the degraded $F(ab')_2$ and Fab fragments may retain some of their neutralizing activity (Akita and Nakai 1993). Therefore, the unfavourable gastric environment should be considered when trying to establish passive immunity via oral administrations (Lee et al. 2012) by better dosage form design, and the stability of IgY to acids and pepsin may be improved by means of dosage form design.

Different embedding materials have been investigated for oral IgY administration including chitosan-alginate microcapsules, hydrogels containing acrylamide and acrylic, acid methacrylic acid copolymers, liposomes, polymeric microspheres and multiple emulsification (Table 12.1). These materials are commonly used to encapsulate and thus protect a molecule of interest from digestive degradation (Holser 2013; Paul et al. 2014; Soumaila Garba et al. 2019). Chitosan is applied as a membrane-coating material to reduce microcapsule swelling, improve encapsulation efficiency and improve stability. The concentration of chitosan in the encapsulation

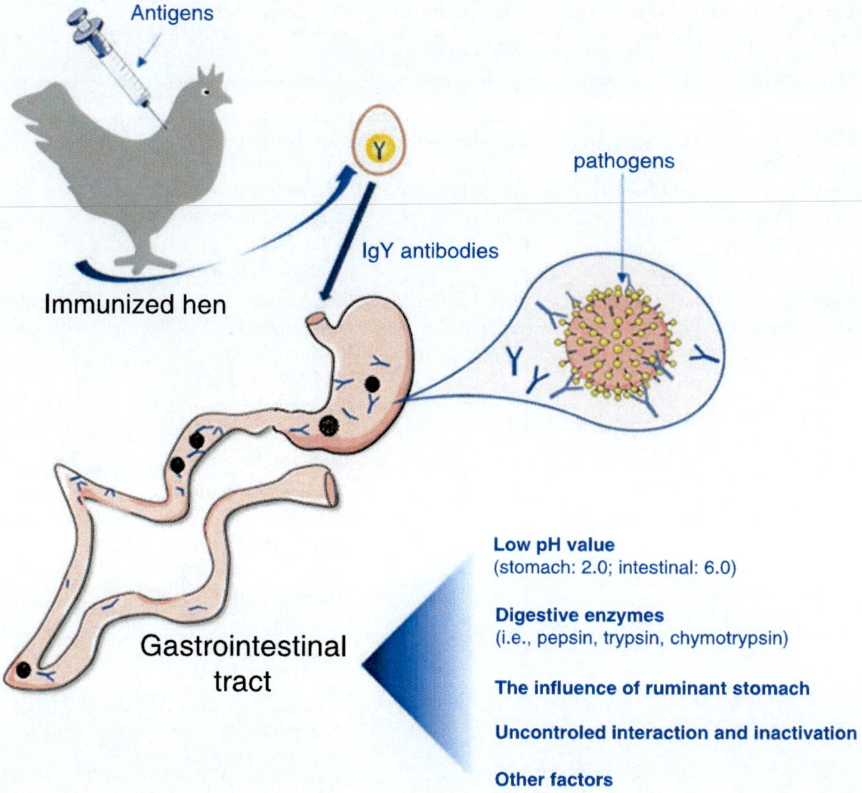

Fig. 12.1 IgY deactivation in the gastrointestinal tract. Notes: Biologically active IgY may be inactivated by the complicated situations of gastrointestinal tract, proper IgY delivery and dosage form design can provide protection to IgY for better bioactivity and targeting. (Source: XYZ-IgY lab)

medium is an important factor which controls the release of various materials from the chitosan-alginate membrane (Li et al. 2009). An effective and feasible prophylaxis was established by feeding chitosan-alginate coated IgY against Edwardsiella tarda infections in the process of turbot farming (Xu et al. 2020) (Chap. 15). Polyacrylamide hydrogels have good versatility and biocompatibility, and non-toxic, can be used as excellent carriers. In vitro study showed that the stability of IgY in simulated gastric fluid was greatly enhanced by encapsulation in hydrogels (Bellingeri et al. 2015). Lipid is a well-tolerated carrier and it can be easily produced in large scale. Most lipids are biodegradable and biocompatible, easy to scale up and sterilize and affordable (Amirijavid and Entezari 2018). Beyond such a good carrier, recently researchers introduced new potential carrier, nanoparticles (NPs; i.e., latex, chitosan, sodium alginate). In recent years, the advantage of NP as a drug delivery system including easy manipulation of particle size and surface characteristics of NPs, controlled release of drug, specific organ distribution, reduction of side effects,

Table 12.1 Design of oral IgY dosage forms for better IgY release and effect

Immunogen	IgY dosage form	Disease or pathological model	Effect of IgY dosage form	References
Clostridium difficile Toxin A and B	Chitosan-Ca pectinate microbeads	Clostridium difficile Infection (rats)	A release of vast (72.6%) biologically active IgY specifically targeting to colon 8 h after oral administration	(Xing et al. 2017)
Clostridium difficile toxin A	Microbeads coated with 35% Eudragit S100	Clostridium difficile Infection (rats)	the loading efficiency = 21%; a release of vast (87.5%) active IgY specifically targeting to colon at 8 h after oral administration	(Zhang et al. 2016a, b)
IgY microspheres	PLGA microspheres	Pigs	The encapsulation efficiency of IgY = 90% (w/w); the positive percentage (total anti-IgY antibody) > 300% versus that of free IgY (≈ 100%) 40 days after intramuscular administration.	(Torché et al. 2006)
Death receptor (DR5) in tumour necrosis factor TRAIL/ Apo-2 L family	Nano liposomal and nanoparticle conjugate	MCF7 cells	IC50 < 0.48 μg/mL versus IC50 = 218.6 μg/mL in the non- conjugated IgY.	(Amirijavid and Entezari 2018)
No immunogen applied	Hydrogels containing acrylamide and acrylic acid	SGF	High IgY activity (87%) compared to non-encapsulated IgY (14.33%).	(Bellingeri et al. 2015)
Edwardsiella tarda	Microcapsules with chitosan-alginate	Edwardsiellosis	Survival rates >20% versus zero survivability of non-specific IgY groups on the tenth day post infection	(Xu et al. 2020)
Vibrio Anguilla rum	Microcapsules with β-cyclodextrin	Vibrio Anguilla rum infection (half-smooth tongue sole)	Sustained release and activity of IgY in intestinal tract up to 24 h.	(Gao et al. 2016)
Enterotoxin genic Escherichia coli (ETEC)	Microcapsules with chitosan wrapped	Porcine neonatal diarrhoea (piglets)	No diarrhoea episode; no immune or inflammatory response.	(Alustiza et al. 2016)

(continued)

Table 12.1 (continued)

Immunogen	IgY dosage form	Disease or pathological model	Effect of IgY dosage form	References
	carbon nanotubes			
Rotavirus	Microcapsules	BALB/c mice	The activity of encapsulated IgY > non-encapsulated one after 3.5 h in the stomach.	(Lee et al. 2012)
ETEC	Microcapsules with chitosan-alginate	SGF	About 70% activity after 2 h in SGF under varying pH (from pH 3 to 6); sustained release (50% of the IgY) within 6 h in SIF.	(Li et al. 2007)
Recombinant HRV subunit protein, VP8	Granules with methacrylic acid copolymer	SGF and SIF	95% activity after 6 h in SGF, 80% activity after 8 h in SIF.	(Kovacs-Nolan and Mine 2005)
Rotavirus	Microcapsules containing whey protein	Diarrhoea (artificial intestinal juice)	74% (70 °C, 15 min) and 80% (pH 1.2) of the reserved activity in the artificial intestinal juice versus 16% and 20% in the non-microencapsulated IgY	(Cho et al. 2006)
Helicobacter pylori urease	Microcapsules with 10% or 20% β-cyclodextrin and Arabic gum	Helicobacter pylori infection	OD405 = 1.03 versus OD405 = 0.90 in the non-encapsulated IgY	(Chang et al. 2002)
No immunogen applied	Microcapsules containing lecithin/ cholesterol liposomes	SGF	Almost no activity loss in SGF (pH 1.8) for 3 h.	(Shimizu et al. 1993)

Notes: *PLGA* poly (D, L-lactide-co-glycolide); *CMC* carboxy methyl cellulose; *HRV* human rotavirus; *IgY-HpUc* helicobacter pylori UreC-specific IgY; *NEYP* nonimmunized egg yolk powder; *PGCR* polyglyceryl condensed ricinolate; *SGF* simulated gastric fluid; *SIF* simulated intestinal fluid; *UreC* an antibody against *Helicobacter pylori* antigens

and site-specific targeting (De Jong and Borm 2008). According to Amirijavid's research, in contrast to normal IgYs that required more than 218.6 μg/mL to kill 50% of cancer cells (MCF7 cells, human breast cancer cell line), IgYs with nano-liposomal reduced their cytotoxic dose to just 0.00056–0.48 μg/mL. The result of significantly reduced IC50 value showed that nanoliposomes played an active role in promoting the function of IgY (Amirijavid and Entezari 2018). The anionic methacrylic acid-ethyl acrylate copolymer is insoluble in acid medium but will begin to dissolve above pH 5.5, and IgY coated with this enteric polymer is resistant to gastric conditions and will dissolve readily in the neutral to slightly alkaline environment of the small intestine (Kovacs-Nolan and Mine 2005). Multiple emulsions (water-in-oil-in-water type emulsion; W/O/W emulsion) are defined as hierarchically dispersed systems having the structure of droplets in a drop. The hierarchical structure of multiple emulsions has the potential to encapsulate different ingredients (i.e., antibody drugs, living cells, incompatible therapeutic molecules) and to protect them from the environment, and release those ingredients in a controlled manner (Dluska et al. 2017). IgY can be encapsulated into a W/O/W emulsion without a loss of activity, the W/O/W emulsion will provide a useful food microcapsule protecting IgY from gastric conditions. Despite of the progress of multiple emulsion technology, the encapsulation of proteins into the inner phase of the W/O/W emulsion has not been fully studied, particularly for encapsulation of IgY into multiple emulsions (Shimizu and Nakane 1995).

It is not yet well studied whether the full-length IgY or functional part of IgY can be absorbed through the gastrointestinal tract, and whether such absorption could lead to systemic effects, especially given the further consideration that IgY cannot be mediated by the human Fc receptor (Chap. 5), and probably also by all mammalian Fc receptors. However, from the dosage form design point of view, it is favourite to reach such effects for oral delivery of proteins, and numerous approaches have been developed to enhance protein's absorption without altering their biological activity, by using such as enzyme inhibitor (protease inhibitor), absorption enhancer, mucoadhesive polymers, cell-penetration peptide, and derivatizations or chemical modification of proteins, and prodrug design strategies (Muheem et al. 2016).

12.3 Parenteral Administrations

Parenteral administrations are another area worthwhile for investigations, particularly for the topical applications of IgY. In previous studies, there have been attempts to deliver IgY by the mouth, nasal cavity, and skin (Table 12.2). For example, patients used *toothpaste* containing anti-gingivitis IgY showed significant differences ($p < 0.05$) in bleeding on probing (BOP) and gingival index compared with the control group (Zhang et al. 2016a, b). The percentage of *Streptococcus mutans* in dental plaque decreased significantly (from 20% to 5%) following a short term (three weeks) application of IgY mouth *spray*, and low level of *S. mutans* persisted for at least 5 weeks after withdrawal of IgY (Zhou et al. 2003). Specially designed functional drinking yogurt containing anti-urease of Helicobacter pylori IgY have been applied commercially; the IgY was stabile in the product up to 7 d, and decreased to 85% after 3 weeks of storage. The relevant clinical study confirmed suppression of *H. pylori* infection in humans by consumption of anti-urease IgY fortified yogurt (Horie et al. 2004). The mouthwash containing anti-*Streptococcus mutans* IgY can be used efficiently to control the colonization of *S. mutans* in the oral

Table 12.2 Design of IgY dosage forms for parenteral administrations

Immunogen	IgY dosage form	Disease or pathological model	Effect of IgY dosage form	References
Compound strain with Porphyromonas gingivalis and nucleated shuttle Bacillus	Toothpaste	Gingivitis	12 weeks later, significant differences ($p < 0.05$) in BOP and gingival index.	(Zhang et al. 2016a, b)
Streptococcus mutans	Toothpaste	Caries	Quick response (at 1 day after toothpaste used) and long duration (> 14 d).	(Chi et al. 2004)
Aeromonas salmonicida	Immersion bath	Ulcer disease (Koi carp)	lower mortality (< 70%) of high and median-dose (≥ 0.2 g /L) IgY-treated fish versus non-specific IgY groups (> 90%) on the fifth day post infection.	(Gan et al. 2015)
Streptococcus mutans	Spray	Caries	3 weeks later, percentage of S. mutans (5%) < that of non-coated IgY group (20%); long duration (> 3 weeks).	(Zhou et al. 2003)
Helicobacter pylori	Functional yogurt containing IgY-Urease	Volunteers who tested positive for Helicobacter pylori	The reserved activity = 85% after storage at 4 °C for 3 weeks; 39.33% reduction in UBT values after 4 weeks.	(Horie et al. 2004)
Streptococcus mutans	Gel (water-based CMC gel that was mixed with water-soluble fraction IgY powder)	Dental caries (Sprague Dawley rats)	S. mutans colony counts (20 CFU/mL) < non-coated IgY group (2287 CFU/mL) in the tooth surface of rats.	(Bachtiar et al. 2016)
Pseudomonas aeruginosa	Gargle	Cystic fibrosis	The slopes of the IgY treated curves > those of the control curves in Kaplan-Meier plot;	(Nilsson et al. 2008)

(continued)

Table 12.2 (continued)

Immunogen	IgY dosage form	Disease or pathological model	Effect of IgY dosage form	References
			Normal bacterial floras of the IgY -treated patients' sputa.	
Porphyromonas gingivalis	Ointment (hydrocarbon gel, sucrose esters of fatty acids, hydroxypropyl methylcellulose 2208)	Periodontitis	Significant reductions ($p < 0.05$) in probing depth; long duration (> 4 weeks).	(Yokoyama et al. 2007)
Streptococcus mutans	Mouth rinse containing IgY-Urease	Caries	59.2% rate of inhibiting S. mutans adherence versus that of the control IgY (8.2%) after 1 h at 25 °C.	(Hatta et al. 1997)

Notes: *CMC* carboxy methyl cellulose; *HRV* human rotavirus; *PGCR* polyglycerol condensed ricinolate; *UBT* Helikit13C-urea breath test

cavity of humans (Hatta et al. 1997). In veterinary application, passive immunization to fish by immersing specific IgY into aquarium water can prevent diseases caused by pathogens that invade the skin and gills (Gan et al. 2015).

12.4 IgY in Food and Feed Use

It is worthy to mention that apart from novel dosage forms, the use of egg yolk powder (EYP) is another simple, economical and practical "dosage form" for IgY, particularly in veterinary applications. IgY bioactivity in the yolk or yolk powder is better preserved, easily absorbed by the host, and avoids gastrointestinal deactivation better than extracted IgY (Ge et al. 2020). Some Food ingredients, such as trehalose and cyclodextrin, could be used as IgY protectants, while sucrose, lactose and dextran are not effective. Infant formula and egg yolk could recover IgY activity (34% and 40%) after pepsin treatment, as compared to sugars or complex carbohydrates which have no protective effect (Jaradat and Marquardt 2000). According to a recent study, tablets containing a commercial available source of IgY formulated to alleviate infections of main gastro intestinal pathogens in dogs (such as parvovirus, *E. coli*, *Salmonella spp.* and *Clostridium spp.*) were administered to the healthy dogs twice per day, combined with basic commercial diet, it is validated that the addition of IgY in the diet maintains the microbial balance and has a positive effect on microbial metabolites, which offers a prophylactic strategy to companion animals (Scheraiber et al. 2019).

12.5 Perspective

The study of IgY delivery and dosage form design is still at early stage of development. With the rapid increase in biopharmaceuticals and biological products, novel and non-invasive dosage forms, materials and technologies have been applied in antibody and biological product delivery. These delivery methods include microneedle based transdermal, transdermal patch, inhalation, aerosol, film based buccal targeting on oral mucosal surface, nasal sprays, as well as self-administration allowed subcutaneous delivery, such as NanoPass MicronJet technology and iontophoresis technology (Anselmo et al. 2019). There are still areas worth exploring to find the most suitable protectant, carrier and delivery systems for IgY. The generation of monoclonal IgY has been successful and practical (Chap. 13), which has increased the value and potential of IgY antibody for broad and precise biomedical applications, and thereby will increase the need of rational and accurate dosage form design.

From the perspective of pharmaceutical sciences, there are still large unknown areas which may be essential to promote rational IgY formulation design and application, such as the pharmaceutical kinetic of IgY in human and animal hosts (Zhou et al. 2018), the regular patterns of IgY absorption and transportation in the hosts, and the possible interactions of IgY to other drug, food ingredients, endogenous components.

References

Akita EM, Nakai S (1993) Production and purification of Fab' fragments from chicken egg yolk immunoglobulin Y (IgY). J Immunol Methods 162(2):155–164

Alustiza F, Bellingeri R, Picco N, Motta C, Grosso MC, Barbero CA, Acevedo DF, Vivas A (2016) IgY against enterotoxigenic Escherichia coli administered by hydrogel-carbon nanotubes composites to prevent neonatal diarrhoea in experimentally challenged piglets. Vaccine 34 (28):3291–3297

Amirijavid S, Entezari M (2018) Comparison of the effects of three kinds of IgYs, (normal, nanoliposomal and nanoparticle conjugated), which are produced against the small domains of DR5 protein on cancer cells. IET Nanobiotechnol 12(4):436–440

Anselmo AC, Gokarn Y, Mitragotri S (2019) Non-invasive delivery strategies for biologics. Nat Rev Drug Discov 18(1):19–40

Bachtiar EW, Afdhal A, Meidyawati R, Soejoedono RD, Poerwaningsih E (2016) Effect of topical anti-Streptococcus mutans IgY gel on quantity of S. mutans on rats' tooth surface. Acta Microbiol Immunol Hung 63(2):159–169

Bellingeri RV, Picco NY, Alustiza FE, Canova JV, Molina MA, Acevedo DF, Barbero C, Vivas AB (2015) pH-responsive hydrogels to protect IgY from gastric conditions: in vitro evaluation. J Food Sci Technol 52(5):3117–3122

Chang HM, Lee YC, Chen CC, Tu YY (2002) Microencapsulation Protects Immunoglobulin in Yolk (IgY) Specific against Helicobacter pylori Urease. J Food Sci 67:15–20

Chi ZB, Gao YX, Pan Y, Zhang B, Feng XP (2004) The inhibitive effect of IgY toothpaste against oral Streptococcus mutans. Shanghai kou qiang yi xue = Shanghai Journal of Stomatology 13 (4):256–258

Cho Y-H, Lee J-J, Park I-B, Huh C-S, Baek Y-J, Park J (2006) Protective effect of microencapsulation consisting of multiple emulsification and heat gelation processes on immunoglobulin in Yolk. J Food Sci 70:E148–E151

De Jong WH, Borm PJ (2008) Drug delivery and nanoparticles: applications and hazards. Int J Nanomedicine 3(2):133–149

Dluska E, Markowska-Radomska A, Metera A, Tudek B, Kosicki K (2017) Multiple emulsions as effective platforms for controlled anti-cancer drug delivery. Nanomedicine (Lond) 12 (18):2183–2197

Gan H, He H, Sato A, Hatta H, Nakao M, Somamoto T (2015) Ulcer disease prophylaxis in koi carp by bath immersion with chicken egg yolk containing anti-Aeromonas salmonicida IgY. Res Vet Sci 99:82–86

Gao X, Zhang X, Sun J, Du X, Li X, Zhang Y, Lin L (2016) Passive protection effect of anti-Vibrio anguillarum IgY-encapsulated feed on half-smooth tongue sole (Cynoglossus semilaevi) against V. anguillarum. Fish Shellfish Immunol 56:483–488

Ge S, Yang Y, Chelliappan B, Michael A, Zhong F, Zhang XY (2020) Evaluation of different igy preparation methods and storage stability as potential animal feed supplement. Pak J Zool

Hatta H, Tsuda K, Ozeki M, Kim M, Yamamoto T, Otake S, Hirasawa M, Katz J, Childers NK, Michalek SM (1997) Passive immunization against dental plaque formation in humans: effect of a mouth rinse containing egg yolk antibodies (IgY) specific to Streptococcus mutans. Caries Res 31(4):268–274

Holser R (2013) Lipid encapsulated phenolic compounds by fluidization. J Encapsulation Adsorption Sci 03:13–15

Horie K, Horie N, Abdou AM, Yang JO, Yun SS, Chun HN, Park CK, Kim M, Hatta H (2004) Suppressive effect of functional drinking yogurt containing specific egg yolk immunoglobulin on Helicobacter pylori in humans. J Dairy Sci 87(12):4073–4079

Jaradat Z, Marquardt R (2000) Studies on the stability of chicken igy in different sugars, complex carbohydrates and food materials. Food Agric Immunol 12:263–272

Kovacs-Nolan J, Mine Y (2005) Microencapsulation for the gastric passage and controlled intestinal release of immunoglobulin Y. J Immunol Methods 296(1–2):199–209

Lee J, Kang HE, Woo HJ (2012) Stability of orally administered immunoglobulin in the gastrointestinal tract. J Immunol Methods 384(1–2):143–147

Li XY, Jin LJ, Lu YN, Zhen YH, Li SY, Wang LH, Xu YP (2009) Chitosan-alginate microcapsules for oral delivery of egg yolk immunoglobulin (IgY): effects of chitosan concentration. Appl Biochem Biotechnol 159(3):778–787

Li XY, Jin LJ, McAllister TA, Stanford K, Xu JY, Lu YN, Zhen YH, Sun YX, Xu YP (2007) Chitosan-alginate microcapsules for oral delivery of egg yolk immunoglobulin (IgY). J Agric Food Chem 55(8):2911–2917

Mine Y, Kovacs-Nolan J (2002) Chicken egg yolk antibodies as therapeutics in enteric infectious disease: a review. J Med Food 5(3):159–169

Muheem A, Shakeel F, Jahangir MA, Anwar M, Mallick N, Jain GK, Warsi MH, Ahmad FJ (2016) A review on the strategies for oral delivery of proteins and peptides and their clinical perspectives. Saudi Pharmaceutical J 24(4):413–428

Nilsson E, Larsson A, Olesen HV, Wejåker PE, Kollberg H (2008) Good effect of IgY against Pseudomonas aeruginosa infections in cystic fibrosis patients. Pediatr Pulmonol 43(9):892–899

Paul NC, Al-Adwani S, Crespo R, Shah DH (2014) Evaluation of passive immunotherapeutic efficacy of hyperimmunized egg yolk powder against intestinal colonization of Campylobacter jejuni in chickens. Poult Sci 93(11):2779–2787

Pérez de la Lastra JM, Baca-González V, Asensio-Calavia P, González-Acosta S, Morales-delaNuez A (2020) Can Immunization of Hens Provide Oral-Based Therapeutics against COVID-19? Vaccine 8(3)

Rahman S, Higo-Moriguchi K, Htun KW, Taniguchi K, Icatlo FC Jr, Tsuji T, Kodama Y, Van Nguyen S, Umeda K, Oo HN, Myint YY, Htut T, Myint SS, Thura K, Thu HM, Fatmawati NN, Oguma K (2012) Randomized placebo-controlled clinical trial of immunoglobulin Y as adjunct to standard supportive therapy for rotavirus-associated diarrhea among pediatric patients. Vaccine 30(31):4661–4669

Scheraiber M, Grześkowiak Ł, Zentek J, Barbosa FF, Félix AP, da Silva AVF (2019) Inclusion of IgY in a dog's diet has moderate impact on the intestinal microbial fermentation. J Appl Microbiol 127(4):996–1003

Shimizu M, Miwa Y, Hashimoto K, Goto A (1993) Encapsulation of chicken egg yolk immunoglobulin G (IgY) by liposomes. Biosci Biotechnol Biochem 57(9):1445–1449

Shimizu M, Nakane Y (1995) Encapsulation of biologically active proteins in a multiple emulsion. Biosci Biotechnol Biochem 59(3):492–496

Soumaila Garba A, Thibodeau A, Perron A, Laurent-Lewandowski S, Letellier A, Fravalo P (2019) In vitro efficacy of potentiated egg yolk powder against Campylobacter jejuni does not correlate with in vitro efficacy. PLoS One 14(3):e0212946

Torché AM, Le Dimna M, Le Corre P, Mesplède A, Le Gal S, Cariolet R, Le Potier MF (2006) Immune responses after local administration of IgY loaded-PLGA microspheres in gut-associated lymphoid tissue in pigs. Vet Immunol Immunopathol 109(3–4):209–217

Xing P, Shi Y, Dong C, Liu H, Cheng Y, Sun J, Li D, Li M, Sun K, Feng D (2017) Colon-Targeted Delivery of IgY Against Clostridium difficile Toxin A and B by Encapsulation in Chitosan-Ca Pectinate Microbeads. AAPS Pharm Sci Tech 18(4):1095–1103

Xu L, Che J, Xu Y, Chen Y, Li Y, Murtaza B, Wang L, Zhang M, Li X (2020) Oral administration of microencapsulated egg yolk immunoglobulin (IgY) in turbot (Scophthalmus maximus) to combat against Edwardsiella tarda 2CDM001 infections. Fish Shellfish Immunol 106:609–620

Yokoyama K, Sugano N, Shimada T, Shofiqur RA, el SM I, Isoda R, Umeda K, Sa NV, Kodama Y, Ito K (2007) Effects of egg yolk antibody against Porphyromonas gingivalis gingipains in periodontitis patients. J Oral Sci 49(3):201–206

Zhang W, Feng XP, Tao DY, Chen JF (2016a) Clinical study on the effect of anti-gingivitis IgY toothpaste in control of gingivitis and dental plaque. Shanghai kou qiang yi xue = Shanghai Journal of Stomatology 25(4):453–455

Zhang S, Xing P, Guo G, Liu H, Lin D, Dong C, Li M, Feng D (2016b) Development of microbeads of chicken yolk antibodies against Clostridium difficile toxin A for colonic-specific delivery. Drug Deliv 23(6):1940–1947

Zhou Z, Wang J, Gao S, Ren H, Xie W, Thirumalai D, Zhang XY (2018) Comparative study on Pharmacokinetic Profiles of Chicken IgY to Horse IgG in Rabbits. J Hellenic Vet Med Soc 68:405

Zhou Z, Zhou R, Tang Z (2003) Effects of topical application of immunoglobulin yolk on mutans streptococci in dental plaque. Hua xi kou qiang yi xue za zhi = Huaxi kouqiang yixue zazhi = West China Journal of Stomatology 21(4):295–297

Monoclonal IgY Antibodies 13

Chiuan Herng Leow, Long Xu, Carol A. Harley, Ricardo S. Vieira-Pires, and Xiaoying Zhang

Abstract

Monoclonal antibodies (mAbs) have the advantages of high specificity and stable expression. Monoclonal IgY (mIgY) is a new antibody development that combines advantages of IgY and mAbs. mIgY have great potential for immunological detection and diagnosis, for screening and validating biomarkers and drug targets, and for the production of antibodies against conserved mammalian proteins. mIgY can be developed into small molecular weight antibody fragments (particularly scFv), chimeric and humanized antibodies to address the many challenges lying ahead to develop mAb for various biomedical applications, especially those for which conventional antibodies are ineffective. Hybridoma

C. H. Leow (✉)
Institute for Research in Molecular Medicine (INFORMM), Universiti Sains Malaysia, Pulau Pinang, Malaysia
e-mail: herng.leow@usm.my

L. Xu
Shaanxi University of Technology, Hanzhong, China

University of Minho, Braga, Portugal

C. A. Harley
i3S–Instituto de Investigação e Inovação em Saúde and IBMC–Instituto do Biologia Molecular e Celular, University of Porto, Porto, Portugal
e-mail: harleyca@ibmc.up.pt

R. S. Vieira-Pires
CNC–Center for Neuroscience and Cell Biology, University of Coimbra, Coimbra, Portugal
e-mail: ricardo.pires@biocant.pt

X. Zhang
College of Biological Science and Engineering, Shaanxi University of Technology, Hanzhong, China
e-mail: zhxying@snut.edu.cn; xzhang67@uoguelph.ca; zhang@bio.uminho.pt

X.-Y. Zhang et al. (eds.), *IgY-Technology: Production and Application of Egg Yolk Antibodies*, https://doi.org/10.1007/978-3-030-72688-1_13

technology or display technology, including phage display, yeast surface display and ribosomal display, can be applied in mIgY generation. However, hybridoma technology is very inefficient in the chicken system due to the lack of a robust avian immortal cell line for hybridoma cell generation. New technologies and different types of antibodies are likely to broaden the application of avian antibodies.

Keywords

Monoclonal IgY (mIgY) · Genetic engineering technology · Functional antibody fragments · IgY single chain antibody fragment (IgY-scFv) · Chimeric antibody · Humanized antibody · Hybridoma technology · Phage display technology

13.1 Introduction

As described earlier (Chap. 2) most antigens offer multiple epitopes which give rise to a variety of B-cell clones and thus a mixture of antibodies; a polyclonal response. Monoclonal antibodies (mAbs) on the other hand are derived from a single B cell clone and are selected to be specific for a single epitope. mAbs are valuable tools in basic and applied research and are important components for medical diagnosis and therapy. Large-scale fermentation production of mAbs is no longer difficult. However, there are still limitations to mouse mAbs development either by hybridoma techniques or by antibody engineering technologies. In particular, it is difficult to induce specific antibodies to some proteins (antigens) that have highly conserved epitopes. In closely related species, highly conserved antigens such as prions (Matsushita et al. 1998) are non-immunogenic. Mice do not always provide a high-affinity response to small hapten molecules. Mouse mAbs also interact with a plethora of mammalian and bacterial proteins, thereby hampering their use in immunodiagnostic approaches (Greunke et al. 2008). Therefore, new antibody technologies need to be developed in order to break through these technical limitations.

13.1.1 Monoclonal Antibody and Functional Antibody Fragments

Mouse mAbs are important tools in various research, diagnostic and clinical applications. Clinically, however, mouse mAbs induce anti-murine antibody response in humans (HAMA), and thus therapeutic effects of mouse mAbs are limited (Tateishi et al. 2008). To reduce the immunogenicity of mouse or chicken antibodies in humans, chimeric and humanized antibodies (Fig. 13.1) have been successfully developed.

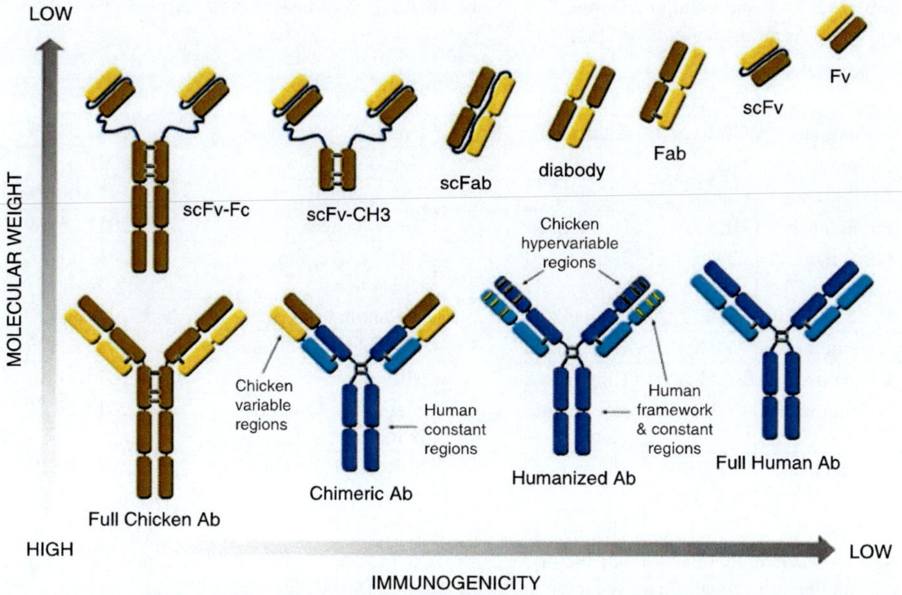

Fig. 13.1 Different types of functional chicken antibodies. Schematic representation of different chicken antibody formats, where different structures enable different functional outcomes. The figure was prepared base on the guidelines of immunogenicity. The specific circumstances shall be subject to the experiment. Chicken H and L chains are shown in brown and yellow, respectively and human H and L chains are shown I dark and light blue

Numerous mouse-human chimeric mAbs (such as butuximab, dituximab, and basiliximab) and humanized mouse mAbs (such as patoizumab, otozumab, daremumab, elotuzumab and perizumab) have been marketed and are now clinically used (Table 13.1).

It is difficult to produce therapeutic mouse mAbs against highly conserved biological molecular targets. A large number of biological molecules are conserved in mammalian species, and therefore induce a limited immune response in rodent hosts due to immunotolerance. Therefore, chickens standout as an useful alternative host for the development of specific antibodies against mammalian-conserved biological molecules. As a result of their phylogenic distance towards mammals (Chap. 4), chickens have a higher potential for antibody production for these conserved molecules. Furthermore, the molecular diversification of immunoglobulins and the organization of antibody genes in the chicken differs from that in mammals (Chap. 5). Indeed, unlike with mammalian immune repertoires, it is possible to amplify avian V-gene repertoires with a single pair of primers, significantly facilitating the molecular cloning of antibody libraries.

Because chicken antibodies are immunogenic in mammals, their clinical applications are limited (Tateishi et al. 2008). In particular, the constant region contributes a significant component to immunogenicity (Morrison et al. 1984). To

Table 13.1 Representative examples of monoclonal antibody treatments approved by the Food and Drug Administration

Name	Target	mAb type	Indication	Approval date
Butuximab	CD30	Chimeric	Hodgkin lymphoma	2011.08.19
Dituximab	GD2	Chimeric	Neuroblastoma	2015.03.10
Basiliximab	CD25	Chimeric	Transplanted liver	1999.03.09
Patoizumab	HER-2	Humanized	Breast cancer	2012.06.08
Otozumab	CD20	Humanized	Chronic lymphocytic leukaemia	2013.11.01
Perizumab	PD-1	Humanized	Melanoma	2014.09.04
Daremumab	CD38	Humanized	Multiple myeloma	2015.11.06
Elotuzumab	SLAMF7	Humanized	Multiple myeloma	2015.11.30
Nesimumab	EGFR	Recombinant human	Metastatic squamous NSCLC	2015.11.24
Olarezumab	PDGFR-α	Human	Soft-tissue sarcoma	2016.10.19
Avelumab	PD-L1	Human	Merck cells cancer	2017.03.23

Notes: *EGFR* epidermal growth factor receptor, *NSCLC* non-small-cell lung cancer, *PDGFR-α* platelet-derived growth factor receptor-α, *PD-L1* programmed death ligand 1

overcome this problem, genetic engineering techniques have been applied to produce chimeric, humanized, and recombinant antibody formats of chicken antibodies (Fig. 13.1), similarly to any other mammalian derived antibody. Chimeric and humanized antibodies result in reduced immunogenicity (Tateishi et al. 2008) and longer half-lives (Nesspor and Scallon 2014), which are important for the pharmaceutical application of mIgY. With the construction of chimeric antibodies or humanized antibodies, more clinical applications of IgY, such as diagnosis or therapy to important disease targets, can be further developed.

13.2 Genetically Engineered Chicken Antibodies

13.2.1 Recombinant IgY Fragments

In view of the fact that the antigen binding site of antibodies is restricted to the variable region, one can generate different formats of antibodies fragments (Fig. 13.1) such as the antigen-binding fragment (Fab), the single-chain variable fragment (scFv), diabody or multivalent fragment, by genetic engineering technology. In many, but not all cases, the variants retain similar affinity and the same antigen-binding specificity when compared with the parent antibodies (Zhen et al. 2009). Of the various recombinant antibody formats, scFv is the most commonly used because it can be rapidly constructed and well expressed in *E. coli*, while retaining a high affinity for the target and in general good stability (Bird et al. 1988).

A chicken sourced single-chain variable fragment antibody (IgY-scFv) was constructed by using a phage display technique against the N-terminal of the mammalian prion protein (PrP) (Nakamura et al. 2000). Western blot and ELISA analysis demonstrated that IgY-scFv was as reactive to PrP from mouse brains as the previously obtained mIgY, derived from chicken hybridoma technology. Indeed the work was particularly relevant to validate the use of recombinant scFv, in order to improve the chronical low levels of mIgY expression from chicken hybridoma system.

On the other hand, generation of IgY-scFv libraries raises a number of technical limitations, including molecular cloning bottlenecks, such as inefficient overlap PCR steps for (VH-VL) genetic fusions, or limitations on downstream expression, detection and purification steps of the scFv final product; for all these different solutions have been worked out.

In the example above, Nakamura et al. used a mouse mAb expression vector pPDS, originally developed by Yamanaka et al. (Yamanaka et al. 1995), in order to obtain IgY-scFv for PrP detection. However, pPDS vectors contain a mouse Cκ sequence as a detection tag, restricting the combined use of a mouse and chicken mAbs derived from such an expression set. To circumvent this problem, Nakamura et al. designed two novel expression vectors pCPDS and His-pCPDS (Nakamura et al. 2003), which contained a chicken Cλ and FLAG with or without 6 × histidine sequences in the 3′ terminus of the scFv, to serve as detection and purification tags. The novel recombinant chicken scFv antibodies showed similar antigen-binding specificity to PrP as the original mAbs.

A number of other approaches have been developed to overcome scFv production shortcomings. For example, pComb3 phage display vector series (Barbas III et al. 1991; Andris-Widhopf et al. 2000) were developed to generate tagged scFv, diabody and Fab fragments of mIgY, that are easier to detect and purify by affinity methods. On the other hand, a new phagemid vector, pCANTAB-link that encoded a polypeptide linker region $(Gly_4Ser)_3$ flanked by two multiple cloning regions, was constructed to overcome the overlap extension PCR step (Sapats et al. 2003), recognized as the most inefficient and problematic stage in constructing scFv libraries (McCafferty and Johnson 1996). This significantly increased the efficiency of generating scFv libraries which increases the library size 500 fold.

13.2.2 Chimeric Antibodies

Currently, mIgY are potentially useful for diagnostic research and have clinical applications, but the applications of chicken antibodies are limited because of their immunogenicity in mammals. To overcome this problem, chimeric antibodies are constructed.

A scFv and a chicken-human chimeric antibody Fab against fluorescein hapten were prepared by engineering the phage vectors pComb3H and pComb3X (Andris-Widhopf et al. 2000). The mIgY prepared by the phage display technology is easily detected and purified by adding an affinity tag. A chicken-human chimeric antibody against human PrP-N-terminal amino peptide still retains similar binding affinity as the parent chicken antibody (Nishibori et al. 2004). A chicken-mouse chimeric antibody against containing chicken variable regions and the murine constant region was constructed (Tateishi et al. 2008). Administration of this chimeric antibody in mice confirmed that the chimeric antibody was less immunogenic than the recombinant IgY, which provides a good tool for a mouse pre-experiment before human pre-clinical and application of mIgY.

13.2.3 Humanized Antibody

In order to further reduce the immunogenicity of antibodies for applications in human clinical therapy, mIgY needs to be humanized. Antibody humanization is typically performed by transplanting the six complementarity-determining regions (CDRs) of the parent non-human Abs onto human framework (FR) regions using CDR-grafting technology (Fig. 13.1) (Jones et al. 1986). However, antibody humanization by simple grafting of CDRs often results in significant reductions in antigen affinity (Foote and Winter 1992). This is because certain FR residues are important to maintain CDR loop structure and affinity (Foote and Winter 1992; Studnicka et al. 1994). Therefore, further engineering of humanized antibodies, which is achieved by replacing the key residues of the FR regions of humanized antibodies to those of the parent antibody, is required to restore affinity. This is a time-consuming process that requires creating a set of humanized variants, characterizing them and re-correcting them if necessary. This is partly because it is difficult to predict which FR residues should be replaced, and partly because minimum replacement is desirable to decrease the risk of immunogenicity. A humanized mIgY was constructed by transferring their CDRs and several framework amino acids onto human acceptor variable regions (Tsurushita et al. 2004). The affinities of humanized mIgY to human and mouse IL-12 measured by competitive binding was nearly identical to those of chicken–human chimeric mIgY. Tsai et al. (Tsai et al. 2019) generated a chicken-derived humanized antibody targeting a novel epitope F2pep of Fibroblast Growth Factor Receptor 2, which had an excellent binding ability with F2pep and had high potential in cancer therapy. A humanized mIgY was constructed by using a phage-display combinatorial library (Nishibori et al. 2006). The humanized Fab variants retained similar binding affinities as the parent Abs, which makes mIgY application *in vivo* possible.

13.3 Biomolecular Methods for Monoclonal IgY Generation

The combination of IgY technology and monoclonal antibody engineering as well as antibody library generation and selection approaches, will likely contribute to a significant expansion of the outcomes and applications of chicken-derived mAbs.

13.3.1 Hybridoma Technology

Since the introduction of hybridoma technology (Köhler and Milstein 1975) attempts have been made to obtain chicken hybrid cell lines to establish stable hybridomas producing specific mIgY. Nishinaka et al. developed the first mIgY using the hybridoma technique (Nishinaka et al. 1989). Several fusion partners cell lines have been developed for the production of chicken mAbs (Table 13.2, (Nishinaka et al. 1989, 1991, 1996).

Although chicken hybridomas have been established, the yield of mAb-producing clones ($>10^6$) is less than the mouse system (around 10^7) in many cases (Pink 1986; Nishinaka et al. 1991, 1996; Matsushita et al. 1998; Matsuda et al. 1999). Furthermore, since these fusion partner lines were derived from the outbred chicken line which might cause mismatching and mis-recombining between the fusion partner and immunized spleen cell (Nishinaka et al. 1996), the chicken hybridomas could not make ascites (Nakamura et al. 2000) and in some cases, specific antibodies were not obtained. Different physiological features of chickens also play a role, for example, the chicken hybridomas are cultivated at 38.5 °C rather than 37 °C as the chicken has a body temperature 38.54 °C ± 0.96 °C at rest (Chap. 2). Furthermore, one-step purification of mIgY developed by hybridoma technology is a difficult process, as mIgY cannot bind to the most commonly used commercial affinity chromatography sources, protein A or protein G (Nakamura et al. 2004). The process of obtaining mIgY by hybridoma technology is thus more complex, time-consuming and more difficult compared to obtaining murine-mAb and all of these factors limit the application of hybridoma technology for mIgY production.

13.3.2 DT40 Cell Line

An alternative to the use of hybridoma is DT40, a chicken lymphoma cell line, which produces IgM-type antibodies and undergoes gene conversion and somatic mutations in the variable region of immunoglobulin (IgV) gene during culture. The hypermutation of IgV can increase the diversity of IgMs, and make DT40 cells a natural antibody library for monoclonal antibody screening (Cumbers et al. 2002; Seo et al. 2005). The fast growth of DT40 cells enables rapid antibody acquisition compared to that in hybridoma and phage display systems (Seo et al. 2006).

An autonomously diversifying library (ADLib) (Fig. 13.2) was developed to rapidly generate specific monoclonal antibodies (Seo et al. 2006), which is generated on the basis of that gene conversion in DT40 cells was enhanced by treatment of the cells with a histone deacetylase inhibitor, trichostatin A (TSA). DT40 clones expressing antigen-specific mAb are isolated from ADLib using antigen-conjugated magnetic beads. The ADLib system has been used to the direct generation of chicken/human chimeric antibody by inserting a DNA segment coding for the constant region of human IgG into the chicken IgM heavy-chain locus of DT40

Table 13.2 Characteristics of three fusion partner cell lines: HU3R27, R27H4 and MuH1

Cell line	Origin	HAT-sensitivity	Ig-expression	Fusion efficiency (%)	Type of specific hybridomas	Stability of specificity hybridomas	Clone efficiency of hybridomas (%)
HU3R27	HU3	+	–	45.8	IgM	Low	0
R27H4	HU3R27	+	IgM	26.4	IgM, IgG	High	9.5
MuH1	R27H1	+	–	14.49–19.4	IgG(0.24–5.22 mg), IgM (50.32 ng)	High	20–25

Notes: *HU3R27* TK-deficient; *R27H4* Resistant to trifluorothymidine (TFT) and TK-deficient; *MuH1* Resistant to Ouabain and TK-deficient; Results referenced from (Nishinaka et al. 1989, 1991, 1996)

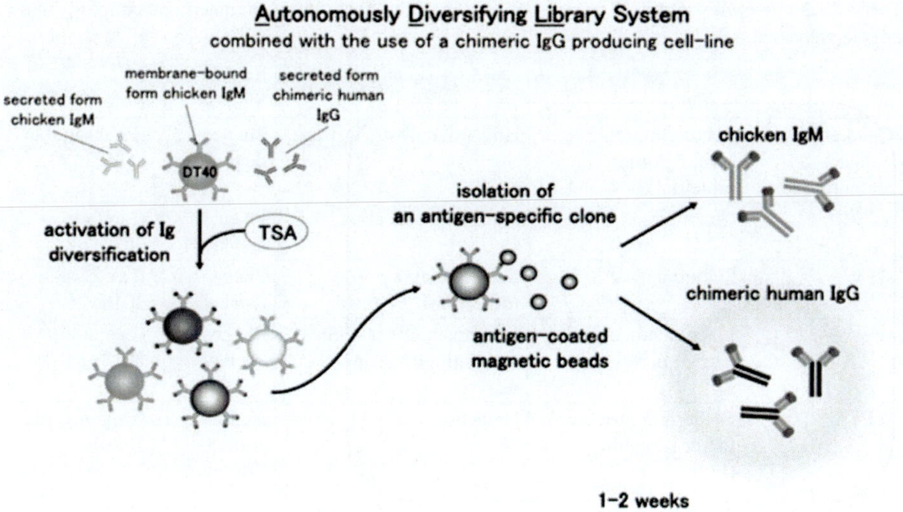

Fig. 13.2 ADLib system using the DT40 cells expressing chicken/human chimeric IgG. The DT40 cells producing antigen-specific antibodies are screened using antibody-conjugated magnetic beads which are recognized by the membrane-bound form chicken IgM, and then the simultaneously expressed chimeric human IgGs can be purified from the culture supernatants of the selected clones. Adapted from (Kurosawa et al. 2014)

cells by homologous gene targeting (Kurosawa et al. 2014). The ADLib system has the potential to be an excellent alternative to the conventional systems for producing mAb reagents and the screening process of drug discovery and design.

DT-40 cells have also been used to generate chimeric IgM containing human Ig light and heavy chain loci with chicken constant regions (Schusser et al. 2013). This provided a model to study how DT-40 cells modified the human variable region by gene conversion and somatic hypermutation. This was also an important initial step in the development transgenic chickens expressing antibodies (Ching et al. 2018).

13.3.3 Antibody Display Technologies

There are a number of display methods available to screen and affinity mature a monoclonal antibody against a particular target of interest. These techniques all rely on the linkage of phenotype with genotype (either DNA or RNA) for selection and can be broadly categorized as *in vivo* display: performing multiple rounds of selection in cells such as bacteriophage and yeast display or *in vitro*: where the selection and subsequent maturation steps do not rely on the transformation of cells such as ribosomal or RNA display. In this section, using the scFv as an example, we will give a general overview of current phage, yeast, and ribosomal display technologies, highlighting key points to consider before starting a screening or affinity maturation campaign. Some representative examples of mIgY developed

Table 13.3 Overview of mIgY preparations using different antibody engineering technology and / or selection methods

mIgY type	Antigen	Use	Eng./Sel. Method	Reference
Chicken/ human chimeric fab	Recombinant hemagglutinin protein	H5N1 diagnostic testing	Phage display	(Pitaksajjakul et al. 2010)
scFv	Gentamicin	Antibiotic detection	Phage display	(Li et al. 2016)
scFv	Salbutamol	Sensitive assay (detection)	Phage display	(Lee et al. 2018a)
scFv	Inactivated cobra venom proteins	Diagnosis of snakebites and antibody affinity test	Phage display	(Lee et al. 2018b)
scFv	Epidermal growth factor receptor (EGFR)	Detection	Yeast surface display	(Bogen et al. 2020)

by genetic engineering technology using the different selection methods are shown (Table 13.3).

13.3.3.1 Phage Display

Phage display relies on the use of viruses (phage is short for bacteriophage) which consist of a genome and a protein coat. Genetic engineering is used to join antibody genes with the gene that encodes one of the coat proteins expressed on its surface. The antibody is then produced linked to the coat protein and a large library can be constructed and the phage amplified by infection of *E. coli* (Smith 1985). The resulting phage is then screened using an immobilized antigen. This technology was first introduced in 1985 by George Smith in which he showed that exogenous DNA could be inserted into the pIII gene, encoding the minor coat protein, of filamentous phage (Smith 1985). This fusion protein is incorporated into the virion, displayed in an accessible form on the particle surface and retains infectivity of E. coli. In fact all the coat proteins of filamentous phage can be fused to proteins with varying degrees of success, however, the most commonly used are N terminal fusions to all or part of the pIII and pVIII coat proteins (Webster 2001). The minor pIII coat protein is present at 5 copies per virion of which all can be expressed as a fusion protein and the major pVIII coat protein is present at 2700 copies per virion of which 10% can be expressed as a fusion protein (Sidhu 2001; Webster 2001). This allows for a low valency expression in pIII systems versus a high valency expression in pVIII systems to select for low-affinity interactions. Antibody phage display library preparations from an immunized chicken involve the isolation of RNA which is converted to cDNA by reverse transcription. A set of defined primers based on avian germline sequences is used to PCR amplify the VL and VH fragments which are then cloned using overlap extension PCR into a phagemid plasmid, such as pHEN (Hoogenboom et al. 1991), pComb3 based vectors, as an N terminal fusion to a pIII gene fragment (αα230–406) (Andris-Widhopf et al. 2000)

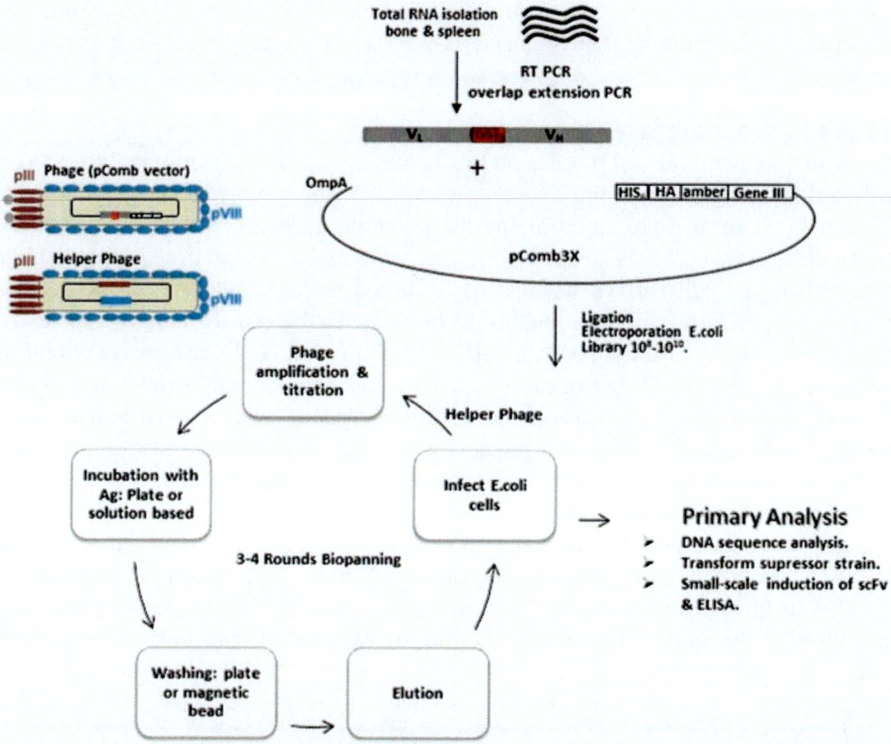

Fig. 13.3 scFv antibody selection by phage display. RNA is isolated from immunized chickens and converted into cDNA from which the VH and VL genes are amplified using defined PCR primers. These are cloned into phagemid vectors such as pComb3X to be displayed on phage and screened through a number of stringent rounds of biopanning. Enriched clones are sequenced then expressed and affinity purified from a suppressor *E. coli* strain for further downstream analysis

and pCANTAB (Amersham Pharmacia Biotech) vector systems which use the entire pIII as the fusion partner. This phagemid plasmid in addition to the plasmid's origin of replication has a phage-derived origin of replication. Propagation of the phagemid in E. coli superinfected with wild-type or helper phage allows packaging of the phagemid DNA identical to the phage DNA and encodes the library to be displayed (Vieira and Messing 1987; Barbas III et al. 1991) (Fig. 13.3). Diverse antibody libraries are produced on the transformation of non-suppressor E. coli strains such as XL1-Blue resulting in libraries between 10^8–10^{10} in size. After amplification and titration, successive rounds of selection and biopanning are performed both in a plate-based ELISA or solution magnetic bead format (Fig. 13.3). A unique property of the pComb3 vector system is the addition of 6 × Histidine and HA tags and an amber suppressor stop codon upstream of gene III. Enriched clones can be transformed into a suppressor E. coli strain and expressed in the periplasm for affinity isolation of the antibody and further characterization. This robust display technology has been successfully used to isolate avian derived scFv molecules against diverse antigen targets such as biomarkers (Drees et al. 2014), antibiotics (Li et al. 2016) and virus (Ge et al. 2020).

13.3.3.2 Yeast Surface Display

Yeast surface display (YSD) relies on the display of an antibody fragment fused to a cell surface A-agglutinin-binding subunit (Aga2p) which projects the protein away from the cell surface thus minimizing the potential interactions with other surface molecules (Boder and Wittrup 1997). Aga2p contains a signal sequence which directs it to the cell surface where it is anchored via disulfide bonding to Aga1p (Fig. 13.4). A limiting factor in this technology is the transformation efficiency which only allows a library size of 10^7–10^9. However, YSD has been extremely successfully for protein engineering of scFv molecules identified following phage display screening. A eukaryotic expression system allows post-translational modifications of scFv molecules, such as glycosylation and proper folding that are more problematic in bacterially propagated display libraries. On the other hand, differential glycosylation patterns in yeast and mammals needs to be taken in account, especially when displaying other e.g. human glycosylated protein scaffolds in YSD. The addition of epitope tags such as HA and c-myc allows the coupling of YSD with flow cytometry technology using fluorescently labeled antibodies. This has enabled the quantitative measurements of equilibrium constants, dissociation constants, stability, and protein expression levels. The engineering of scFv

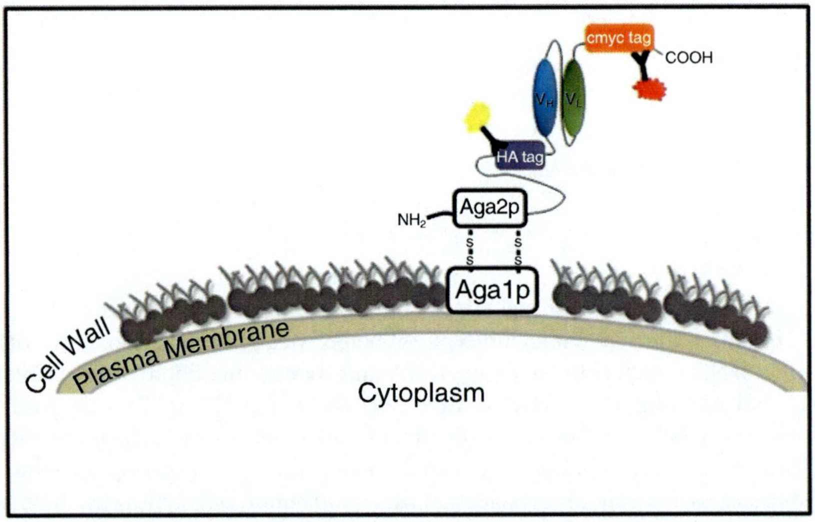

Fig. 13.4 Yeast surface display of scFv antibody for directed evolution. The scFv of interest is cloned into a YSD vector in-frame with Aga2 and HA tag genes and a c-myc C terminal tag. This plasmid can be subjected to error-prone PCR before being transformed into the *Saccharomyces cerevisiae* strain EBY100 which has Aga1p integrated under the control of the GAL1 promoter. After growth of cells in galactose media detection of surface displayed protein can be monitored by flow cytometry analysis using fluorescently labeled antibodies that recognize the HA or c-myc tags

molecules for particular properties can be accomplished by molecular evolution using error-prone PCR and after transformation of yeast cells and induction of expression under the GAL promoter the variants are displayed on the cell surface. The library can then be screened for cells having high expression and higher affinity, which recognize sequentially lower concentrations of the ligand by flow cytometry and then sorted using fluorescent activated cell sorting (FACS) (Bogen et al. 2020).

13.3.3.3 Ribosomal Display

Ribosomal display, an *in vitro* evolution technology for proteins, was first introduced in 1997 by Hanes and Plückthun (Hanes and Plückthun 1997; Qi et al. 2009). It is based on in vitro translation, but prevents the newly synthesized protein and the mRNA encoding it from leaving the ribosome. It thereby couples phenotype and genotype. By incorporating a linker between the peptide and ribosome protein folding can occur correctly. Since cells do not need to be transformed, very large libraries (up to 10^{14}) can be used directly in selections, and the in vitro amplification provides a very convenient integration of random mutagenesis (Plückthun 2012). The main steps of the ribosomal display are shown in Fig. 13.5. Successive rounds using error-prone PCR can increase the diversity or affinity maturation of the selected molecule. Ribosome display can be carried out using either prokaryotic gene expression machinery or more recently eukaryotic systems (He and Taussig 2007) which allows direct cDNA synthesis from the ribosome-mRNA-antibody complex (Hanes et al. 1998). Unfortunately, many biochemistry laboratories are not experienced in RNA stability and recovery, which has limited its use. Ribosomal display has not yet been applied for the generation of IgY fragments, but it is of interest to explore the feasibility and efficiency of such an application.

13.3.4 Gel Encapsulated Microenvironment (GEM) Screening Technology

The most recent development in chicken mAb production is the gel encapsulated microenvironment (GEM) antibody-secreting single B-cell screening (Fig. 13.6) (Izquierdo et al. 2016). The technology is based on single-cell screening from entire B-cell repertoires by diluting the B-cells in a similar manner to the hybridoma method of “limiting dilution” to achieve a statistical probability of a single cell in a GEM. The GEM is an encapsulated agarose droplet containing a single antigen-specific antibody-secreting B-cell and multiple “reporters” (usually 2 or 3) which may be microbeads coated with an antigen of interest or whole cells expressing surface antigens. The GEM is then cultured under standard tissue culture environments to allow antibody-secretion by the B-cell and binding to the “reporter(s)” which is then detected using fluorophore-conjugated secondary antibodies. The GEM is then visualized by fluorescent microscopy. Labeled GEMs are then picked from the culture using a micropipette and the gel matrix is dissolved to obtain antigen-specific B-cells. The V-genes from the isolated B-cells

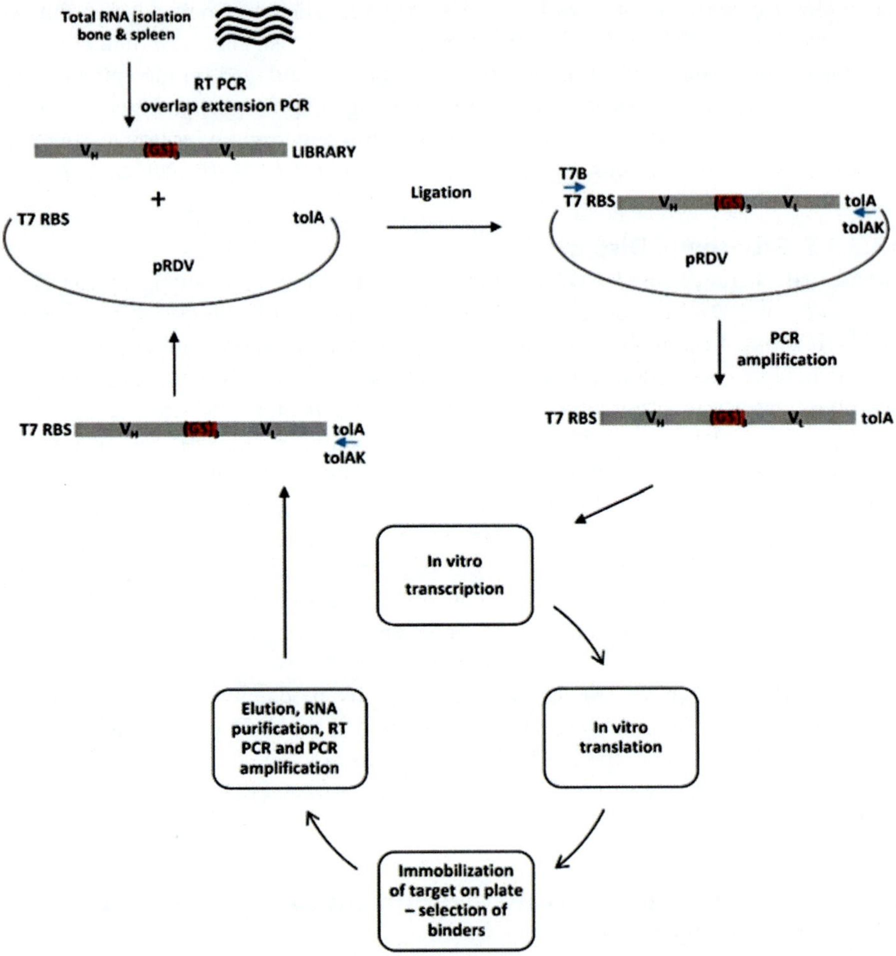

Fig. 13.5 scFv antibody selection by ribosomal display. Total RNA is isolated and cDNA corresponding to VH/VL domains is amplified. The PCR products are ligated into the Ribosomal Display vector (pRDV) to incorporate 5'and 3'required regions. After transcription and translation, the absence of a stop codon gives an RNA-Ribosome-Antibody complex which is subjected to multiple rounds of panning and enrichment for binding to the protein of interest

are then amplified and assembled into full-length scFvs for downstream cloning and expression via the standard molecular biology procedures.

The technology allows almost all B-cells in the initial repertoire (up to 100 million; (Izquierdo et al. 2016) to be screened for desirable binding characteristics without loss of diversity as with cell-surface display and is useful for isolating rare antigen-specific V-genes which may otherwise be lost during molecular cloning procedures. The use of multiple antigen "reporters" also allows for high-throughput screening where multiple binding interactions can be determined. However, the

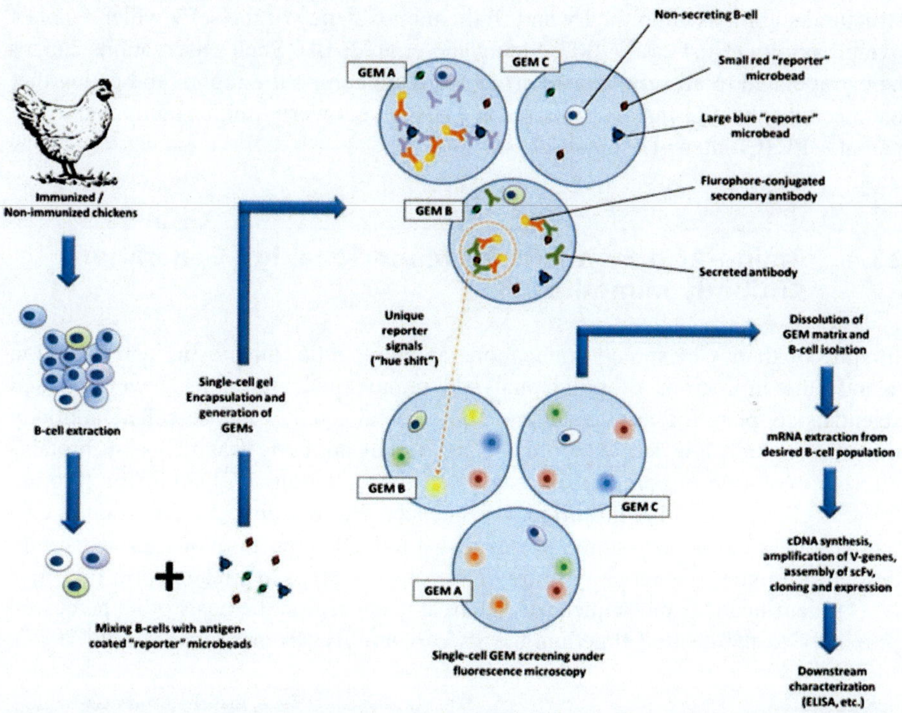

Fig. 13.6 Overview of the GEM single chicken B-cell screening technology. This technology allows each "reporter" to be identified based on microscopic size or "color/hue shifts" generated upon binding of the fluorescently-labeled secondary antibodies to the "reporter" after primary antibody binding. For example, binding of an orange, fluorescent secondary antibody to a red-colored small bead may create a visibly yellow reporter (e.g. GEM B in figure) indicating a specific antibody-antigen interaction. Non-secreting B-cells (such as in GEM C) produce no detectable reporter signals in the GEM. Refer to (Lee et al. 2017) with the permission of the Journal of immunological methods

drawbacks of the technology include the need for high-investment infrastructures such as fluorescent microscopes. Additionally, the technology cannot be used concurrently with FACS since interrogating each GEM is not only fluorescent quantitation-dependent but also relies on identifying the morphology (e.g. size of microbeads) of the "reporter". This limits the operation of this technology to skilled and experienced personnel capable of differentiating or identifying parameters correlating to individual antibody-antigen interactions in a GEM using microscopy. Furthermore, single B-cells within a GEM have to be monitored for their viability during the prolonged culture conditions to achieve detectable amounts of secreted antibody.

It is noteworthy that generating mAbs in their recombinant scFv formats are not without drawbacks. Even if scFvs are successfully expressed using display systems or GEMs, some scFvs have shown to have reduced affinity towards the target when compared to the parent Ig due to: i) lack of the constant regions which influences the structural stability within the Fv and ii) the monovalency of the scFv which reduces overall avidity (Hust et al. 2007; Steinwand et al. 2014). Such observations cannot be extrapolated to all scFv constructs as there are conflicting reports indicating that engineered scFv-CH fusions resembling Fabs have lower affinities compared to the parent scFv (Quintero-Hernández et al. 2007).

13.4 Amino-Acid Synthetic for Monoclonal IgY Generation Antibody Mimetics

In spite of their widespread applications as therapeutic, diagnostic, and detection agents, the limitations of polyclonal and monoclonal antibodies have enthused scientists to plan for next-generation biomedical agents, the so-called antibody mimetics, which are derived from the functional antibody fragments by protein-directed evolution or complementarity determining regions' (CDR) fusion through framework regions (FRs) in different sequences. Protein-directed evolution is currently employed to harness the power of natural selection to evolve proteins with desired properties. Generically, it involves four key steps as illustrated in Fig. 13.7 A: (1) Identification: the sequence of interest is chosen on the basis of its perceived proximity to the desired function and its evolvability (Romero and Arnold 2009);

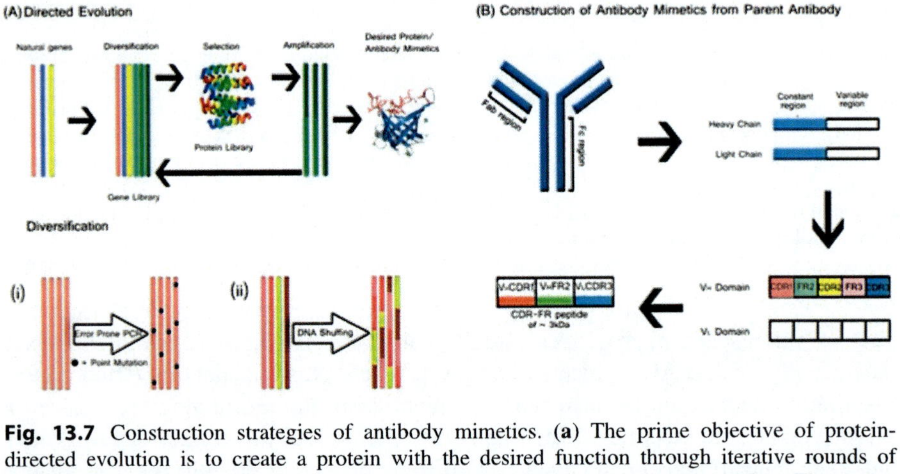

Fig. 13.7 Construction strategies of antibody mimetics. (**a**) The prime objective of protein-directed evolution is to create a protein with the desired function through iterative rounds of mutation and selection. Natural genes are selected that are amenable to mutation. The parent sequence is then mutated to form a huge library by either: (i) error-prone PCR or (ii) DNA shuffling. The mutated sequences are evaluated by using high-throughput screening or selection. The fittest sequence/s is/are used as parent sequence/s for the next round and the process is iterated until the objectives are met (usually after 5 to 10 generations). (**b**). CDR-FR peptides of 3 kDa were synthesized that mimic HB-168 IgG, a mAb IgG against EBV gp350/220 envelope glycoprotein. In CDR-FR peptides, the complementarity-determining regions are linked by a framework region. The structure of a minimal 28-residue mimetic comprising VHCDR1, VHFR2 and VLCDR3 domains of Fab is shown above. Refer to (Baloch et al. 2014) with the permission of the Critical Reviews in Biotechnology

(2) Diversification: the parent sequence is subjected to diversification by error-prone PCR and DNA shuffling; (3) Selection: the screening or selection is used to test the presence of mutants or variants in the generated library; and (4) Amplification: the variants are screened, selected and replicated many-fold to harvest a variant with the desired functional properties. Massive combinatorial libraries have been constructed by randomizing amino acid positions in structurally variable loops of proteins (Schöenfeld et al. 2009) or by exon shuffling and phage display (Silverman et al. 2005) (Fig. 13.7).

CDR-framework region (FR) peptides are constructed by fusing two CDRs through a cognate FR (Baloch et al. 2014). Protein antigens are usually recognized by all six CDRs from both the VL and VH domains of the intact antibody combining site. The CDRH3 loop is considered the most indispensable part of the mimetic, as it is often the most accessible of the CDR loops, the one with the highest diversity and thus almost always involved in antigen binding (Chap. 5). The C-terminus of the chosen CDR1 or CDR2 loop and the N-terminus of the selected CDRH3 loop are joined with a FR selected from VH or VL (Qiu et al. 2007) (Fig. 13.7b). On the basis of these principles, numerous antibody mimetics against certain pharmacologically relevant targets have been developed. Antibody mimetics derived from chicken antibodies have not yet been developed, but it is of interest to explore them in therapeutic, diagnostic, and detection application.

CDR and framework region (VHFR2) sequences of IgY-scFv against the VLP of CPV-VP2 (Ge et al. 2020) were used to create an antibody mimetic peptide comprising two interacting VHCDR1 and VLCDR3 derived CDRs. There was no difference in the detection performance of peptide and IgY-scFv on CPV clinical fecal samples. The mimetic peptide showed no cross-reactivity with canine distemper virus (CDV) and canine coronavirus (CCV) and had excellent protective properties for cells. The mimetic was the world's first peptide derived from chicken IgY, revealed the feasibility of a novel functional antibody fragment development strategy for both detection and therapeutic purposes in veterinary medicine.

13.5 Conclusion and Future Prospect

Chicken-derived recombinant antibodies share similar limitations with the broadly explored mammalian counterparts. Indeed, most of the recombinant antibody development shortcomings described in this Chapter are extended to all eukaryotes including chickens.

Cellular-based approaches, such the use of immortalized DT-40 cell line or chicken hybridomas are preferred when proper folding and secretion of antibodies needs to be mediated by pre-existing intracellular eukaryotic chaperones and post-translational modifications (PTM). Moreover these approaches also retain the native heavy-light chain pairing, sometimes critical for the antibody specificity, unlike with other library generation display technologies, where pairing is lost due to inherent heavy-light chain shuffling during *in vitro* library cloning. Nevertheless, the versatility of molecular biology techniques has offset many of the associated shortcomings. Cloning of selected scFv genes into mammalian expression vectors for expression in eukaryotic host cells for example, may increase production costs; but combining bacterial- and mammalian-based expression methods seems to be the most cost efficient option.

Taken together, each method established for the generation of chicken mAbs has to be considered carefully especially depending on the envisaged final use, being it a research and development, diagnostic or therapeutic outcome; it is also reasonable to assume that the versatile combination of cellular and molecular approaches is the basis for successful chicken mAbs development.

Chicken-derived monoclonal antibodies and related recombinant formats have unique features and thus enormous prospects for the development of new antibody-based platforms contributing to broaden IgY applications and overall rediscover of the IgY technology field.

References

Andris-Widhopf J, Rader C, Steinberger P, Fuller R, Barbas CF (2000) Methods for the generation of chicken monoclonal antibody fragments by phage display. J Immunol Methods 242 (1–2):159–181

Baloch AR, Baloch AW, Sutton BJ, Zhang X (2014) Antibody mimetics: promising complementary agents to animal-sourced antibodies. Crit Rev Biotechnol 36(2):268–275

Barbas CF III, Kang AS, Lerner RA, Benkovic SJ (1991) Assembly of combinatorial antibody libraries on phage surfaces: the gene III site. Proc Natl Acad Sci U S A 88(18):7978–7982

Bird RE, Hardman KD, Jacobson JW, Johnson S, Kaufman BM, Lee SM, Lee T, Pope SH, Riordan GS, Whitlow M (1988) Single-chain antigen-binding proteins. Science 242(4877):423–426

Boder ET, Wittrup KD (1997) Yeast surface display for screening combinatorial polypeptide libraries. Nat Biotechnol 15(6):553–557

Bogen JP, Grzeschik J, Krah S, Zielonka S, Kolmar H (2020) Rapid generation of chicken immune libraries for yeast surface display. Methods Mol Biol. (Clifton, N.J.) 2070:289–302

Ching KH, Collarini EJ, Abdiche YN, Bedinger D, Pedersen D, Izquierdo S, Harriman R, Zhu L, Etches RJ, van de Lavoir MC, Harriman WD, Leighton PA (2018) Chickens with humanized immunoglobulin genes generate antibodies with high affinity and broad epitope coverage to conserved targets. MAbs 10(1):71–80

Cumbers SJ, Williams GT, Davies SL, Grenfell RL, Takeda S, Batista FD, Sale JE, Neuberger MS (2002) Generation and iterative affinity maturation of antibodies in vitro using hypermutating B-cell lines. Nat Biotechnol 20(11):1129–1134

Drees JJ, Augustin LB, Mertensotto MJ, Schottel JL, Leonard AS, Saltzman DA (2014) Soluble production of a biologically active single-chain antibody against murine PD-L1 in Escherichia coli. Protein Expr Purif 94:60–66

Foote J, Winter G (1992) Antibody framework residues affecting the conformation of the hypervariable loops. J Mol Biol 224(2):487–499

Ge S, Xu L, Li B, Zhong F, Liu X, Zhang X (2020) Canine parvovirus is efficiently diagnosed and it neutralized by chicken IgY-scFv generated against the virus capsid protein. Vet Res 51(1):110

Greunke K, Braren I, Alpers I, Blank S, Sodenkamp J, Bredehorst R, Spillner E (2008) Recombinant IgY for improvement of immunoglobulin-based analytical applications. Clin Biochem 41 (14–15):1237–1244

Hanes J, Jermutus L, Weber-Bornhauser S, Bosshard HR, Plückthun A (1998) Ribosome display efficiently selects and evolves high-affinity antibodies in vitro from immune libraries. Proc Natl Acad Sci U S A 95(24):14130–14135

Hanes J, Plückthun A (1997) In vitro selection and evolution of functional proteins by using ribosome display. Proc Natl Acad Sci U S A 94(10):4937–4942

He M, Taussig MJ (2007) Eukaryotic ribosome display with in situ DNA recovery. Nat Methods 4 (3):281–288

Hoogenboom HR, Griffiths AD, Johnson KS, Chiswell DJ, Hudson P, G. (Winter 1991) Multi-subunit proteins on the surface of filamentous phage: methodologies for displaying antibody (fab) heavy and light chains. Nucleic Acids Res 19(15):4133–4137

Hust M, Jostock T, Menzel C, Voedisch B, Mohr A, Brenneis M, Kirsch MI, Meier D, Dübel S (2007) Single chain fab (scFab) fragment. BMC Biotechnol 7:14

Izquierdo SM, Varela S, Park M, Collarini EJ, Lu D, Pramanick S, Rucker J, Lopalco L, Etches R, Harriman W (2016) High-efficiency antibody discovery achieved with multiplexed microscopy. Microscopy (Oxford, England) 65(4):341–352

Jones PT, Dear PH, Foote J, Neuberger MS, G. (Winter 1986) Replacing the complementarity-determining regions in a human antibody with those from a mouse. Nature 321(6069):522–525

Köhler G, Milstein C (1975) Continuous cultures of fused cells secreting antibody of predefined specificity. Nature 256(5517):495–497

Kurosawa K, Lin W, Ohta K (2014) Chimeric antibodies. In: Steinitz M (ed) Human monoclonal antibodies: methods and protocols, vol 1060. Springer Science+Business Media, New York, pp 139–148

Lee W, Ali SA, Leow CY, Tan SC, Leow CH (2018a) Isolation and characterization of a novel anti-salbutamol chicken scFv for human doping urinalysis. Anal Biochem 555:81–93

Lee W, Atif AS, Tan SC, Leow CH (2017) Insights into the chicken IgY with emphasis on the generation and applications of chicken recombinant monoclonal antibodies. J Immunol Methods 447:71–85

Lee CH, Leu SJ, Lee YC, Liu CI, Lin LT, Mwale PF, Chiang JR, Tsai BY, Chen CC, Hung CS, Yang YY (2018b) Characterization of chicken-derived single chain antibody fragments against venom of Naja Naja Atra. Toxins 10(10):383

Li C, He J, Ren H, Zhang X, Du E, Li X (2016) Preparation of a chicken scFv to analyze gentamicin residue in animal derived food products. Anal Chem 88(7):4092–4098

Matsuda H, Mitsuda H, Nakamura N, Furusawa S, Mohri S, Kitamoto T (1999) A chicken monoclonal antibody with specificity for the N-terminal of human prion protein. FEMS Immunol Med Microbiol 23(3):189–194

Matsushita K, Horiuchi H, Furusawa S, Horiuchi M, Shinagawa M, Matsuda H (1998) Chicken monoclonal antibodies against synthetic bovine prion protein peptide. J Vet Med Sci 60 (6):777–779

McCafferty J, Johnson KS (1996) Construction and screening of antibody display libraries. In: Kay BK, Winter J, McCafferty J (eds) Phage display of peptides and proteins: a laboratory manual. Academic, San Diego, pp 79–111

Morrison SL, Johnson MJ, Herzenberg LA, Oi VT (1984) Chimeric human antibody molecules: mouse antigen-binding domains with human constant region domains. Proc Natl Acad Sci U S A 81(21):6851–6855

Nakamura N, Aoki Y, Horiuchi H, Furusawa S, Yamanaka HI, Kitamoto T, Matsuda H (2000) Construction of recombinant monoclonal antibodies from a chicken hybridoma line secreting specific antibody. Cytotechnology 32(3):191–198

Nakamura N, Shimokawa M, Miyamoto K, Hojyo S, Horiuchi H, Furusawa S, Matsuda H (2003) Two expression vectors for the phage-displayed chicken monoclonal antibody. J Immunol Methods 280(1–2):157–164

Nakamura N, Shuyama A, Hojyo S, Shimokawa M, Miyamoto K, Kawashima T, Aosasa M, Horiuchi H, Furusawa S, Matsuda H (2004) Establishment of a chicken monoclonal antibody panel against mammalian prion protein. J Vet Med Sci 66(7):807–814

Nesspor TC, Scallon B (2014) Chimeric antibodies with extended half-life in ferrets. Influenza Other Respir Viruses 8(5):596–604

Nishibori N, Horiuchi H, Furusawa S, Matsuda H (2006) Humanization of chicken monoclonal antibody using phage-display system. Mol Immunol 43(6):634–642
Nishibori N, Shimamoto T, Nakamura N, Shimokawa M, Horiuchi H, Furusawa S, Matsuda H (2004) Expression vectors for chicken-human chimeric antibodies. Biologicals 32(4):213–218
Nishinaka S, Akiba H, Nakamura M, Suzuki K, Suzuki T, Tsubokura K, Horiuchi H, Furusawa S, Matsuda H (1996) Two chicken B cell lines resistant to ouabain for the production of chicken monoclonal antibodies. J Vet Med Sci 58(11):1053–1056
Nishinaka S, Matsuda H, Murata M (1989) Establishment of a chicken X chicken hybridoma secreting specific antibody. Int Arch Allergy Appl Immunol 89(4):416–419
Nishinaka S, Suzuki T, Matsuda H, Murata M (1991) A new cell line for the production of chicken monoclonal antibody by hybridoma technology. J Immunol Methods 139(2):217–222
Pink JRL (1986) Counting components of the chicken's B cell system. Immunol Rev 91(1):115–128
Pitaksajjakul P, Lekcharoensuk P, Upragarin N, Barbas CF III, Ibrahim MS, Ikuta K, Ramasoota P (2010) Fab MAbs specific to HA of influenza virus with H5N1 neutralizing activity selected from immunized chicken phage library. Biochem Biophys Res Commun 395(4):496–501
Plückthun A (2012) Ribosome display: a perspective. In: Douthwaite J, Jackson R (eds) Ribosome display and related technologies. Methods in molecular biology (methods and protocols), vol 805. Springer, New York
Qi Y, Wu C, Zhang S, Wang Z, Huang S, Dai L, Wang S, Xia L, Wen K, Cao X, Wu Y, Shen J (2009) Selection of anti-sulfadimidine specific ScFvs from a hybridoma cell by eukaryotic ribosome display. PLoS One 4(7):e6427
Qiu XQ, Wang H, Cai B, Wang LL, Yue ST (2007) Small antibody mimetics comprising two complementarity-determining regions and a framework region for tumor targeting. Nat Biotechnol 25(8):921–929
Quintero-Hernández V, Juárez-González VR, Ortíz-León M, Sánchez R, Possani LD, Becerril B (2007) The change of the scFv into the fab format improves the stability and in vivo toxin neutralization capacity of recombinant antibodies. Mol Immunol 44(6):1307–1315
Romero PA, Arnold FH (2009) Exploring protein fitness landscapes by directed evolution. Nat Rev Mol Cell Biol 10(12):866–876
Sapats SI, Heine HG, Trinidad L, Gould GJ, Foord AJ, Doolan SG, Prowse S, Ignjatovic J (2003) Generation of chicken single chain antibody variable fragments (scFv) that differentiate and neutralize infectious bursal disease virus (IBDV). Arch Virol 148(3):497–515
Schöenfeld D, Matschiner G, Chatwell L, Trentmann S, Gille H, Hülsmeyer M, Brown N, Kaye PM, Schlehuber S, Hohlbaum AM, Skerra A (2009) An engineered lipocalin specific for CTLA-4 reveals a combining site with structural and conformational features similar to antibodies. Proc Natl Acad Sci U S A 106(20):8198–8203
Schusser B, Yi H, Collarini EJ, Izquierdo SM, Harriman WD, Etches RJ, Leighton PA (2013) Harnessing gene conversion in chicken B cells to create a human antibody sequence repertoire. PLoS One 8(11):e80108
Seo H, Hashimoto S, Tsuchiya K, Lin W, Shibata T, Ohta K (2006) An ex vivo method for rapid generation of monoclonal antibodies (ADLib system). Nat Protoc 1(3):1502–1506
Seo H, Masuoka M, Murofushi H, Takeda S, Shibata T, Ohta K (2005) Rapid generation of specific antibodies by enhanced homologous recombination. Nat Biotechnol 23(6):731–735
Sidhu SS (2001) Engineering M13 for phage display. Biomol Eng 18(2):57–63
Silverman J, Liu Q, Bakker A, To W, Duguay A, Alba BM, Smith R, Rivas A, Li P, Le H, Whitehorn E, Moore KW, Swimmer C, Perlroth V, Vogt M, Kolkman J, Stemmer WPC (2005) Multivalent avimer proteins evolved by exon shuffling of a family of human receptor domains. Nat Biotechnol 23(12):1556–1561
Smith GP (1985) Filamentous fusion phage: novel expression vectors that display cloned antigens on the virion surface. Science 228(4705):1315–1317
Steinwand M, Droste P, Frenzel A, Hust M, Dübel S, Schirrmann T (2014) The influence of antibody fragment format on phage display based affinity maturation of IgG. MAbs 6 (1):204–218

Studnicka GM, Soares S, Better M, Williams RE, Nadell R, Horwitz AH (1994) Human-engineered monoclonal antibodies retain full specific binding activity by preserving non-CDR complementarity-modulating residues. Protein Eng 7(6):805–814

Tateishi Y, Nishimichi N, Horiuchi H, Furusawa S, Matsuda H (2008) Construction of chicken-mouse chimeric antibody and immunogenicity in mice. J Vet Med Sci 70(4):397–400

Tsai KC, Chang CD, Cheng MH, Lin TY, Lo YN, Yang TW, Chang FL, Chiang CW, Lee YC, Yen Y (2019) Chicken-derived humanized antibody targeting a novel epitope F2pep of fibroblast growth factor receptor 2: potential cancer therapeutic agent. ACS Omega 4(1):2387–2397

Tsurushita N, Park M, Pakabunto K, Ong K, Avdalovic A, Fu H, Jia A, Vásquez M, Kumar S (2004) Humanization of a chicken anti-IL-12 monoclonal antibody. J Immunol Methods 295 (1–2):9–19

Vieira J, Messing J (1987) Production of single-stranded plasmid DNA. Methods Enzymol 153:3–11

Webster R (2001) Filamentoous phage biology. In: Barbas CF III, Burton DR, Scott JK, Silverman GJ (eds) A laboratory manual. Cold Spring Harbor Laboratory Press, New York, pp 1–37

Yamanaka HI, Kirii Y, Ohmoto H (1995) An improved phage display antibody cloning system using newly designed PCR primers optimized for Pfu DNA polymerase. J Biochem 117 (6):1218–1227

Zhen ZP, Zhang J, Zhang SY (2009) Development of a novel small antibody that retains specificity for tumor targeting. J Exp Clin Cancer Res 28(1):59

Protein Production in Transgenic Chickens 14

Greg Papadakos

Abstract

There is a growing need for resource-sustainable and cost-efficient technologies in protein production. Transgenic animals have drawn considerable attention as protein makers owing to the high quality of their products. The white of chicken eggs is particularly attractive for this purpose and a series of therapeutic proteins have been produced in it. An overview of the advantages of chickens as bioreactors and the challenges in their transgenesis are presented here.

Keywords

Chicken transgenesis · Egg white · Protein therapeutics

14.1 Introduction

Biotechnology provides a multitude of protein-based tools to the life sciences and healthcare sectors. The demand for proteins used as therapeutics or as imaging, diagnostics or cell culture agents is soaring due to their high efficiency. It has been almost forty years since the approval for therapeutic use of the recombinant human insulin, the first protein pharmaceutical (Johnson 1983). Since then, the repertoire of protein targets has been continuously expanding due to the emergence of new genetic information and the uncovering of subcellular pathways and disease processes. Combined with the great advances in protein production methods this has led to a burst in the development of therapeutic proteins. Nowadays proteins of diverse biological activities are applied as antigens, hormones, cytokines, enzymes and monoclonal antibodies in the prevention or treatment of conditions in oncology,

G. Papadakos (✉)
Roslin Technologies Ltd, Edinburgh, UK
e-mail: grigorios.papadakos@roslintech.com

X.-Y. Zhang et al. (eds.), *IgY-Technology: Production and Application of Egg Yolk Antibodies*, https://doi.org/10.1007/978-3-030-72688-1_14

haematology, autoimmune and inflammatory diseases, or as mediators for transplantation and tissue reconstitution. With their low off-target toxicity and high specificity and efficacy relative to chemical therapeutics, protein-based biologics have become invaluable in the medical armoury.

14.2 Protein Production Methods

Protein therapeutics are produced by transgenesis where the gene of interest is isolated from human cells, modified and inserted into prokaryotic or eukaryotic cells or the genome of an animal or plant. The protein is then expressed either by the transformed cells, plant or within an animal product (e.g., milk from farm livestock mammals or eggs from poultry). Although bacteria have been a common protein expression host, a wide range of recombinant proteins have been produced in yeasts, insect and mammalian cell lines such as Chinese hamster ovary (CHO) cells, murine myeloma (NS0) cells and human embryo kidney (HEK-93) cells (Tripathi and Shrivastava 2019).

Bacteria, yeast and insect cells offer the advantage of simple, scalable and fast growth with high protein yields at low cost. However, the biological activity of the produced proteins is affected by either the lack of necessary post-translational modifications (PTMs) especially glycosylation, or by bearing modifications different from the human ones that in turn make the molecules immunogenic. Mammalian cells are more suitable with respect to the correct addition of PTMs (Hossler et al. 2009) although their growth is significantly slower and comes at high costs that make scaling-up unfavourable.

On the other hand, and despite the laborious transgenesis process associated with their production, transgenic animals are highly efficient protein producers. Milk from cow, sheep or goat is a great depository of high-quality protein which however comes with a strong lipid component that hampers the purification process and adds to the high animal maintenance cost (Harvey and Ivarie 2003). Combined with the long mammalian gestation time and the limited number of offspring, this cost prohibits the scale-up of protein production from milk.

14.3 Advantages of Protein Production in Chicken Egg White

In contrast to milk, chicken eggs provide several advantages that make them attractive for protein production in their egg white. Breeding, flock maintenance and egg production in poultry are well-developed and low-cost processes. A single transgenic chicken can be bred under controllable conditions to quickly produce a large egg-producing flock due to the short timespan (–20 weeks) towards sexual maturity (Lillico et al. 2005) (Chap. 2). Ideally for hosting protein expression, the oviduct secretory cells have a strong protein synthetic capacity which is demonstrated by the natively high protein content of approximately 4 g of protein per –30 ml egg white (1 egg). Natural protease inhibitors and ovalbumin confer stabilization to heterologous proteins while the natively low protein complexity allows the simplification of their purification process (Ivarie 2006; Park et al. 2020). Hens annually produce 300–330 eggs (Chap. 2) and protein expression can be optimized up to 2 mg/ml of egg white thus potentially yielding tens of protein grams per annum. Additionally, the egg collection and manipulation can be easily scaled up, automated and regulated according to good manufacturing practice (Chap. 8), the basis of which has already been established by the wide production of vaccines in eggs (World Health Organization 2016). Furthermore, chicken-produced human proteins benefit from the similar glycosylation patterns between the two species thus presenting low immunogenicity and equivalent biological activity to the mammalian-produced versions (Raju et al. 2000; Zhu et al. 2005; Kodama et al. 2008).

14.4 Overview of Chicken Transgenesis

In mammalian cells the transfer of genetic material to the host organism is mostly achieved by pronuclear microinjection to a single cell zygote (Rülicke and Hübscher 2000). In the chicken, the isolation and manipulation of the fertilized oocyte from the reproductive system are technically challenging (Lee et al. 2020). Consequently, genome editing in chickens is conducted by microinjection of the transgene to the ~60,000 blastodermal cells of the embryo in a freshly laid egg right after oviposition, termed stage X (Ivarie 2006). The transgene is delivered to the blastoderm by injection of viral particles, or gene-edited germline competent cells (either embryonic stem cells, ESC or primordial germline cells, PGC) (Lee et al. 2020). The delivery methods have evolved over the years leading to improved germline transmission frequencies. Additionally, the application of a series of promoters has been driving the tissue-specificity and yield of the transgene expression.

14.4.1 Evolution of Delivery Methods and Germline Transmission Frequencies

The injection of low-titre replication-defective avian retroviruses to blastodermal cells pioneered the recombinant protein production in chickens (Harvey et al. 2002; Mozdziak et al. 2003; Rapp et al. 2003). However, the method has proven inefficient in germline transmission of the transgene due to host silencing of the viral sequences (McGrew et al. 2004; Bahrami et al. 2020). A significant transmission increase has been achieved by use of non-avian retroviruses as in the application of concentrated murine retrovirus in the embryo heart at 50–60 h of incubation (Kamihira et al. 2005) or directly to the blastoderm (Kwon et al. 2008; Koo et al. 2010; Lee et al. 2013). Improved transmission rates without host silencing have been observed from high-titre application of lentiviruses, another subclass of retroviruses that can typically deliver transgenes up to ~8 kb (Lillico et al. 2005). Lentiviruses maintain the transgene expression in a tissue-specific manner after germline transmission (Sang

2004) and have the unique ability to reach the host genome and integrate the viral RNA without the need of compromising the nuclear membrane (i.e. in non-dividing cells) (Naldini 1998). Stable germline transmission has been reported for several proteins produced by application of equine lentivirus (McGrew et al. 2004; Lillico et al. 2007; Herron et al. 2018), a replication-defective HIV-1 lentiviral vector (Chapman et al. 2005), the pWPXL lentiviral vector (Cao et al. 2015) the feline immunodeficiency virus (Kwon et al. 2018) and the pLenti6/V5-DEST vector (Byun et al. 2011).

Gene delivery to the chicken embryo can also be achieved by injection of gene edited ESC or PGC cells. ESC are developed by culturing blastodermal cells at stage IX-XI (Zhang et al. 2018) while PGC are obtained by culturing blood isolated from the embryonic vasculature system at stage 15-16HH (Macdonald et al. 2010). To date, *in vitro* gene editing of ES or PGC cells for protein expression has been performed by random integration following electroporation (Zhu et al. 2005; van de Lavoir et al. 2006), viral transduction (Motono et al. 2010), transposons (Macdonald et al. 2012; Park and Han 2012) and CRISPR/Cas9 (Dimitrov et al. 2016; Oishi et al. 2018), all resulting in higher germline transmission frequencies than the direct embryonic injections of retroviruses. Notably, CRISPR/Cas9 and transposons provide considerably higher site-specificity in transgene integration than viral gene delivery thus avoiding a major risk for insertional mutagenesis and potentially harmful health effects (Woodfint et al. 2018).

14.4.2 The Effects of Promoters in Protein Expression

The transgene expression performed in chickens to date has been either ubiquitous or tissue-specific. The constitutive Cytomegalovirus (CMV), β-actin, phosphoglycerate kinase (PGK) and Rous Sarcoma virus (RSV) promoters have been applied in the ubiquitous expression of protein therapeutics in the serum and egg (white and yolk) with significant yields. However, the lack of tissue-specificity is not desirable when the protein is active in chicken. Such activity was observed in the ectopic expression and serum circulation of the human parathyroid hormone under the RSV promoter (Lee et al. 2007) and human erythropoietin under the CMV promoter (Koo et al. 2017) both of which had detrimental effects on the chicken's health leading to death and lack of progeny.

The only tissue-specific promoter reported in chicken protein production to date is the ovalbumin promoter (OVA) (Woodfint et al. 2018). Ovalbumin is the most abundant protein in egg white (54%) (Stevens 1991) which manifests the strong expression potential by the OVA promoter whose regulatory elements inhibit protein expression in tissues other than the oviduct (Lillico et al. 2005). The OVA promoter has been extensively studied and transcription factor sites that enhance protein expression such as the oestrogen response element (ERE) have been revealed. Multiples of ERE and other parts of the promoter have been utilized in constructs in order to increase the protein yield (Lillico et al. 2007; Cao et al. 2015; Herron et al. 2018; Park et al. 2020).

Using these promoters, cytokines, growth factors, antibodies, enzymes and an antimicrobial peptide have been produced in chicken egg white with considerable yield variation in the low ng-to-mg range per ml egg white (Woodfint et al. 2018). For example, the growth factor erythropoietin has been produced at 40–55μg/ml in egg white (Kwon et al. 2018) while the anti-PMSA antibody yielded up to 0.15μg/ml (Zhu et al. 2005) and the anti-prion-scFv-hIgG1-Fc antibody up to 1.5 mg/ml (Kamihira et al. 2005). Cytokines such as human IFNα-2a has been produced at 0.15 mg/ml and human CSF1-Fc fusion at 1 mg/ml (Herron et al. 2018). Although further optimization of the expression yields is always desirable, the ones achieved so far might be sufficient for commercialization depending on the potency and required dose of the therapeutic. Additionally, moderate yields might also be preferable so that the protein does not interfere with embryo development thus allowing the continuation of progeny production. Further optimization of the OVA promoter and exploration of the promoters of other oviduct-specific genes will furnish further flexibility in modulating the yield and safety of expression in the egg white.

14.5 Closing Remarks

From concept to materialization, the expression of a protein in chicken egg white requires a bit over a year involving the transgene preparation and injection into eggs followed by breeding up to the production of heterozygous egg-laying G1 birds. Six months are further required to produce homozygous G2 birds. The biological cycle of the chicken is the bottleneck of the process and cannot be altered, although a complementary system for screening protein constructs and yields before the actual bird production would be desirable as well as any acceleration to the transgene preparation (e.g. by CRISPR/Cas9). Addressing these timescale improvements while optimizing the yield and ensuring animal safety, will make protein production in transgenic chickens greatly appealing to investment for commercial purposes.

References

Bahrami S, Amiri-Yekta A, Daneshipour A, Jazayeri SH, Mozdziak PE, Sanati MH, Gourabi H (2020) Designing a transgenic chicken: applying new approaches toward a promising bioreactor. Cell J 22(2):133–139

Byun SJ, Kim SW, Kim KW, Kim JS, Hwang IS, Chung HK, Kan IS, Jeon IS, Chang WK, Park SB, Yoo JG (2011) Oviduct-specific enhanced green fluorescent protein expression in transgenic chickens. Biosci Biotechnol Biochem 75(4):646–649

Cao D, Wu H, Li Q, Sun Y, Liu T, Fei J, Zhao Y, Wu S, Hu X, Li N (2015) Expression of recombinant human lysozyme in egg whites of transgenic hens. PLoS One 10(2):e0118626

Chapman SC, Lawson A, Macarthur WC, Wiese RJ, Loechel RH, Burgos-Trinidad M, Wakefield JK, Ramabhadran R, Mauch TJ, Schoenwolf GC (2005) Ubiquitous GFP expression in transgenic chickens using a lentiviral vector. Development 132(5):935–940

Dimitrov L, Pedersen D, Ching KH, Yi H, Collarini EJ, Izquierdo S, van de Lavoir MC, Leighton PA (2016) Germline gene editing in chickens by efficient CRISPR-mediated homologous recombination in primordial germ cells. PLoS One 11(4):e0154303

Harvey AJ, Ivarie R (2003) Validating the hen as a bioreactor for the production of exogenous proteins in egg white. Poult Sci 82(6):927–930

Harvey AJ, Speksnijder G, Baugh LR, Morris JA, Ivarie R (2002) Expression of exogenous protein in the egg white of transgenic chickens. Nat Biotechnol 20(4):396–399

Herron LR, Pridans C, Turnbull ML, Smith N, Lillico S, Sherman A, Gilhooley HJ, Wear M, Kurian D, Papadakos G, Digard P, Hume DA, Gill AC, Sang HM (2018) A chicken bioreactor for efficient production of functional cytokines. BMC Biotechnol 18(1):82

Hossler P, Khattak SF, Li ZJ (2009) Optimal and consistent protein glycosylation in mammalian cell culture. Glycobiology 19(9):936–949

Ivarie R (2006) Competitive bioreactor hens on the horizon. Trends Biotechnol 24(3):99–101

Johnson IS (1983) Human insulin from recombinant DNA technology. Science (New York, NY) 219(4585):632–637

Kamihira M, Ono K, Esaka K, Nishijima K, Kigaku R, Komatsu H, Yamashita T, Kyogoku K, Iijima S (2005) High-level expression of single-chain Fv-fc fusion protein in serum and egg white of genetically manipulated chickens by using a retroviral vector. J Virol 79 (17):10864–10874

Kodama D, Nishimiya D, Iwata K, Yamaguchi K, Yoshida K, Kawabe Y, Motono M, Watanabe H, Yamashita T, Nishijima K, Kamihira M, Iijima S (2008) Production of human erythropoietin by chimeric chickens. Biochem Biophys Res Commun 367(4):834–839

Koo B, Kwon M, Kim D, Kim S, Kim N-H, Kim T (2017) Production of transgenic chickens constitutively expressing human erythropoietin (hEPO): problems with uncontrollable overexpression of hEPO gene. Biotechnol Bioprocess Eng 22:22–29

Koo BC, Kwon MS, Lee H, Kim M, Kim D, Roh JY, Park YY, Cui XS, Kim NH, Byun SJ, Kim T (2010) Tetracycline-dependent expression of the human erythropoietin gene in transgenic chickens. Transgenic Res 19(3):437–447

Kwon MS, Koo BC, Choi BR, Park YY, Lee YM, Suh HS, Park YS, Lee HT, Kim JH, Roh JY, Kim NH, Kim T (2008) Generation of transgenic chickens that produce bioactive human granulocyte-colony stimulating factor. Mol Reprod Dev 75(7):1120–1126

Kwon MS, Koo BC, Kim D, Nam YH, Cui XS, Kim NH, Kim T (2018) Generation of transgenic chickens expressing the human erythropoietin (hEPO) gene in an oviduct-specific manner: production of transgenic chicken eggs containing human erythropoietin in egg whites. PLoS One 13(5):e0194721

Lee SH, Gupta MK, Han DW, Han SY, Uhm SJ, Kim T, Lee HT (2007) Development of transgenic chickens expressing human parathormone under the control of a ubiquitous promoter by using a retrovirus vector system. Poult Sci 86(10):2221–2227

Lee SH, Gupta MK, Ho YT, Kim T, Lee HT (2013) Transgenic chickens expressing human urokinase-type plasminogen activator. Poult Sci 92(9):2396–2403

Lee J, Kim DH, Lee K (2020) Current approaches and applications in avian genome editing. Int J Mol Sci 21(11)

Lillico SG, McGrew MJ, Sherman A, Sang HM (2005) Transgenic chickens as bioreactors for protein-based drugs. Drug Discov Today 10(3):191–196

Lillico SG, Sherman A, McGrew MJ, Robertson CD, Smith J, Haslam C, Barnard P, Radcliffe PA, Mitrophanous KA, Elliot EA, Sang HM (2007) Oviduct-specific expression of two therapeutic proteins in transgenic hens. Proc Natl Acad Sci U S A 104(6):1771–1776

Macdonald J, Glover JD, Taylor L, Sang HM, McGrew MJ (2010) Characterisation and germline transmission of cultured avian primordial germ cells. PLoS One 5(11):e15518

Macdonald J, Taylor L, Sherman A, Kawakami K, Takahashi Y, Sang HM, McGrew MJ (2012) Efficient genetic modification and germ-line transmission of primordial germ cells using piggyBac and Tol2 transposons. Proc Natl Acad Sci U S A 109(23):E1466–E1472

McGrew MJ, Sherman A, Ellard FM, Lillico SG, Gilhooley HJ, Kingsman AJ, Mitrophanous KA, Sang H (2004) Efficient production of germline transgenic chickens using lentiviral vectors. EMBO Rep 5(7):728–733

Motono M, Yamada Y, Hattori Y, Nakagawa R, Nishijima K, Iijima S (2010) Production of transgenic chickens from purified primordial germ cells infected with a lentiviral vector. J Biosci Bioeng 109(4):315–321

Mozdziak PE, Borwornpinyo S, McCoy DW, Petitte JN (2003) Development of transgenic chickens expressing bacterial beta-galactosidase. Dev Dynam 226(3):439–445

Naldini L (1998) Lentiviruses as gene transfer agents for delivery to non-dividing cells. Curr Opin Biotechnol 9(5):457–463

Oishi I, Yoshii K, Miyahara D, Tagami T (2018) Efficient production of human interferon beta in the white of eggs from ovalbumin gene-targeted hens. Sci Rep 8(1):10203

Park TS, Han JY (2012) piggyBac transposition into primordial germ cells is an efficient tool for transgenesis in chickens. Proc Natl Acad Sci U S A 109(24):9337–9341

Park JS, Lee KY, Han JY (2020) Precise genome editing in poultry and its application to industries. Genes 11(10)

Raju T, Briggs J, Borge S, Jones A (2000) Species-specific variation in glycosylation of IgG: evidence for the species-specific sialylation and branch-specific galactosylation and importance for engineering recombinant glycoprotein therapeutics. Glycobiology 10:477–486

Rapp JC, Harvey AJ, Speksnijder GL, Hu W, Ivarie R (2003) Biologically active human interferon alpha-2b produced in the egg white of transgenic hens. Transgenic Res 12(5):569–575

Rülicke T, Hübscher U (2000) Germ line transformation of mammals by pronuclear microinjection. Exp Physiol 85(6):589–601

Sang H (2004) Prospects for transgenesis in the chick. Mech Dev 121(9):1179–1186

Stevens L (1991) Egg white proteins. Comp Biochem Physiol B 100(1):1–9

Tripathi NK, Shrivastava A (2019) Recent developments in bioprocessing of recombinant proteins: expression hosts and process development. Front Bioeng Biotechnol 7:420

van de Lavoir MC, Diamond JH, Leighton PA, Mather-Love C, Heyer BS, Bradshaw R, Kerchner A, Hooi LT, Gessaro TM, Swanberg SE, Delany ME, Etches RJ (2006) Germline transmission of genetically modified primordial germ cells. Nature 441(7094):766–769

Woodfint RM, Hamlin E, Lee K (2018) Avian bioreactor systems: a review. Mol Biotechnol 60 (12):975–983

World Health Organization (2016) WHO Technical Report Series No. 999: good manufacturing practices for biological products. WHO Expert Committee on Biological Standardization

Zhang L, Wu Y, Li X, Wei S, Xing Y, Lian Z, Han H (2018) An alternative method for long-term culture of chicken embryonic stem cell in vitro. Stem Cells Int 2018:2157451

Zhu L, van de Lavoir MC, Albanese J, Beenhouwer DO, Cardarelli PM, Cuison S, Deng DF, Deshpande S, Diamond JH, Green L, Halk EL, Heyer BS, Kay RM, Kerchner A, Leighton PA, Mather CM, Morrison SL, Nikolov ZL, Passmore DB, Pradas-Monne A, Preston BT, Rangan VS, Shi M, Srinivasan M, White SG, Winters-Digiacinto P, Wong S, Zhou W, Etches RJ (2005) Production of human monoclonal antibody in eggs of chimeric chickens. Nat Biotechnol 23 (9):1159–1169

Part III

Applications of IgY Technology

Applications of IgY in Veterinary Medicine 15

Brindha Chelliappan, Antonysamy Michael, Celina G. Vega, Fagang Zhong, Xiaoying Zhang, and Patricia M. Morgan

Abstract

Animal health and welfare is under constant threat from pathogenic organisms and there is need to move from the use of antibiotics as a first line of defence. The oral administration of IgY antibodies constitutes a powerful passive immune strategy and has been used in terrestrial and aquatic animals. The prevention and treatment of gastrointestinal diseases, particularly diarrhoea from different infectious agents in livestock and dogs, is one of the most studied uses of IgY for passive protection. The use of IgY for the treatment of calves, adult cattle, pigs and poultry is described. One of the major factors impacting on the productivity of aquaculture is the outbreak of disease and there is a need for development of

B. Chelliappan · A. Michael
Department of Microbiology, PSG College of Arts & Science, Bharathiar University, Coimbatore, Tamil Nadu, India

C. G. Vega
Centro de Investigación en Ciencias Veterinarias y Agronómicas, INTA, Buenos Aires, Argentina
e-mail: vega.celina@inta.gob.ar

F. Zhong
State Key Laboratory for Sheep Genetic Improvement and Healthy Production, Xinjiang Academy of Agricultural and Reclamation Science, Shihezi, China

X. Zhang
College of Biological Science and Engineering, Shaanxi University of Technology, Hanzhong, China
e-mail: zhxying@snut.edu.cn; xzhang67@uoguelph.ca; zhang@bio.uminho.pt

P. M. Morgan (✉)
Emerita Lecturer, School of Natural Sciences, National University of Ireland Galway, Galway, Ireland
e-mail: pat.morgan@nuigalway.ie

X.-Y. Zhang et al. (eds.), *IgY-Technology: Production and Application of Egg Yolk Antibodies*, https://doi.org/10.1007/978-3-030-72688-1_15

treatments suitable for an aqueous environment. IgY used in the treatment of dogs is also expanding. Finally, the use of IgY in the detection of disease is outlined.

Keywords

Veterinary medicine · Aquaculture · Prophylaxis · Therapy · IgY product

15.1 Introduction

The focus of this chapter is on the use of IgY for passive immunization of food producing livestock, poultry and companion animals such as dogs. Currently, it is evident that a pandemic such as corona virus has the potential to disrupt agriculture and food supplies (Jámbor et al. 2020); therefore the control of disease in animals and fish assumes even greater importance and the role of companion animals in human welfare is recognized.

There are two main drivers for higher production of food: the increase in the global population which is predicted to reach 9.7 billion by 2050; and the increase in per capita income in developing countries which is matched with an increased demand for protein (Davis and White 2020). Meeting these demands will require increased productivity and a reduction of losses while meeting the demands of climate change and limiting resources (Crute and Muir 2011). Together with the livestock and poultry, aquaculture is also likely to contribute to a greater share of food production (Bostock 2011). The relationship between companion animals, such as dogs and cats and human wellbeing is well documented. It is estimated that 11.5 million dogs are owned as pets in UK and that there are similar levels of ownership in Europe, China, Japan and Australia (Brooks et al. 2018); there were notable increases in dog ownership during the pandemic of 2020 and a positive impact on loneliness during the lockdown (Oliva and Johnston 2020).

Intensive animal rearing systems are open to cross-infections with subsequent negative impacts on growth or even large-scale death of infected animals. The use of antibiotics which are relatively cheap and easy to administer has been implicated in the development of antibiotic resistance in humans (Barton 2000). Vaccination plays a major role in animal protection against a range of pathogens including bacteria, viruses and parasites (Hedegaard and Heegaard 2016). However, vaccines are seen to be ineffective in very young animals due to interference from maternally-derived antibodies and for protection from enteric infections (Hedegaard and Heegaard 2016). An alternative to both the use of antibiotics and vaccination, is passive immunization where immunoglobulins (Ig) from a donor are administered to a recipient to provide immediate immunity and short-term protection (Fig. 15.1).

The source of Ig is not important and as reviewed by Hedegaard and Heegaard 2016 (Hedegaard and Heegaard 2016), bovine colostrum, horse serum and IgY have been licensed for use in ruminants, horses and pigs. Several strategies have been

Fig. 15.1 Generation of IgY for passive immunization. Hens are immunized with a relevant antigen and antigen-specific antibodies are extracted from the egg yolk. The treated animals are provided with immediate immunity and short-term protection

developed to protect IgY from inactivation in the intestine, as IgY is unstable below a pH of 3.5 (Chap. 12). These include, for example, chitosan/alginate-based (Li et al. 2007) or cholesterol/lecithin-based microencapsulation methods (Rahman et al. 2013). An example of a disease which is one of the most common in the new-born of many animal species is diarrhoea. Diarrhoea associated with infectious agents leads to a deteriorating condition in neonates and will directly hamper the overall performance of the animal in early as well as later stage of their life (USDA and APHIS 2014). The clinical presentation can range from mild diarrhoea without systemic disease to profuse acute diarrhoea associated with rapid dehydration, severe disturbance of acid-base and electrolyte balance, and death, sometimes in as few as 12 hours (Grünberg 2014). A systematic review has shown that IgY treatment was effective in controlling and preventing diarrhoea in domesticated animals including calves, poultry, piglets, and mice (Diraviyam et al. 2014). The current chapter reviews passive immunization with IgY as an effective treatment for a range of diseases (bacterial, viral and parasitic) in important food producing species such as cattle, pigs, poultry and a variety of fish species. Dogs, which are an important companion animal, are also treated with IgY. Finally, the in vitro use of IgY in immunoassays is discussed.

15.2 Cattle

World milk production is projected to grow at 1.6% p.a. (to 997 million tonnes by 2029); this is faster than most other main agricultural commodities (FAO and OECD 2020). Globally, cattle meat production has more than doubled since 1961—increasing from 28 million tonnes per year to 68 million tonnes in 2014 (Ritchie 2017). The impacts of cattle diseases can have wide ranging economic effects including loss of livelihood, disruption of trade, loss of access to international markets and indeed threats to human health (FAO 2016).

15.2.1 Diarrhoea in Calves

Neonatal calf diarrhoea is a common disease affecting new-born calves and is the leading cause of calf mortality before weaning in both beef and dairy calves (Cho and Yoon 2014). Diarrhoea is more common between 1 and 21 days of age, with a peak incidence at two weeks, but it can extend up to 30–45 days of age (Bartels et al. 2010). Infected calves show watery faeces of increased frequency. Fluid loss from the vascular compartment leads to hypovolemia (dehydration), hypotension, and shock (Heller and Chigerwe 2018). Calves develop metabolic acidosis, which results in death in less than 2 days, if not treated promptly (Foster and Smith 2009).

Most of the time, diarrhoea is caused by a combination of agents which may result in more severe disease than a single agent alone (Hoet et al. 2003; Izzo et al. 2011; Cho et al. 2013; Ferragut et al. 2016). Historically, calf diarrhoea has been attributed to bovine rotavirus group A (RVA), bovine coronavirus (BCoV), *Salmonella spp.*, *Escherichia coli* strains with virulent factors and *Cryptosporidium parvum* as the main aetiological agents (Foster and Smith 2009; Heller and Chigerwe 2018). However other viral agents have also been identified and proven to cause diarrhoea in calves including bovine parvovirus, astrovirus, torovirus (Breda virus), enteric calicivirus and nebovirus (Saif and Thiel 1990; Hoet et al. 2003; Garaicoechea et al. 2006; Otto et al. 2011; Badaracco et al. 2012; Bok et al. 2015; Ferragut et al. 2016; Langel et al. 2016). Enteric diseases are also common in adult cattle. As diagnostic testing is often not available, these treatment protocols are generally based on knowledge of the most likely pathogen and the veterinarian's clinical experience.

Antibodies from bovine colostrum protect against RVA-associated diarrhoea in calves (Parreño et al. 2004; Parreño et al. 2010). However, this strategy is industrially unfeasible at a large scale and several bovine infectious agents could be spread as adventitious virus/bacteria, like bovine leukaemia virus. A promising, economically feasible and practical strategy which has been explored is the supplementation of the milk diet of calves with specific IgY antibodies from egg yolk (Kuroki et al. 1994, 1997; Mine and Kovacs-Nolan 2002; Vega et al. 2011, 2015). The production of IgY against specific pathogens has been described in Chap. 10.

It has been shown that IgY antibodies can resist digestion in the gastrointestinal tract of calves, remaining biologically active (Vega et al. 2011, 2015). This is critical as most of the infectious agents causing diarrhoea infect cells from the intestinal mucosa. Furthermore, IgY antibodies can neutralize virus and bacteria being shed in faeces, preventing environmental contamination and dissemination of this agents. In a recent review of the effect of IgY in the treatment of rotavirus infection (Thu et al. 2017), it is speculated that its mode of action involves either blocking the entry of the rotavirus into the host cells and/or minimizing the cell to cell spread of the virus. IgY antibodies can retain activity through various manufacturing processes and dried IgY preparations are stable for several years without a significant loss of their biological activity (Larsson et al. 1993; Vega et al. 2015).

The oral administration of egg-derived preparations to new born calves has not only shown protection against diarrhoea but also an increase in weight gain and an improvement in growth performance (DiLorenzo et al. 2006; Vega et al. 2015). A number of commercial products have been developed to treat calf diarrhoea (Chap. 17).

15.2.2 Bovine Viral Diarrhoea Virus (BVDV): Prevention and Treatment

Bovine viral diarrhoea-mucosal disease virus (BVDV) is one of the most significant animal diseases in the livestock industry worldwide (Houe 2003). Bovine diarrhoea virus (BDV), including BDV-1, BDV-2 and BDV-3, are a members of the genus Pestivirus, belonging to the family Flaviviridae (Zhong et al. 2011). BVDV causes high mortality of new born calves due to acute infections and also results in the development of persistently infected calves which shed large viral loads throughout their life and are a source of infection for the entire herd. Despite ongoing research (Bollini et al. 2018), there are no effective treatments nor preventive drugs for this disease. At present, there are a variety of attenuated vaccines and inactivated vaccines (Newcomer et al. 2017). However, due to the rapid variation and diversity of RNA viruses, the use of vaccines is not fully protective for the animal, and the occurrence of bovine viral diarrhoea cannot be eliminated. Live attenuated vaccines also present a risk of causing host immune suppression.

The first anti-BVDV IgY antibody was reported by Zhang (Zhang et al. 2016), using the Xinjiang BVDV strain (purified E2 protein) to immunize chickens (Zhong et al. 2011). The peak titre was obtained after the third booster injection, and the binding activity of the IgY to the E2 protein was confirmed by Western blot analysis. In order to develop an assay for BVDV diagnosis, indirect ELISA and immunochromatographic assays (ICA) were performed to evaluate their efficacy. The specificity of the IgY was confirmed using different strains of viruses and bacteria (BVDV1, BADV-1, BADV-3, EPEC, *Salmonella enteritidis*, *Staphylococcus aureus,* and *Clostridium Difficile*). ELISA and ICA results showed that the IgY and BVDV-E2 and BVDV1 bind strongly and do not cross-react with other strains of virus and bacterial species.

15.2.3 Mastitis

Worldwide bovine mastitis continues to be a costly disease for the dairy industry and the control of this disease is necessary to ensure the sustainability of dairy farming and the production of milk which meets high quality global standards (Ruegg 2017). The disease can occur in either a clinical form easily detected by abnormalities in the milk, udder or other clinical signs; or subclinical mastitis which is often undetected as it relies on indirect measurements such as increased somatic cell count but has the greatest economic impact due to lower milk yields (Ruegg 2017). The incidence of clinical mastitis in a Canadian study was reported as 19 cases/100 cow-year (Aghamohammadi et al. 2018) and on large Chinese dairy farms was 3.3 cases per 100 cows per month (range = 1.7 to 8.1) (Gao et al. 2017). A number of different pathogens cause mastitis. The pathogens are generally classified as contagious (primarily *Staphylococcus aureus (S. aureus)*) or environmental (primarily *E. coli*) (Riffon et al. 2001). While *Staphylococus* spp. continues to be one of the most prevalent causes of intramammary infections (Monistero et al. 2018), and the causes of clinical mastitis vary across regions and the most frequently isolated pathogens reported by Gao and references therein are as follows: in Scandinavia it was *S. aureus*; in the Netherlands it was *E. coli*; in the United States, *Klebsiella* and *E. coli* mastitis were of equal importance; in China *E. coli*, and *Klebsiella spp* had the highest incidence (Gao et al. 2017). Staphylococci can be coagulase-positive (*S. aureus*) or coagulase negative based on the ability to coagulate rabbit plasma (Xu et al. 2015); coagulase negative species were shown to be the dominant pathogens of subclinical or mild clinical infections. *S. aureus* can express a wide range virulence factors helping it to attach and infect the host (Zhen et al. 2009; Xu et al. 2015; Monistero et al. 2018). The most common treatment for mastitis is the administration of antibiotics directly into the mammary tissue but the cure rates can be low and antibiotic resistant strains have emerged (Tenhagen et al. 2006). Alternative treatments include the use of caprylic acid and monocaprylin which have been shown to be effective *in vitro* (Nair et al. 2005); and vaccination (Ma et al. 2004). A recent review of various vaccine treatments has concluded that they are not necessarily effective or economical (Ismail 2017). The previously stated advantages of IgY technology (Chap. 5) have led to the development of specific IgY treatments for the main causative agents of mastitis namely, *S. aureus* and *E. coli*. An initial *in vitro* assay of IgY produced by hens immunized with *E. coli* O111 showed a dose-dependent inhibitory effect on growth of *E. coli* 0111 and five other mastitis causing strains of *E. coli* (Zhen et al. 2008). The specific IgY also enhanced the uptake of *E. coli* by milk macrophages or leukocytes suggesting it increased phagocytic activity. A further study of IgY antibodies raised against *E. coli* showed concentrations of IgY up 7.1 mg/mL of yolk and inhibition of growth of *E. coli* in the presence of the specific IgY (Meenatchisundaram et al. 2011). Specific IgY to *S. aureus* was shown to reduce clinical and experimental mastitis during a 6-day intramammary 10 mL infusion of IgY morning and evening at a concentration of 100 mg/mL (Zhen et al. 2009). *S. aureus* is not only present in milk but is internalized in mammary epithelial cells thus avoiding host defence and antibiotic treatment and leading to longer-term persistence (Bayles et al. 1998). Specific IgY against encapsulated type 5 (IgY-T5) and type 8 (IgY-T8) and non-encapsulated type 336 (IgY-T336) *S. aureus* strains (at 5 mg/mL) significantly blocked the internalization of their homologous strains by bovine mammary epithelial cells (MAC-T cells) within 6 h (Wang et al. 2011). These results suggest that IgY acts to control mastitis by prevention of uptake rather than by impacting on growth of the pathogen.

15.3 Pigs

Pork is the world's most consumed meat from terrestrial animals (USDA 2019). Major improvements in pig breeding technologies in recent decades have turned commercial pig production into a high-input, high-output industry (FAO 2009; FAO 2016).

Increased productivity in the porcine industry relies on improved animal health. Even though there have been major achievements in disease control and prevention, the pig production sector continues to be threatened by emerging trans-boundary diseases (Farms 2017). Systemic risks are emerging owing to the combination of rapid structural changes in the livestock sector, geographical clustering of intensive livestock production facilities near urban population centres and the movement of animals, people and pathogens between intensive and traditional production systems (Patra et al. 2014). The economic impact of diseases and the cost of control measures are high and becoming higher (FAO 2009).

15.3.1 Diarrhoea in Pigs

Neonatal diarrhoea is a frequent cause of heavy economic losses for the porcine industry, increasing significantly pre-weaning mortality and piglets weaned below the target weight (Rhouma et al. 2017). Enteric diseases are often of endemic presentation but may also occur as outbreaks with high morbidity and mortality. The aetiology of diarrhoea in specific herds may differ and is often incompletely diagnosed. This complexity was not of major concern when the pig industry relied on vaccines and metaphylactic use of antibiotics (Kongsted et al. 2014). Vaccines are usually administered to the pregnant sows and passive immunity is transferred to the piglets during the first weeks of life. However, there is an increasing demand for cost effective, safe and sustainable treatments to control diarrhoea in young pigs (particularly neonatal and early-weaned piglets) (Li et al. 2015). Stress associated with weaning, changes in social relationships and changes in diet (including lack of sow milk) are critical factors that influence the piglets' susceptibility to infection (Stefaniak 2006). There are several infectious agents associated with diarrhoea in piglets. Enterotoxigenic *Escherichia coli* (ETEC) and Enteropathogenic *E. coli* (EPEC), *Clostridium perfringens* type A and C, Transmissible gastroenteritis virus (TGEV), porcine epidemic diarrhoea (PEDV) and Rotavirus A, B and C being the most common ones. In most of the cases, more than one enteric pathogen is present (Mesonero-Escuredo et al. 2018).

IgY antibodies specific against gastrointestinal pathogens can be orally administered to piglets during the first weeks of life as a passive treatment against infectious diarrhoea and are summarized in Table 15.1. The effect is usually reported as a dose-dependent one (Stefaniak et al. 2014).

The protective effect of IgY antibodies in piglets appears to be effective beyond their specific role. A proteomic analysis carried out on pigs infected with *E. coli* and *S. Typhimurium* and fed a diet supplemented with natural herbs and IgY showed that

Table 15.1 Use of IgY antibodies to control enteric diseases in piglets

Pathogen	Effects of IgY	Reference
Escherichia coli	Prevention of K88+, K99+, 987P+ ETEC infection in neonatal piglets	(Yokoyama et al. 1992)
	Inhibits the adhesion of K88+ ETEC to piglet intestinal mucosa	(Jin et al. 1998)
	Prevention of K88+ ETEC infection in neonatal and early weaned piglets	(Marquardt et al. 1999)
	Fast protection from diarrhoea in piglets orally treated with anti-K88+ ETEC IgY encapsulated on chitosan-alginate microparticles	(Li et al. 2007)
	Protection of ETEC infected piglets from diarrhoea with IgY delivered in hydrogel-carbon nanotubes composites	(Alustiza et al. 2016)
Porcine epidemic Diarrhoea virus (PEDV)	Partial protection of piglets against PEDV associated mortality	(Kweon et al. 2000)
	Protection of neonatal piglets against PEDV by oral administration of anti-PEDV S1 IgY	(Lee et al. 2015)
Porcine transmissible gastroenteritis virus (TGEV)	Prophylactic administration: Increase in piglets' survival rate after challenge Therapeutic administration: Reduction in mortality	(Zuo et al. 2009)
Rotavirus group A (RVA)	Protection of gnotobiotic piglets from human RVA-associated diarrhoea	(Vega et al. 2012)

oral egg yolk administration containing IgY could regulate the immune system and reduce the stress of microbial infections (Park et al. 2011). IgY immunoglobulins from chicken egg yolk are attractive not only because of its large scale production but also for their relatively resistance to porcine gastrointestinal proteases (Vega et al. 2012) and their positive impact on growth rate and on the stabilization and enhancement of the protective barriers against gastrointestinal tract disturbances (Vega et al. 2012, Stefaniak et al. 2014). IgY preparations for treatment of piglets are commercially available in many countries (Chap. 17).

15.4 Poultry

Based on estimates from the US Food and Agricultural Organization, the number of hens worldwide is on the increase (Shahbandeh 2020). Poultry meat production reached almost 125 million metric tons in 2018 and is forecast to increase on an annual basis. The countries who are the main poultry meat producers are the USA, China and Brazil. Egg production has also increased by nearly 25% in the last decade and accounted for 80.1 million metric tons of eggs in 2017. China is the top egg-producing country, producing nearly 529 billion eggs annually. The protection of the health of these large number of animals is paramount for global food production. The primary objective of the poultry industry is to convert feed into meat (in broilers) and eggs (in layers). The physical, chemical or biological disturbances of the digestive system of the birds can result in various gastrointestinal disorders (enteric diseases). The severity and course of the disease are influenced by multiple factors including management practices and the microbial agent (s) involved. In the poultry sector, enteric diseases are one of the major groups of diseases tending to cause significant economic losses due to poor performance (decreased body weight gain or egg production), high mortality and increased treatment costs (Gadde et al. 2015). Various pathogens (bacteria, viruses and parasites) cause enteric infections either alone (mono-causal) or synergistically with other pathogens (multi-casual) (Hafez 2011). Passive immunization with IgY is a promising alternative approach to combat the emergence of new and drug resistance pathogens; as well as diseases that are unresponsive to drug therapy (including viral infections) (Kovacs-Nolan and Mine 2012).

15.4.1 IgY Treatment of Bacterial Infections of Poultry

Bacterial infections are a problem of economic concern in all phases of poultry husbandry and also the source of transmission of food borne diseases to humans. The presence of *Escherichia coli, Campylobacter and Salmonella* in poultry meat have placed a considerable burden on public health. *Campylobacter jejuni* is present in the large intestine of chickens and contamination by undercooked meat is one of the most common sources of acute gastrointestinal infections in humans (Igwaran and Okoh 2019). *Salmonella* species cause disease in a wide range of animals and zoonotic transfer is a major concern (Stevens et al. 2009). Colibacillosis is also a cause of morbidity and mortality in poultry; this is partly or entirely caused by *E. coli* infection (Barnes et al. 2003). The effects of chicken IgY used to treat a range of bacterial infections in chickens are outlined in Table 15.2. A number of studies have shown IgY to be effective in the control of *E. coli* infection. In 2004 it was shown that treatment with IgY was protective against an intra-air sac challenge of *E. coli* (Kariyawasam et al. 2004) and a later study showed that the dietary administration of IgY was shown to enhance the intestinal health and growth performance of broiler chickens (Mahdavi et al. 2010a, b). As early as 1997, Tsubokura et al. evaluated the prophylactic and therapeutic property of IgY against *Campylobacter jejuni* in broilers (Tsubokura et al. 1997). This report was the starting point for the approach to treat *Campylobacter jejuni* infections (Tsubokura et al. 1997, Hermans et al. 2014, Vandeputte et al. 2019). Treatment for Salmonella infections by specific IgY was shown to be effective, particularly if protected from gastrointestinal degradation (Table 15.2).

Increased incidences of disease, mortality and economic losses due to infection with *Gallibacterium anatis* suggest that it may be an emerging pathogen in poultry (Zhang et al. 2019). Due to its antigenic diversity and multidrug resistance, an alternate treatment strategy such as IgY treatment is suggested. *Gallibacterium* toxin A (GtxA) is a member of RTX toxins (contain repeat sequences of nine amino acids), secreted by *G. anatis*. The GtxA has both leukotoxic and haemolytic actions. In this study, chickens were immunized with N-terminus recombinant GtxA

Table 15.2 Application of IgY against bacterial infections of poultry

Pathogen	Poultry	IgY form	Outcomes	Reference
E. coli	11 day old broilers	100 mg i.m.	Protection from homologous challenge by *E. coli* 078	(Kariyawasam et al. 2004)
E. coli	Day old chicks	IgY 3 mL orally	Reduced symptoms, lesions	(Tamilzarasan et al. 2009)
E. coli	7 day broiler chickens	Lyophilized IgY powder in diet (0.1–0.4%)	Decreased ileal *E. coli* counts and the circulating heterophil lymphocyte ratio	(Mahdavi et al. 2010a, b)
C. Jejuni	14–34 day old chicks	IgY preparation oral	Prophylactic (99%) and therapeutic treatments (80%–95%) reduction in bacteria	(Tsubokura et al. 1997)
C. Jejuni	6 day old broiler chicks	Egg yolk 5% (w/w) in feed	Significantly reduced bacterial cell counts	(Hermans et al. 2014)
C. Jejuni	10 day old broilers	Egg yolk 5% (w/w) in feed	Significantly reduced bacterial cell counts when administered prophylactically and therapeutically.	(Vandeputte et al. 2019)
Salmonella enterica subsp. enterica	Salmonella-free White Leghorn hens	Egg yolk powder or capsulated immune yolk	Capsulated specific IgY reduced the colonization of salmonella in chicken's caeca and liver. Capsulation may be necessary to protect from gastrointestinal degradation	(Hatamzade Isfahani et al. 2020)
Salmonella enteritidis and salmonella typhimurium	16-week-old salmonella free ISA Brown hen	Freeze-dried egg yolk powder	Egg yolk powder, other than specific IgY, has a beneficial effect on growth performance. Specific IgY in feed may be denatured and degraded in chicken gastrointestinal tract and did not have any effect on growth performance and caecal colonization of salmonella.	(Chalghoumi et al. 2009)
Salmonella enteritidis	Ross 308-day old broiler	Extracted IgY solution	Specific IgY and IgY-probiotic co-treatment significantly decreased faecal shedding and S. enteritidis colonization in the major organs.	(Rahimi et al. 2007)

(continued)

Table 15.2 (continued)

Pathogen	Poultry	IgY form	Outcomes	Reference
Salmonella enteritidis	SPF laying hens	Whole egg powder	IgY egg powder as a feed additive reduced the production of salmonella-positive eggs (13.3%) as compared to the control group (29.4%).	(Gürtler et al. 2004)
Salmonella enteritidis	Commercial Pekin ducklings	Egg yolk liquid	Oral IgY and probiotics act synergistically to prevent salmonella infection	(Fulton et al. 2002)
Gallibacterium anatis	4-week-old chicks	Extracted IgY	Intramuscular. Injection (10 mg/ml IgY) protected chickens against challenge	(Zhang et al. 2019)

to produce anti-GtxA-N IgY antibodies. In the challenge test, after contamination of chickens with *G. anatis,* the anti-GtxA-N IgY was injected into chickens. The results showed that anti-GtxA-N IgY protected the chickens against *G. anatis* infection and lower lesion scores were seen in the peritoneum, liver, and duodenum of chickens (Zhang et al. 2019). There is a commercially available product which contains IgY antibodies raised against *E. coli* and Salmonella (Chap. 17). Overall these results suggest that IgY is a promising alternative to sub-therapeutic and therapeutic usage of antibiotics in the poultry industry.

15.4.2 IgY Treatment of Viral Infections of Poultry

Infectious bursal disease virus (IBDV) is a highly contagious immunosuppressive disease of young chicks, infecting the bursa of Fabricius (Chap. 3) (Dey et al. 2019). The effect of specific IgY against IBDV both in the prevention and therapeutic treatment of the disease are detailed in Table 15.3.

Numerous studies have been carried out to develop IgY preparations against duck hepatitis virus. A related veterinary drug has been licenced in China for protection of ducklings against duck hepatitis by prophylactic subcutaneous or intramuscular injection (Gui-rong and Yun-ying 2011).

15.4.3 IgY Treatment of Protozoan Infections of Poultry

The reader is referred to a recent review of the use of IgY as both a diagnostic and therapeutic treatment of parasitic infections (Thirumalai et al. 2019). Avian coccidiosis is an intestinal disease caused by infection with a protozoan parasite of the genus *Eimeria* (coccidia). There are several distinct species which infect chickens including *E. acervuline*, *E. maxima*, *E. tenella*, *E. praecox* and *E. necatrix*. Vaccination

Table 15.3 Role of IgY antibodies in treating viral infections in poultry

Pathogen	Poultry	Prevention or therapeutic	IgY form	Outcomes	Reference
IBDV	Broiler	Therapeutic	Diluted yolk with antibodies	Birds recovered	(Muhammad et al. 2001)
IBDV	Mixed breeds of IBDV infected broilers and layers	Therapeutic	IgY solution injected ip	92% recovery rate	(Malik et al. 2006)
IBDV	Day old Hubbard broiler chickens	Prevention	IgY solution orally 0.5 mL/ bird	Decreased morbidity, mortality, lesions	(El-Ghany 2011)
IBDV	14 day old Golden birds	Prevention	Diluted yolk with antibodies	No mortality or morbidity	(Farooq et al. 2012)

with live or attenuated parasites is routinely used to confer active immunity to the pathogen (Gadde et al. 2015). There are two reports on the use of specific IgY powder (Supracox®) in feed which prevented infection by *E. acervulina* (Lee et al. 2009a, b), *E. maxima* and *E. tenella* (Lee et al. 2009a, b). In 2013 the protective efficacy against experimental *E. tenella* infection in chickens was assessed (Xu et al. 2013). The results of these investigations represent a promising strategy to prevent avian coccidiosis by supplementing with IgY.

15.5 Applications of IgY in Aquaculture

Globally fish production has been estimated to be about 179 million tonnes in 2018 of which aquaculture accounted for 46% (82 million tonnes); aquaculture production is projected to reach 109 million tonnes in 2030 (FAO 2020). The leading producers of farmed fish are China, Egypt, Chile, India, Indonesia, Viet Nam, Bangladesh and Norway. Of the major finfish species produced in 2018, carp accounted for 33%, salmon for 4.5% and rainbow trout for 1.6% of the total production; of the world production of crustaceans, white shrimp accounted for 53% (FAO 2020). One of the major factors impacting on the productivity of the aquaculture sector is the outbreak of diseases; the widespread use of antimicrobials is problematic and can lead to the development of antibiotic resistance (Baloch et al. 2014). While finfish have innate and adaptive immune systems (Sunyer 2013), crustaceans depend completely on the innate immune system for defence and thus vaccination is ineffective (Vazquez et al. 2009). The use of passive immunization in aquaculture poses a number of specific challenges which need to be considered. For routine application of antibodies oral administration would be preferred (Winkelbach et al. 2015a, b) but may be species specific.

Administration of IgY through oral and anal routes was not detectable in gastric rainbow trout probably impacted by pepsin activity in the gut (Winkelbach et al. 2015a, b); in contrast uptake in the gastric carp (lacking acidic stomach enzymes) demonstrated an efficient uptake of IgY (Winkelbach et al. 2015a, b) and probably increased IgY transcytosis in carp compared to trout. Other considerations are the period of immunization which is most effective for uptake and the duration for which passively transmitted IgY are present in the different fish species; continuous administration may be necessary in some cases (Rajan et al. 2017). Despite these challenges, advances have been made in the use of IgY antibodies for the treatment of fish diseases and these are summarized in Table 15.4.

White spot syndrome infection is caused by a pathogen called white spot syndrome virus (WSSV) which is a highly lethal, stress-dependent dsDNA virus with 305,107-bp length, first reported in 1992–1993 in the southern provinces of China and in the northern counties of Taiwan (Leu et al. 2009). WSSV has envelope proteins VP28 and VP19; VP28 interacts with the shrimp receptor PmRab7 forming a PmRab7-VP28 complex (Verma et al. 2013) and the virus can be neutralized by IgY produced against truncated fusion proteins of VP28 and VP19 (Kim et al. 2004). By using specific IgY from the hens that had been immunized with inactivated WSSV to neutralize WSSV, shrimp showed a higher survival rate (73.3%) than those hens that were immunized with DNA vaccine (33.3%) (Lu et al. 2008) (Table 15.4). In another study, the use of IgY antibodies significantly reduced the mortality rate in crayfish against the WSSV when effectiveness of IgY antibodies was investigated using intramuscular injection, oral administration and immersion. It was shown that the crayfish treated with IgY antibodies from inactivated WSSV had 20% mortality, while those that were treated with DNA vaccine had 80% mortality rate (Lu et al. 2009) (Table 15.4).

Vibrio parahaemolyticus is a pathogen associated with acute hepatopancreatic necrosis disease (AHPND) and can cause up to 100% mortality in post-larvae shrimp. IgY has been shown to confer protection to this infection (Gao et al. 2016) (Table 15.4). The Gram-negative bacteria *Yersinia ruckeri* is endemic to the North America and also widespread in several regions of world, such as Germany, France, United Kingdom, Australia, South Africa and Europe, and causes significant losses to salmonid farming industry (Saleh et al. 2008). The infected fish carry *Y. ruckeri* and shed the bacteria in faeces presenting a continuous source of infection to uninfected fish. On the basis of organ and intestine culture, it was found that fish treated with anti-*Y. ruckeri* IgY antibodies had lower levels of infection and carried lower number of *Y. ruckeri* in intestine samples (Lee et al. 2000) (Table 15.4). *Piscirickettsia salmonis* causes the disease piscirickettsiosis in salmon, resulting in significant mortality and economic losses in fish (Oliver et al. 2015). Specific IgY was produced by immunizing hens with *P. salmonis* proteins and was effective against the infection in SHK-1 cells; further studies are necessary to show its efficacy in fish. *Vibrio anguillarum* is an important pathogen of marine and fresh water fish and alternates to vaccination which can be stressful for fish were sought (Arasteh et al. 2004). Oral treatment is preferable and was shown to be as effective in some cases as intraperitoneal injection (Table 15.4). *Edwardsiella tarda*, a Gram-negative

Table 15.4 IgY treatment of important aquaculture species

Pathogen	Species	IgY form	Outcomes	Reference
White spot syndrome virus (WSSV)	Shrimp	Injected I m.	When challenged with virus, 73% survival when IgY prepared with inactivated virus; 33% with IgY prepared from WSSV-DNA	(Lu et al. 2008)
White spot syndrome virus	Shrimp	IgY in feed	Highly resistant To WSSV challenge	(Kumaran et al. 2010)
White spot syndrome virus (WSSV)	Crayfish	Injected i.m., feed or immersion in IgY solution	All methods provided some protection of crayfish against infection.	(Lu et al. 2009)
V. harveyi and V. parahaemolyticus	Shrimp	β-Cyclodextrin encapsulated egg yolk powder	Protected from infection	(Gao et al. 2016)
Vibrio parahaemolyticus	Shrimp	Whole egg powder with specific IgY fed at 20%	Survival of 86% when treated with recombinant PirA-like toxin induced IgY	(Nakamura et al. 2019)
Yersinia ruckeri	Rainbow trout	Microencapsulated IgY	Lower mortalities than control fish	(Lee et al. 2000)
Piscirickettsia salmonis	Salmon	Chitosan-alginate micro encapsulated IgY	Encapsulated IgY can resist degradation and is absorbed into the bloodstream	(Oliver et al. 2015)
Vibrio anguillarum	Rainbow trout	intraperitoneal (IP) injection, oral intubation, or feeding	Injection conferred protection for 14 days. Oral intubation and feeding gave comparable results in some cases	(Arasteh et al. 2004)
Edwardsiella tarda	Turbot	chitosan-alginate encapsulated IgY at 1, 3 and 5% in feed	survival rates of the 1%, 3% and 5% micro-encapsulated specific IgY groups were 20% 56.7% 63.3%, on the tenth day post infection with 10^7 CFU *E. tarda*	(Xu et al. 2020)
V. Alginolyticus	Abalone	5 or 10% alginate encapsulated IgY in feed	Survival rates of fed with 5 or 10% anti-vibrio IgY egg	(Wu et al. 2011)

(continued)

Table 15.4 (continued)

Pathogen	Species	IgY form	Outcomes	Reference
			powders ranged from 65–70% 14 days post-*V. alginolyticus* challenge (1x10^6c.f.u.)	
Shewanella marisflavi	Sea cucumber	Oral yolk powder at 25, 5 and 1 mg/ml	25, 5, and 1 mg/mL anti-*S. marisflavi* AP629 IgY gave 77.5%, 50%, and 22.5% survival rates at day 12, respectively, when challenged with 4.2 × 10^6 CFU S. marisflavi	(Xu et al. 2019)

bacterium belonging to the family *Enterobacteriaceae*, is considered a common pathogen infecting mainly economically important fish species such as Japanese eel (*Anguilla japonica*), red sea bream (*Pagrus major*), yellowtail (*Seriola quinqueradiata*), channel catfish (*Ictalurus punctatus*) and turbot (*Scophthalmus maximus*) (Park et al. 2012). The use of encapsulated IgY shows promise for the treatment of turbot (Xu et al. 2020) (Table 15.4). *Vibrio alginolyticus* infections, which cause mass death of small abalone and thus great economic losses have also been treated effectively with IgY (Wu et al. 2011). In 2016, approximately 218,038 tons of sea cucumber *Apostichopus japonicus* were produced in China with an economic value of almost four billion dollars (Xu et al. 2019) and references therein. Sea cucumbers depend on their innate immune system, as they lack an adaptive immune response (Chap. 3). Treatment with IgY raised against *Shewanella marisflavi* AP629 was shown to be protective against the pathogen and specific IgY was shown to stimulate the immune system of sea cucumbers to eliminate the pathogen (Table 15.4).

15.6 Dogs

As companion animals, dogs are very important for human wellbeing and have been kept as pets for over 14 centuries; increases in dog ownership was evident during the pandemic of 2020 (Oliva and Johnston 2020). IgY antibodies have been used for the diagnosis, prophylaxis and treatment of diseases of dogs.

15.6.1 Canine Parvovirus

Canine parvovirus (CPV) infection is a fatal disease of dogs, believed to have originated from cats and parvoviruses of carnivores including bush dogs, cats, coyotes, bears and wolves, causes haemorrhagic diarrhoea and myocarditis resulting from severe gastroenteritis. The disease emerged during 1970's and currently the prevalence is worldwide (Appel et al. 1979, Burtonboy et al. 1979, Gerlach et al. 2020).

The aetiological agent Canine parvovirus type 2 (CPV-2) is a single stranded, non-enveloped DNA virus with an icosahedral capsid (Nandi and Kumar 2010). The virus has undergone several genetic mutations and antigenic changes to evolve as CPV-2a, CPV-2b, New CPV-2a, New CPV-2b and CPV-2c in the past three decades (Parrish et al. 1991, Buonavoglia et al. 2001, Nakamura et al. 2004, Kapil et al. 2007). CPV capsid consists of about 90% of the capsid proteins VP2 and copies of a combination of 60 viral coat proteins VP1 (82 kDa), VP2 (67 kDa) and VP3 (63.5 kDa) (Han et al. 2012). Mutations in the VP2 gene of the virus resulted in the evolution of the different antigenic variants (Raj et al. 2010).

The infection spreads from infected dogs or from their faeces to healthy ones which later show symptoms of pyrexia, vomiting, anorexia, and dysentery (Hoskins et al. 1988). CPV attacks the rapidly diving cells of the bone marrow and the small intestine (Guimaraes et al. 2008). The pandemics of CPV infection are attributed to the fact that the virus is highly stable and has the ability to survive in harsh environmental conditions for extended periods of time (Hoelzer and Parrish 2008).

In spite of the availability of live modified vaccines to control the infection, the disease still remains as a major problem in the canine population worldwide, especially impacting on young puppies (Chinchkar et al. 2006, Qi et al. 2020, Kelman et al. 2020). Interference of maternal antibodies transferred to the puppies through milk, the timing of the vaccination, and also the emergence of antigenic variations between the currently available vaccine strain and prevalent virus types accounts for the failure of current vaccines (Decaro et al. 2008, Spibey et al. 2008, Chinchkar et al. 2014, Akila et al. 2016, Kelman et al. 2020). As the maternal antibody level reduces, and their own immune system is not mature enough to fight the infection, puppies become more susceptible to the infection which can be fatal (Van Nguyen et al. 2006).

Oral passive immunization with IgY antibodies specific to CPV-2 virus, administered as an egg yolk powder to dogs challenged orally with a CPV-2 viral strain for 7 days controlled the infection to a great extent (Van Nguyen et al. 2006). Animal groups treated with 2 g of IgY antibodies in powdered form for 16 days post challenge showed significant weight gain and shorter duration of virus shedding than the control group. Apart from oral administration of antibodies, an intravenous route of anti CPV-2 IgY treatment has been used to protect dogs from CPV infection. Experimental infection induced by an oral challenge of dogs with CPV, after treatment with 1000 and 10,000 PD (Protective Doses) of Anti CPV IgY antibodies showed recovery rates of 25% and 100%, respectively. Higher doses produce increased antibody titres which suppressed the viral load and minimized the excretion of virus from stool samples (Suartini et al. 2014). Therefore, in addition to oral therapy, intravenous immunization with antibodies also proves to be an efficient passive therapy method. The kinetics of anti CPV IgY antibodies was studied by regression analysis to determine the level of IgY in serum from the time of injection (Suartini et al. 2016).

The use of monoclonal IgY antibodies and scFv to treat disease has become more prominent including the treatment of CPV (Chap. 13). Oral passive immunotherapy can be extended to the development of supplementary oral products to protect dogs from parvoviral infections (Chap.17).

15.6.2 Canine Morbillivirus

Canine morbillivirus (formerly termed Canine distemper virus, CDV) is an extremely contagious immunosuppressive disease which affects dogs. Morbilliviruses belong to the family *Paramyxoviridae* and the genus *Morbillivirus* and cause moderate to severe immunosuppressive, respiratory, gastrointestinal and neurological diseases in a variety of hosts from humans (measles) to canines; and present an interesting model to research inter-species jumping (Quintero-Gil et al. 2019).

The early symptoms are similar to a cold, followed by bronchitis, catarrhal pneumonia, and gastroenteritis. In the later stages of the disease, there are neurological symptoms such as convulsions. In some cases, there may be a high degree of keratosis of the nose and hardening of the skin of the footpads of the paws (hard pad disease). Dogs are vaccinated to protect from the disease, but there are geographical genetic variations in the virus (Pratelli 2011).

Due to the broad clinical symptoms, laboratory tests are required to confirm the disease and a range of biological samples have been used to measure the virus, mainly by PCR but also by immunochromatography, immunofluorescence and ELISA (Costa et al. 2019).

Currently, there is no specific drug for the treatment of this disease. Antibody based therapy has been considered as a relatively efficient intervention. IgY antibodies have been generated by immunizing laying hens with CDV (Guimaraes et al. 2008, Guimaraes et al. 2009).

15.7 Applications of IgY to Treat Other Animals

Global rabbit production has been estimated at more than one million tonnes per year, according to the FAO (FAOSTAT 2012) and therefore infections of rabbits need to be controlled. Rabbit haemorrhagic disease (RHD) is a contagious and highly lethal viral disease of rabbits. Hens were immunized with the N-terminal of the viral VP-60 capsid protein and IgY treated rabbits showed a significant reduction in the onset, duration and severity of RHD infection (Li et al. 2014).

Allergies to cats are the most common human allergy of animal origin; the major cat allergen is Fel d 1 (Satyaraj et al. 2019). Hens were immunized with this allergen and allergen-specific IgY fed to cats was shown to neutralize Fel d 1 after its production and lessen the release into the environment thus decreasing the allergic response of the owners.

15.8 Application of IgY Antibodies in Detection & Immunoassay

Polyclonal IgY has less cross-reactivity than mammalian IgG as it does not bind to the mammalian Fc receptor and does not cross-react with many mammalian proteins (Chap. 5). IgY does not activate mammalian complement factors, which also helps to reduce the assay interference by these factors in mammalian serum samples, resulting in increased sensitivity as well as decreased background in immunological assays (Mine and Kovacs-Nolan 2002).

The polyclonal nature of IgY antibodies has limited their use as the development and application of immunoassays require IgY to be extracted and purified (Chap. 11), to yield a homogenous antibody preparation. Avian hybridoma techniques are not well established for IgY generation (Chap. 13). Recombinant IgY can be efficiently generated by phage display (Chap. 13) which makes it possible to exploit the advantageous properties of both IgY and monoclonal antibodies. However, the nature of scFv antibodies can result in VH and VL domains with sub-optimal biophysical properties, such as reduced thermodynamic stability and enhanced aggregation propensity, which leads to poorer production and more limited applications (Spillner et al. 2012, Zhang et al. 2017, Pereira et al. 2019).

15.8.1 Igy Based Diagnostics of Animal Diseases

The application of IgY antibodies to the detection of various analytes is summarized in (Table 15.5).

The use of IgY antibodies to develop varied rapid diagnostics for CPV-2 detection has gained attention in recent years (He et al. 2015). Recently, CPV IgY-scFvs generated against VLPs (Virus Like Particles) by a T7 phage display system expressed in *E.coli* showed 95% of phage contained the scFv fragments (Table 15.5). IgY-scFv fragments used for the detection of CPV in clinical samples showed good agreement with PCR and no cross reactivity towards canine distemper virus and canine coronavirus (Ge et al. 2020). A latex agglutination with anti CPV polyclonal IgY antibodies has been developed (González-Figueredo et al. 2015) as was an assay to detect the virus strain from infected faeces (Guimaraes et al. 2008). IgY based systems have also be applied to detect Cytochrome P450 in mouse liver tissue using immunomagnetic beads which is suitable for large-scale and rapid screening (Table 15.5). The specific determination of potential food contaminants such as gentamicin, Staphylococcal enterotoxins and fumonisin has also been reported (Table 15.5). A variety of cancer and infection-related antigens have also

Table 15.5 Applications of IgY based diagnostics

Antibody	Antigen	Detection method	Sample	Analytical performance	References
IgY-scFv	VP2 protein of canine parvovirus (CPV)	ELISA	Clinical sample (faeces)	Able to detect VP2 (2 ng/μL) and CPV samples in good agreement with PCR and commercial colloidal gold strip; no cross reactivity with CDV and CCV.	(Ge et al. 2020)
IgY	Cytochrome P450 2E1 (CYP2E1)	ELISA and immunomagnetic beads (IMBs)	Mouse liver tissue	IgY bound to CYP2E1 protein at a dilution of 1:1000 in both methods; no cross-reactivity with other CYP proteins; the sensitivity of IMB is lower than that for ELISA and Western blotting.	(Jiang et al. 2016)
IgY-scFv	Gentamicin	Indirect competitive ELISA	Animal food products (milk, pork, beef, and chicken)	LOD = 0.147 ng/mL, no cross reactivity with kanamycin and amikacin.	(Li et al. 2016)
IgY and ssDNA aptamers	Staphylococcal enterotoxin B (SEB)	Sandwich ELISA	Naturally contaminated food	Specific IgY highly reacted to SEB, with LOD of 50 ng/mL, without cross-reactivity with other toxins such as SEA and SEC.	(Mudili et al. 2015)
IgY	Fumonisin group B (FB)	Competitive lateral flow immunoassay	Maize grains	LOD = 4000μg/kg; no cross-reaction with deoxynivalenol, ochratoxin A, aflatoxin B1 and zearalenone.	(Tran et al. 2019)
IgY	Karilysin	Competitive ELISA	Adolescent saliva	Antibody specific to human MMP-3; the IgY showed affinity in low nanomolar range.	(Skottrup et al. 2019)
IgY- alkaline phosphatase (ALP) bio-conjugate	Influenza A	Sandwich ELISA	Human nasopharyngeal specimens	High consistent between the ELISA and Rt-PCR analysis in 42 nasopharyngeal swab samples. The cut-off value was around 0.26.	(Ozkan et al. 2019)

(continued)

Table 15.5 (continued)

Antibody	Antigen	Detection method	Sample	Analytical performance	References
IgY	Prostate specific antigen (PSA)	Sandwich ELISA	Human serum samples	Good agreement between the IgY-based and commercial IBL ELISA.	(Łupicka-Słowik et al. 2019)
IgY	Human epithelial growth factor family receptor-2(HER2)	Immunodetection with quantum dot fluorophores	Human breast cancer cell lines MCF-7 and human cell lines A549 and IMR90	At most antibody concentrations from 1 to 16 mg/mL, IgY antibody showed significantly higher binding signal than IgG. Quantum dot fluorophore-labeled secondary antibodies were used.	(Xiao et al. 2008)
IgY	Human C-reactive protein (CRP)	Colloidal gold assay	Human pleural fluid	Excellent stabilization in a wide range of concentrations and in various buffer systems.	(Gasparyan 2005)
IgY	HMGP-A	ELISA	Human milk	Higher binding with HMGP-A than HMGP-C.	(Shimizu et al. 1995)

Notes: *CDV* Canine distemper virus; *CCV* coronavirus; *VP2* structural viral proteins of canine parvovirus; *IBV* infectious bronchitis virus; *ALV* Avian leukosis virus; *MMP-3* matrix metalloproteinase-3; *HMGP-A* high-molecular weight mucin-like glycoprotein-A; *HMGP-C* high-molecular weight mucin-like glycoprotein-C; *SEA* Staphylococcal Enterotoxin A; *SEC* Staphylococcal Enterotoxin C been detected in human samples and presented here, though not strictly relevant to this chapter (Table 15.5).

15.8.2 IgY for Detection of Microbial Contamination in Sea Food

IgY antibodies have been successfully used to detect mature stages of the parasite *Loma salmonae* (Microsporidia), a causative agent of microsporidia gill disease (MGD) in various members of the Oncorhynchus genus, in particular chinook salmon (*O. Tshawytscha*) and coho salmon (*O. Kisutch*) (Alcorn and Pascho 2002). The *Loma salmonae* causes respiratory distress, secondary infection and is often fatal (Rodríguez-Tovar et al. 2006). IgY antibodies were developed by immunizing hens with $1x10^6$ purified *L. salmonae* spores by an intramuscular route. The extracted and purified IgY antibodies were used in immunohistochemistry, and the *L. salmonae* were stained strongly in the immunohistochemical preparations of the tissues of sockeye, chinook and pink salmon. IgY antibodies detected mature xenomas from gill, heart and spleen of chinook, and heart tissue of sockeye salmon. These specific IgY antibodies did not show any cross-reactivity with myoprotein parasites *K. thyrsites* or *P. minibicornis*. (Campora et al. 2008) developed a sensitive IgY sandwich ELISA to detect a poisonous organic compound called ciguatoxin in the tissues of fish. This toxin accumulates in fish organs such as skin, head, viscera and roe, and it cannot be destroyed by cooking the food at high temperature. Several diagnostic techniques forming the basis of detection of aquatic animal diseases have been developed (Prasad and Sukumaran 2013), and IgY could be used for the development of immunodiagnostic kits.

15.9 Conclusion and Perspectives

The advantages of passive immunization in the control of diseases in a wide variety of animal species has been demonstrated. There are now also a wide range of veterinary products (Chap. 17) which have been developed based on the research reported in this chapter. Egg yolk is an ideal source of antibodies and in many of the applications reported here egg yolk powder is used as a treatment. The presence of other components of the yolk, in addition to the specific IgY also promoted growth performance. IgY-scFv based genetically engineered antibody products (Chap. 13) have the potential for successful product development in veterinary applications, particularly to pet animals and highly-economically valued animals.

References

Aghamohammadi M, Haine D, Kelton DF, Barkema HW, Hogeveen H, Keefe GP, Dufour S (2018) Herd-level mastitis-associated costs on Canadian dairy Farms. Front Vet Sci 5:100

Akila S, Jyothi VP, Selvan MK, Naidu H, Raghunathan S, Kota S, Sundaram RC, Rana SK, Raj GD, Srinivasan VA, Mohana Subramanian B (2016) Direct typing of canine parvovirus (CPV) from infected dog faeces by rapid mini sequencing technique. J Virol Methods 238:66–69

Alcorn SW, Pascho RJ (2002) Antibody responses by Chinook salmon (Oncorhynchus tshawytscha) and rainbow trout (Oncorhynchus mykiss) to various protein antigens. Fish Shellfish Immunol 13(4):327–333

Alustiza F, Bellingeri R, Picco N, Motta C, Grosso MC, Barbero CA, Acevedo DF, Vivas A (2016) IgY against enterotoxigenic Escherichia coli administered by hydrogel-carbon nanotubes composites to prevent neonatal diarrhoea in experimentally challenged piglets. Vaccine 34 (28):3291–3297

Appel MJ, Scott FW, Carmichael LE (1979) Isolation and immunisation studies of a canine parco-like virus from dogs with haemorrhagic enteritis. Vet Rec 105(8):156–159

Arasteh N, Aminirissehei AH, Yousif AN, Albright LJ, Durance TD (2004) Passive immunization of rainbow trout (Oncorhynchus mykiss) with chicken egg yolk immunoglobulins (IgY). Aquaculture 231(1):23–36

Badaracco A, Garaicoechea L, Rodríguez D, Uriarte EL, Odeón A, Bilbao G, Galarza R, Abdala A, Fernandez F, Parreño V (2012) Bovine rotavirus strains circulating in beef and dairy herds in Argentina from 2004 to 2010. Vet Microbiol 158(3–4):394–399

Baloch A, Zhang XY, Schade R (2014) IgY technology in aquaculture–a review. Rev Aquac 6:1–8

Barnes JH, Vaillancourt JP, Gross WB (2003) Colibacillosis. In: Diseases of poultry, 11th edn. Iowa State University Press, Ames, pp 631–652

Bartels CJ, Holzhauer M, Jorritsma R, Swart WA, Lam TJ (2010) Prevalence, prediction and risk factors of enteropathogens in normal and non-normal faeces of young Dutch dairy calves. Prev Vet Med 93(2–3):162–169

Barton MD (2000) Antibiotic use in animal feed and its impact on human healt. Nutr Res Rev 13 (2):279–299

Bayles KW, Wesson CA, Liou LE, Fox LK, Bohach GA, Trumble WR (1998) Intracellular Staphylococcus aureus escapes the endosome and induces apoptosis in epithelial cells. Infect Immun 66(1):336–342

Bok M, Miño S, Rodriguez D, Badaracco A, Nuñes I, Souza SP, Bilbao G, Louge Uriarte E, Galarza R, Vega C, Odeon A, Saif LJ, Parreño V (2015) Molecular and antigenic characterization of bovine coronavirus circulating in Argentinean cattle during 1994-2010. Vet Microbiol 181(3–4):221–229

Bollini M, Leal ES, Adler NS, Aucar MG, Fernández GA, Pascual MJ, Merwaiss F, Alvarez DE, Cavasotto CN (2018) Discovery of novel bovine viral diarrhea inhibitors using structure-based virtual screening on the envelope protein E2. Front Chem 6:79

Bostock J (2011) The application of science and technology development in shaping current and future aquaculture production systems. J Agric Sci 149(S1):133–141

Brooks HL, Rushton K, Lovell K, Bee P, Walker L, Grant L, Rogers A (2018) The power of support from companion animals for people living with mental health problems: a systematic review and narrative synthesis of the evidence. BMC Psychiatry 18(1):1–12

Buonavoglia C, Martella V, Pratelli A, Tempesta M, Cavalli A, Buonavoglia D, Bozzo G, Elia G, Decaro N, Carmichael L (2001) Evidence for evolution of canine parvovirus type 2 in Italy. J Gen Virol 82(Pt 12):3021–3025

Burtonboy G, Coignoul F, Delferriere N, Pastoret PP (1979) Canine hemorrhagic enteritis: detection of viral particles by electron microscopy. Arch Virol 61(1–2):1–11

Campora CE, Hokama Y, Yabusaki K, Isobe M (2008) Development of an enzyme-linked immunosorbent assay for the detection of ciguatoxin in fish tissue using chicken immunoglobulin Y. J Clin Lab Anal 22(4):239–245

Chalghoumi R, Marcq C, Théwis A, Portetelle D, Beckers Y (2009) Effects of feed supplementation with specific hen egg yolk antibody (immunoglobin Y) on salmonella species cecal colonization and growth performances of challenged broiler chickens. Poult Sci 88 (10):2081–2092

Chinchkar SR, Mohana Subramanian B, Hanumantha Rao N, Rangarajan PN, Thiagarajan D, Srinivasan VA (2006) Analysis of VP2 gene sequences of canine parvovirus isolates in India. Arch Virol 151(9):1881–1887

Chinchkar S, Subramanian B, Naidu H, Dorairajan T, Srinivasan VA (2014) Canine parvovirus isolates of India and the relevance of canine parvovirus Type-2 vaccines. J Adv Vet Res 4 (1):34–41

Cho YI, Han JI, Wang C, Cooper V, Schwartz K, Engelken T, Yoon KJ (2013) Case-control study of microbiological etiology associated with calf diarrhea. Vet Microbiol 166(3–4):375–385

Cho YI, Yoon KJ (2014) An overview of calf diarrhea - infectious etiology, diagnosis, and intervention. J Vet Sci 15(1):1–17

Costa VGD, Saivish MV, Rodrigues RL, Lima Silva RF, Moreli ML, Krüger RH (2019) Molecular and serological surveys of canine distemper virus: a meta-analysis of cross-sectional studies. PLoS One 14(5):e0217594

Crute IR, Muir JF (2011) Improving the productivity and sustainability of terrestrial and aquatic food production systems: future perspectives. J Agric Sci 149(S1):1–7

Davis TC, White RR (2020) Breeding animals to feed people: the many roles of animal reproduction in ensuring global food security. Theriogenology 150:27–33

Decaro N, Desario C, Lucente MS, Amorisco F, Campolo M, Elia G, Cavalli A, Martella V, Buonavoglia C (2008) Specific identification of feline panleukopenia virus and its rapid differentiation from canine parvoviruses using minor groove binder probes. J Virol Methods 147(1):67–71

Dey S, Pathak DC, Ramamurthy N, Maity HK, Chellappa MM (2019) Infectious bursal disease virus in chickens: prevalence, impact, and management strategies. Vet Med (Auckland, NZ) 10 (2019):85–97

DiLorenzo N, Diez-Gonzalez F, DiCostanzo A (2006) Effects of feeding polyclonal antibody preparations on ruminal bacterial populations and ruminal pH of steers fed high-grain diets. J Anim Sci 84(8):2178–2185

Diraviyam T, Zhao B, Wang Y, Schade R, Michael A, Zhang X (2014) Effect of chicken egg yolk antibodies (IgY) against diarrhea in domesticated animals: a systematic review and meta-analysis. PLoS One 9(5):e97716

El-Ghany WA (2011) Comparison between immunoglobulins IgY and the vaccine for prevention of infectious bursal disease in chickens. Global Vet 6:16–24

FAO (2009) The state of food and agriculture. http://www.fao.org/3/i0680e/i0680e.pdf

FAO (2016) Economic analysis of animal diseases. http://www.fao.org/3/a-i5512e.pdf

FAO (2020) The state of world fisheries and aquaculture 2020. http://www.fao.org/publications/sofia/2020/en/

FAO, OECD (2020) Dairy and dairy products. In: OECD-FAO Agricultural Outlook 2020–2029, pp 174–183

FAOSTAT (2012) Food and agricultural organization statistical da-tabase. . http://faostat3.fao.org/faostatgateway/go/to/download/Q/QL/E

Farms T (2017) What are the three biggest challenges facing the pork industry? https://toshfarms.net/what-are-the-three-biggest-challenges-facing-the-pork-industry/#:~:text=Bill%20Even%2C%20CEO%20of%20theand%20consumers'%20fear%20of%20science.&text=A%20foreign%20animal%20disease%20outbreak%20would%20be%20devastating

Farooq A, Rabbani M, Khushi M, Akram Z, Ahad A, Fatima Z, Khurram T, Anwar Z (2012) Passive immunization in infectious bursal disease virus infected birds using chemically purified immune yolk immunoglobulins (IgY). Afr J Microbiol Res 6(12):2993–2998

Ferragut F, Vega CG, Mauroy A, Conceição-Neto N, Zeller M, Heylen E, Uriarte EL, Bilbao G, Bok M, Matthijnssens J, Thiry E, Badaracco A, Parreño V (2016) Molecular detection of bovine noroviruses in Argentinean dairy calves: circulation of a tentative new genotype. Infect Genet Evol 40:144–150

Foster DM, Smith GW (2009) Pathophysiology of diarrhea in calves. Veterinary clinics of North America. Food Anim Pract 25(1):13–36

Fulton RM, Nersessian BN, Reed WM (2002) Prevention of *Salmonella enteritidis* infection in commercial ducklings by oral chicken egg-derived antibody alone or in combination with probiotics. Poult Sci 81(1):34–40

Gadde U, Rathinam T, Lillehoj HS (2015) Passive immunization with hyperimmune egg-yolk IgY as prophylaxis and therapy for poultry diseases--a review. Anim Health Res Rev 16(2):163–176

Gao J, Barkema HW, Zhang L, Liu G, Deng Z, Cai L, Shan R, Zhang S, Zou J, Kastelic JP, Han B (2017) Incidence of clinical mastitis and distribution of pathogens on large Chinese dairy farms. J Dairy Sci 100(6):4797–4806

Gao X, Zhang X, Lin L, Yao D, Sun J, Du X, Li X, Zhang Y (2016) Passive immune-protection of Litopenaeus vannamei against Vibrio harveyi and Vibrio parahaemolyticus infections with anti-vibrio egg yolk (IgY)-encapsulated feed. Int J Mol Sci 17(5):723

Garaicoechea L, Bok K, Jones LR, Combessies G, Odeón A, Fernandez F, Parreño V (2006) Molecular characterization of bovine rotavirus circulating in beef and dairy herds in Argentina during a 10-year period (1994-2003). Vet Microbiol 118(1–2):1–11

Gasparyan VK (2005) Hen egg immunoglobulin Y in colloidal gold agglutination assay: comparison with rabbit immunoglobulin G. J Clin Lab Anal 19(3):124–127

Ge S, Xu L, Li B, Zhong F, Liu X, Zhang X (2020) Canine parvovirus is diagnosed and neutralized by chicken IgY-scFv generated against the virus capsid protein. Vet Res 51(1):110

Gerlach M, Proksch AL, Dörfelt R, Unterer S, Hartmann K (2020) Therapy of canine parvovirus infection - review and current insights. Tierarztliche praxis. Ausgabe K, Kleintiere/Heimtiere 48 (1):26–37

González-Figueredo J, Rajme-Manzur D, Negrin-Thomson G, Dopico-Paz JR (2015) Obtención de anticuerpos policlonales IgY antiparvovirus canino y evaluación en un sistema de aglutinación con látex. VacciMonitor 25(4):451–463

Grünberg W (2014) Diarrhoea in neonatal ruminants. https://www.msdvetmanual.com/digestive-system/intestinal-diseases-in-ruminants/diarrhea-in-neonatal-ruminants

Guimaraes MCC, Amaral LG, Borges FV, Vieira HPL, Matta CGF, Matta MFDR (2009) Characterization of an IgY polyclonal antibodies directed against the canine distemper virus. J Med Biol Sci 8(1):18–25

Guimaraes MCC, Amaral LG, Borges FV, Vieira HPL, Shimoya A, Gomes C, Matta F, Matta R (2008) Production and use of egg-yolk antibody for detection of canine parvovirus in feces. J Med Biol Sci 7(3):241–248

Gui-rong Z, Yun-ying Z (2011) Development of yolk antibodies against young duck viral hepatitis. China Anim Hus Vet Med 38(10):142–144

Gürtler M, Methner U, Kobilke H, Fehlhaber K (2004) Effect of orally administered egg yolk antibodies on *Salmonella enteritidis* contamination of hen's eggs. J Vet Med 51(3):129–134

Hafez H (2011) Enteric diseases of poultry with special attention to Clostridium perfringens. Pak Vet J 31:175–184

Han S, Zhang XY, Zhao J (2012) Production of egg yolk antibody (IgY) against recombinant canine parvovirus VP2 protein. Acta Sci Vet 40(2):1029

Hatamzade IN, Rahimi S, Rasaee MJ, Karimi Torshizi MA, Zahraei Salehi T, Grimes JL (2020) The effect of capsulated and noncapsulated egg-yolk-specific antibody to reduce colonization in the intestine of Salmonella enterica ssp. enterica serovar Infantis-challenged broiler chickens. Poult Sci 99(3):1387–1394

He J, Wang Y, Sun S, Zhang X (2015) Evaluation of chicken IgY generated against canine parvovirus viral-like particles and development of enzyme-linked immunosorbent assay and Immunochromatographic assay for canine parvovirus detection. Viral Immunol 28(9):489–494

Hedegaard CJ, Heegaard PM (2016) Passive immunisation, an old idea revisited: basic principles and application to modern animal production systems. Vet Immunol Immunopathol 174:50–63

Heller MC, Chigerwe M (2018) Diagnosis and treatment of infectious enteritis in neonatal and juvenile ruminants. Veterinary clinics of North America. Food Anim Pract 34(1):101–117

Hermans D, Van Steendam K, Verbrugghe E, Verlinden M, Martel A, Seliwiorstow T, Heyndrickx M, Haesebrouck F, De Zutter L, Deforce D, Pasmans F (2014) Passive immunization to reduce campylobacter jejuni colonization and transmission in broiler chickens. Vet Res 45(1):27

Hoelzer K, Parrish CR (2008) Evolution and variation of the parvoviruses. In: Domingo E, Parrish CR, Holland JJ (eds) Origin and evolution of viruses, 2nd edn. Academic

Hoet AE, Smiley J, Thomas C, Nielsen PR, Wittum TE, Saif LJ (2003) Association of enteric shedding of bovine torovirus (Breda virus) and other enteropathogens with diarrhea in veal calves. Am J Vet Res 64(4):485–490

Hoskins JD, Breitschwerdt EB, Gaunt SD, French TW, Burgdorfer W (1988) Antibodies to Ehrlichia canis, Ehrlichia platys, and spotted fever group rickettsiae in Louisiana dogs. J Vet Intern Med 2(2):55–59

Houe H (2003) Economic impact of BVDV infection in dairies. Biologicals 31(2):137–143

Igwaran A, Okoh AI (2019) Human campylobacteriosis: a public health concern of global importance. Heliyon 5(11):e02814

Ismail ZB (2017) Mastitis vaccines in dairy cows: recent developments and recommendations of application. Veterinary World 10(9):1057–1062

Izzo MM, Kirkland PD, Mohler VL, Perkins NR, Gunn AA, House JK (2011) Prevalence of major enteric pathogens in Australian dairy calves with diarrhoea. Aust Vet J 89(5):167–173

Jámbor A, Czine P, Balogh PJS (2020) The impact of the coronavirus on agriculture: first evidence based on global newspapers. Sustainability 12(11):4535

Jiang Z, Jiang X, Li C, Xue H, Zhang X (2016) Development of an IgY antibody-based immunoassay for the screening of the CYP2E1 inhibitor/enhancer from herbal medicines. Front Pharmacol 7:502

Jin LZ, Baidoo SK, Marquardt RR, Frohlich AA (1998) In vitro inhibition of adhesion of enterotoxigenic Escherichia coli K88 to piglet intestinal mucus by egg-yolk antibodies. FEMS Immunol Med Microbiol 21(4):313–321

Kapil S, Cooper E, Lamm C, Murray B, Rezabek G, Johnston L 3rd, Campbell G, Johnson B (2007) Canine parvovirus types 2c and 2b circulating in north American dogs in 2006 and 2007. J Clin Microbiol 45(12):4044–4047

Kariyawasam S, Wilkie BN, Gyles CL (2004) Resistance of broiler chickens to Escherichia coli respiratory tract infection induced by passively transferred egg-yolk antibodies. Vet Microbiol 98(3–4):273–284

Kelman M, Barrs VR, Norris JM, Ward MP (2020) Canine parvovirus prevention and prevalence: veterinarian perceptions and behaviors. Prev Vet Med 174:104817

Kim DK, Jang IK, Seo HC, Shin SO, Yang SY, Kim JW (2004) Shrimp protected from WSSV disease by treatment with egg yolk antibodies (IgY) against a truncated fusion protein derived from WSSV. Aquaculture 237(1):21–30

Kongsted H, Toft N, Nielsen JP (2014) Risk factors and epidemiological characteristics of new neonatal porcine diarrhoea syndrome in four Danish herds. BMC Vet Res 10:151

Kovacs-Nolan J, Mine Y (2012) Egg yolk antibodies for passive immunity. Annu Rev Food Sci Technol 3:163–182

Kumaran T, Michaelbab M, Thangaswamy S, Albindhas S, Citarasu T (2010) Production of anti WSSV IgY edible antibody using herbal Immunoadjuvant Asparagus racemosus and its immunological influence against WSSV infection in Penaeus monodon. J Aquac Feed Sci Nutr 2 (1):1–5

Kuroki M, Ohta M, Ikemori Y, Icatlo FC, Kobayashi C, Yokoyama H, Kodama Y (1997) Field evaluation of chicken egg yolk immunoglobulins specific for bovine rotavirus in neonatal calves. Arch Virol 142(4):843–851

Kuroki M, Ohta M, Ikemori Y, Peralta RC, Yokoyama H, Kodama Y (1994) Passive protection against bovine rotavirus in calves by specific immunoglobulins from chicken egg yolk. Arch Virol 138(1–2):143–148

Kweon CH, Kwon BJ, Woo SR, Kim JM, Woo GH, Son DH, Hur W, Lee YS (2000) Immunoprophylactic effect of chicken egg yolk immunoglobulin (Ig Y) against porcine epidemic diarrhea virus (PEDV) in piglets. J Vet Med Sci 62(9):961–964

Langel SN, Paim FC, Lager KM, Vlasova AN, Saif LJ (2016) Lactogenic immunity and vaccines for porcine epidemic diarrhea virus (PEDV): historical and current concepts. Virus Res 226:93–107

Larsson A, Bålöw RM, Lindahl TL, Forsberg PO (1993) Chicken antibodies: taking advantage of evolution--a review. Poult Sci 72(10):1807–1812

Lee DH, Jeon YS, Park CK, Kim S, Lee DS, Lee C (2015) Immunoprophylactic effect of chicken egg yolk antibody (IgY) against a recombinant S1 domain of the porcine epidemic diarrhea virus spike protein in piglets. Arch Virol 160(9):2197–2207

Lee SH, Lillehoj HS, Park DW, Jang SI, Morales A, García D, Lucio E, Larios R, Victoria G, Marrufo D, Lillehoj EP (2009a) Induction of passive immunity in broiler chickens against Eimeria acervulina by hyperimmune egg yolk immunoglobulin Y. Poult Sci 88(3):562–566

Lee SH, Lillehoj HS, Park DW, Jang SI, Morales A, García D, Lucio E, Larios R, Victoria G, Marrufo D, Lillehoj EP (2009b) Protective effect of hyperimmune egg yolk IgY antibodies against Eimeria tenella and Eimeria maxima infections. Vet Parasitol 163(1–2):123–126

Lee SB, Mine Y, Stevenson RM (2000) Effects of hen egg yolk immunoglobulin in passive protection of rainbow trout against *Yersinia ruckeri*. J Agric Food Chem 48(1):110–115

Leu JH, Yang F, Zhang X, Xu X, Kou G, Lo CF (2009) Whispovirus. book: lesser known large DsDNA viruses, current topics. In: Microbiolo-gy and immunology, vol 328. Springer, Berlin, Heidelberg, pp 197–227

Li C, He J, Ren H, Zhang X, Du E, Li X (2016) Preparation of a chicken scFv to analyze gentamicin residue in animal derived food products. Anal Chem 88(7):4092–4098

Li ZX, Hu WD, Li BC, Li TY, Zhou XY, Zhang Z (2014) Egg yolk IgY against RHDV capsid protein VP60 promotes rabbit defense against RHDV infection. Vet Immunol Immunopathol 157(1–2):97–104

Li XY, Jin LJ, McAllister TA, Stanford K, Xu JY, Lu YN, Zhen YH, Sun YX, Xu YP (2007) Chitosan-alginate microcapsules for oral delivery of egg yolk immunoglobulin (IgY). J Agric Food Chem 55(8):2911–2917

Li X, Wang L, Zhen Y, Li S, Xu Y (2015) Chicken egg yolk antibodies (IgY) as non-antibiotic production enhancers for use in swine production: a review. J Anim Sci Biotechnol 6(1):40

Lu Y, Liu J, Jin L, Li X, Zhen Y, Xue H, Lin Q, Xu Y (2009) Passive immunization of crayfish (Procambius clarkiaii) with chicken egg yolk immunoglobulin (IgY) against white spot syndrome virus (WSSV). Appl Biochem Biotechnol 159(3):750–758

Lu Y, Liu J, Jin L, Li X, Zhen Y, Xue H, You J, Xu Y (2008) Passive protection of shrimp against white spot syndrome virus (WSSV) using specific antibody from egg yolk of chickens immunized with inactivated virus or a WSSV-DNA vaccine. Fish Shellfish Immunol 25 (5):604–610

Łupicka-Słowik A, Grzywa R, Leporowska E, Procyk D, Oleksyszyn J, Sieńczyk M (2019) Development and evaluation of an immunoglobulin Y-based ELISA for measuring prostate specific antigen in human serum. Ann Lab Med 39(4):373–380

Ma J, Cocchiaro J, Lee JC (2004) Evaluation of serotypes of Staphylococcus aureus strains used in the production of a bovine mastitis bacterin. J Dairy Sci 87(1):178–182

Mahdavi A, Rahmani H, Nili N, Samie A, Soleimania S (2010a) Chicken egg yolk antibody (IgY) powder against Escherichia coli O78:K80. J Anim Vet Adv 9(2):366–373

Mahdavi AH, Rahmani HR, Nili N, Samie AH, Soleimanian-Zad S, Jahanian R (2010b) Effects of dietary egg yolk antibody powder on growth performance, intestinal Escherichia coli colonization, and immunocompetence of challenged broiler chicks. Poult Sci 89(3):484–494

Malik MW, Ayub N, Qureshi IZ (2006) Passive immunization using purified IgYs against infectious bursal disease of chickens in Pakistan. J Vet Sci 7(1):43–46

Marquardt RR, Jin LZ, Kim JW, Fang L, Frohlich AA, Baidoo SK (1999) Passive protective effect of egg-yolk antibodies against enterotoxigenic Escherichia coli K88+ infection in neonatal and early-weaned piglets. FEMS Immunol Med Microbiol 23(4):283–288

Meenatchisundaram S, Michael A, Subbraj T, Diraviam T, Shanmugam V (2011) Isolation, purification and neutralizing potential of chicken egg yolk immunoglobulin (IgY) against mastitis causing Escherichia coli in dairy cows in Coimbatore District. Int J Drug Develop Res 3 (2):147–153

Mesonero-Escuredo S, Strutzberg-Minder K, Casanovas C, Segalés J (2018) Viral and bacterial investigations on the aetiology of recurrent pig neonatal diarrhoea cases in Spain. Porcine Health Manag 4:5

Mine Y, Kovacs-Nolan J (2002) Chicken egg yolk antibodies as therapeutics in enteric infectious disease: a review. J Med Food 5(3):159–169

Monistero V, Graber HU, Pollera C, Cremonesi P, Castiglioni B, Bottini E, Ceballos-Marquez A, Lasso-Rojas L, Kroemker V, Wente N, Petzer IM, Santisteban C, Runyan J, Veiga Dos Santos M, Alves BG, Piccinini R, Bronzo V, Abbassi MS, Said MB, Moroni P (2018) Staphylococcus aureus isolates from bovine mastitis in eight countries: genotypes, detection of genes encoding different toxins and other virulence genes. Toxins (Basel) 10(6):247

Mudili V, Makam SS, Sundararaj N, Siddaiah C, Gupta VK, Rao PVL (2015) A novel IgY-aptamer hybrid system for cost-effective detection of SEB and its evaluation on food and clinical samples. Sci Rep 5:15151

Muhammad K, Rabbi F, Hussain I, Riaz A, Saeed K, Hussain T (2001) Passive immunization against infectious bursal disease in chicks. Int J Agric Biol 3:413–415

Nair MK, Joy J, Vasudevan P, Hinckley L, Hoagland TA, Venkitanarayanan KS (2005) Antibacterial effect of caprylic acid and monocaprylin on major bacterial mastitis pathogens. J Dairy Sci 88(10):3488–3495

Nakamura R, Pedrosa-Gerasmio IR, Alenton RRR, Nozaki R, Kondo H, Hirono I (2019) Anti-PirA-like toxin immunoglobulin (IgY) in feeds passively immunizes shrimp against acute hepatopancreatic necrosis disease. J Fish Dis 42(8):1125–1132

Nakamura M, Tohya Y, Miyazawa T, Mochizuki M, Phung HT, Nguyen NH, Huynh LM, Nguyen LT, Nguyen PN, Nguyen PV, Nguyen NP, Akashi H (2004) A novel antigenic variant of canine parvovirus from a Vietnamese dog. Arch Virol 149(11):2261–2269

Nandi S, Kumar M (2010) Canine parvovirus: current perspective. Ind J Virol 21(1):31–44

Newcomer BW, Chamorro MF, Walz PH (2017) Vaccination of cattle against bovine viral diarrhea virus. Vet Microbiol 206:78–83

Oliva JL, Johnston KL (2020) Puppy love in the time of Corona: dog ownership protects against loneliness for those living alone during the COVID-19 lockdown. Int J Soc Psychiatry 23:1–11

Oliver C, Valenzuela K, Silva H, Haro RE, Cortés M, Sandoval R, Pontigo JP, Álvarez C, Figueroa JE, Avendaño-Herrera R, Troncoso JM, Yáñez AJ (2015) Effectiveness of egg yolk immunoglobulin against the intracellular salmonid pathogen *Piscirickettsia salmonis*. J Appl Microbiol 119(2):365–376

Otto PH, Clarke IN, Lambden PR, Salim O, Reetz J, Liebler-Tenorio EM (2011) Infection of calves with bovine norovirus GIII.1 strain Jena virus: an experimental model to study the pathogenesis of norovirus infection. J Virol 85(22):12013–12021

Ozkan B, Budama-Kilinc Y, Cakir-Koc R, Mese S, Badur S (2019) Application of an immunoglobulin Y-alkaline phosphatase bioconjugate as a diagnostic tool for influenza a virus. Bioengineered 10(1):33–42

Park SB, Aoki T, Jung TS (2012) Pathogenesis of and strategies for preventing Edwardsiella tarda infection in fish. Vet Res 43(1):67

Park HS, Park KI, Nagappan A, Lee DH, Kang SR, Kim JA, Kim EH, Han DY, Kim GS (2011) Proteomic analysis of effects on natural herb additive containing immunoglobulin Yolksac (IgY) in pigs. Am J Chin Med 39(3):477–488

Parreño V, Béjar C, Vagnozzi A, Barrandeguy M, Costantini V, Craig MI, Yuan L, Hodgins D, Saif L, Fernández F (2004) Modulation by colostrum-acquired maternal antibodies of systemic and mucosal antibody responses to rotavirus in calves experimentally challenged with bovine rotavirus. Vet Immunol Immunopathol 100(1–2):7–24

Parreño V, Marcoppido G, Vega C, Garaicoechea L, Rodriguez D, Saif L, Fernández F (2010) Milk supplemented with immune colostrum: protection against rotavirus diarrhea and modulatory effect on the systemic and mucosal antibody responses in calves experimentally challenged with bovine rotavirus. Vet Immunol Immunopathol 136(1–2):12–27

Parrish CR, Aquadro CF, Strassheim ML, Evermann JF, Sgro JY, Mohammed HO (1991) Rapid antigenic-type replacement and DNA sequence evolution of canine parvovirus. J Virol 65 (12):6544–6552

Patra MK, Begum S, Deka B (2014) Problems and prospects of traditional pig farming for tribal livelihood in Nagaland. Ind Res J Exten Edu 14(4):6–11

Pereira EPV, van Tilburg MF, Florean E, Guedes MIF (2019) Egg yolk antibodies (IgY) and their applications in human and veterinary health: a review. Int Immunopharmacol 73:293–303

Prasad K, Sukumaran JK (2013) Advances in aquatic animal health manage-ment. Int J Bio-Tech Res 3:35–44

Pratelli A (2011) Canine distemper virus: the emergence of new variants. Vet J (London, England: 1997) 187(3):290–291

Qi S, Zhao J, Guo D, Sun D (2020) A mini-review on the epidemiology of canine parvovirus in China. Front Vet Sci 7:5

Quintero-Gil C, Rendon-Marin S, Martinez-Gutierrez M, Ruiz-Saenz J (2019) Origin of canine distemper virus: consolidating evidence to understand potential Zoonoses. Front Microbiol 10:1982

Rahimi S, Shiraz Z, Salehi T, Karimi Torshizi MA, Grimes J (2007) Prevention of salmonella infection in poultry by specific egg-derived antibody. Int J Poult Sci 6(4):230–235

Rahman S, Van Nguyen S, Icatlo FC Jr, Umeda K, Kodama Y (2013) Oral passive IgY-based immunotherapeutics: a novel solution for prevention and treatment of alimentary tract diseases. Human Vaccines and Immunotherapeutics 9(5):1039–1048

Raj JM, Mukhopadhyay HK, Thanislass J, Antony PX, Pillai RM (2010) Isolation, molecular characterization and phylogenetic analysis of canine parvovirus. Infect Genet Evol 10 (8):1237–1241

Rajan B, Løkka G, Koppang EO, Austbø L (2017) Passive immunization of farmed fish. J Immunol. (Baltimore, Md. : 1950) 198(11):4195–4202

Rhouma M, Fairbrother JM, Beaudry F, Letellier A (2017) Post weaning diarrhea in pigs: risk factors and non-colistin-based control strategies. Acta Vet Scand 59(1):31

Riffon R, Sayasith K, Khalil H, Dubreuil P, Drolet M, Lagacé J (2001) Development of a rapid and sensitive test for identification of major pathogens in bovine mastitis by PCR. J Clin Microbiol 39(7):2584–2589

Ritchie H (2017) Meat and dairy production. Published online at OurWorldInData.org. https://ourworldindata.org/meat-production

Rodríguez-Tovar LE, Becker JA, Markham RJ, Speare DJ (2006) Induction time for resistance to microsporidial gill disease caused by Loma salmonae following vaccination of rainbow trout (Oncorhynchus mykiss) with a spore-based vaccine. Fish Shellfish Immunol 21(2):170–175

Ruegg PL (2017) A 100-year review: mastitis detection, management, and prevention. J Dairy Sci 100(12):10381–10397

Saif LJ, Thiel K (1990) Viral diarrheas of man and animals. CRC Press, Boca Raton, Florida

Saleh M, Soliman H, El-Matbouli M (2008) Loop-mediated isothermal amplification as an emerging technology for detection of *Yersinia ruckeri* the causative agent of enteric red mouth disease in fish. BMC Vet Res 4:31

Satyaraj E, Wedner HJ, Bousquet J (2019) Keep the cat, change the care pathway: a transformational approach to managing Fel d 1, the major cat allergen. Allergy 74(107):5–17

Shahbandeh M (2020) Egg production worldwide 1990–2018. https://www.statista.com/statistics/263972/egg-production-worldwide-since-1990/

Shimizu M, Watanabe A, Tanaka A (1995) Detection of high-molecular weight mucin-like glycoprotein-a (HMGP-A) of human milk by chicken egg yolk antibody. Biosci Biotechnol Biochem 59(1):138–139

Skottrup PD, López R, Ksiazek M, Højrup P, Baelum V, Potempa J, Kaczmarek JZ (2019) An IgY-based immunoassay to evaluate the biomarker potential of the Tannerella forsythia virulence factor karilysin in human saliva. J Immunol Methods 469:26–32

Spibey N, Greenwood NM, Sutton D, Chalmers WS, Tarpey I (2008) Canine parvovirus type 2 vaccine protects against virulent challenge with type 2c virus. Vet Microbiol 128(1–2):48–55

Spillner E, Braren I, Greunke K, Seismann H, Blank S, du Plessis D (2012) Avian IgY antibodies and their recombinant equivalents in research, diagnostics and therapy. Biologicals 40 (5):313–322

Stefaniak T (2006) Chapter 8 control of intestinal diseases by dietary supplementation with antibodies. In: Mosenthin R, Zentek J, Żebrowska T (eds) Biology of growing animals, vol 4. Elsevier, pp 285–309

Stefaniak T, Rzasa A, Jawor P, Zyzak A, Niemczuk W, Kuczaj M, Popławski M, Borkowski J (2014) Field application of egg yolk immunoglobulin as the feed additive in prophylaxis of diseases in weaned piglets. Med Weter 70(9):553–557

Stevens MP, Humphrey TJ, Maskell DJ (2009) Molecular insights into farm animal and zoonotic salmonella infections. Philosophical transactions of the Royal Society of London. Series B, Biol Sci 364(1530):2709–2723

Suartini IGAA, Sendow I, Agustini NLP, Suprayogi A, Wibawan IWT, Mahardika IGNK (2016) Kinetika Immunoglobulin Kuning Telur Antiparvovirus Anjing Pada Anjing (KINETICS OF ANTICANINE PARVOVIRUS YOLK IMMUNOGLOBULIN IN DOGS). Jurnal Veteriner 17 (2):292–299

Suartini GAA, Suprayogi A, Wibawan WT, Sendow I, Mahardika G (2014) Intravenous administration of chicken immunoglobulin has a curative effect in experimental infection of canine parvovirus. Global Vet 13:801–808

Sunyer JO (2013) Fishing for mammalian paradigms in the teleost immune system. Nat Immunol 14 (4):320–326

Tamilzarasan KB, Dinakaran AM, Selvaraju G, Dorairajan N (2009) Efficacy of egg yolk immunoglobulins (IgY) against enteric pathogens in poultry. Tamilnadu J Vet Sci 5(6):264–268

Tenhagen BA, Köster G, Wallmann J, Heuwieser W (2006) Prevalence of mastitis pathogens and their resistance against antimicrobial agents in dairy cows in Brandenburg, Germany. J Dairy Sci 89(7):2542–2551

Thirumalai D, Visaga Ambi S, Vieira-Pires RS, Xiaoying Z, Sekaran S, Krishnan U (2019) Chicken egg yolk antibody (IgY) as diagnostics and therapeutics in parasitic infections - a review. Int J Biol Macromol 136:755–763

Thu HM, Myat TW, Win MM, Thant KZ, Rahman S, Umeda K, Nguyen SV, Icatlo FC Jr, Higo-Moriguchi K, Taniguchi K, Tsuji T, Oguma K, Kim SJ, Bae HS, Choi HJ (2017) Chicken egg yolk antibodies (IgY) for prophylaxis and treatment of rotavirus diarrhea in human and animal neonates: a concise review. Korean J Food Sci Anim Resour 37(1):1–9

Tran TV, Do BN, Nguyen TPT, Tran TT, Tran SC, Nguyen BV, Nguyen CV, Le HQ (2019) Development of an IgY-based lateral flow immunoassay for detection of fumonisin B in maize. F1000Research 8:1042

Tsubokura K, Berndtson E, Bogstedt A, Kaijser B, Kim M, Ozeki M, Hammarström L (1997) Oral administration of antibodies as prophylaxis and therapy in campylobacter jejuni-infected chickens. Clin Exp Immunol 108(3):451–455

USDA (2019) What is the most consumed meat in the world? https://ask.usda.gov/s/article/What-is-the-most-consumed-meat-in-the-world#:~:text=Information&text=According%20to%20the%20United%20Nationsgoats%2Fsheep%20(5%25)

USDA, APHIS (2014) Veterinary Services, Health and management practices on U.S dairy operations. In. Fort Collins, CO, USA

Van Nguyen S, Umeda K, Yokoyama H, Tohya Y, Kodama Y (2006) Passive protection of dogs against clinical disease due to canine parvovirus-2 by specific antibody from chicken egg yolk. Canadian J Vet Res 70(1):62–64

Vandeputte J, Martel A, Canessa S, Van Rysselberghe N, De Zutter L, Heyndrickx M, Haesebrouck F, Pasmans F, Garmyn A (2019) Reducing campylobacter jejuni colonization in broiler chickens by in-feed supplementation with hyperimmune egg yolk antibodies. Sci Rep 9 (1):8931

Vazquez L, Alpuche J, Maldonado G, Agundis C, Pereyra-Morales A, Zenteno E (2009) Review: immunity mechanisms in crustaceans. Innate Immun 15(3):179–188

Vega C, Bok M, Chacana P, Saif L, Fernandez F, Parreño V (2011) Egg yolk IgY: protection against rotavirus induced diarrhea and modulatory effect on the systemic and mucosal antibody responses in newborn calves. Vet Immunol Immunopathol 142(3–4):156–169

Vega C, Bok M, Saif L, Fernandez F, Parreño V (2015) Egg yolk IgY antibodies: a therapeutic intervention against group a rotavirus in calves. Res Vet Sci 103:1–10

Vega CG, Bok M, Vlasova AN, Chattha KS, Fernández FM, Wigdorovitz A, Parreño VG, Saif LJ (2012) IgY antibodies protect against human rotavirus induced diarrhea in the neonatal gnotobiotic piglet disease model. PLoS One 7(8):e42788

Verma AK, Gupta S, Verma S, Mishra A, Nagpure NS, Singh SP, Pathak AK, Sarkar UK, Singh SP, Singh M, Seth PK (2013) Interaction between shrimp and white spot syndrome virus through PmRab7-VP28 complex: an insight using simulation and docking studies. J Mol Model 19(3):1285–1294

Wang LH, Li XY, Jin LJ, You JS, Zhou Y, Li SY, Xu YP (2011) Characterization of chicken egg yolk immunoglobulins (IgYs) specific for the most prevalent capsular serotypes of mastitis-causing Staphylococcus aureus. Vet Microbiol 149(3–4):415–421

Winkelbach A, Günzel D, Schulz C, Wuertz S (2015a) Differences in IgY gut absorption in gastric rainbow trout (Oncorhynchus mykiss) and agastric common carp (Cyprinus carpio) assessed in vivo and in vitro. Comp Biochem Physiol C Toxicol Pharmacol 167:58–64

Winkelbach A, Schade R, Schulz C, Wuertz S (2015b) Comparison of oral, rectal and intraperitoneal administration of IgY antibodies in passive immunization of rainbow trout (Oncorhynchus mykiss). Aquaculture Int 23(2):427–438

Wu CJ, Wang H, Chan YL, Li TL (2011) Passive immune-protection of small abalone against vibrio alginolyticus infection by anti-vibrio IgY-encapsulated feed. Fish Shellfish Immunol 30 (4–5):1042–1048

Xiao Y, Gao X, Gannot G, Emmert-Buck MR, Srivastava S, Wagner PD, Amos MD, Barker PE (2008) Quantitation of HER2 and telomerase biomarkers in solid tumors with IgY antibodies and nanocrystal detection. Int J Cancer 122(10):2178–2186

Xu L, Che J, Xu Y, Chen Y, Li Y, Murtaza B, Wang L, Zhang M, Li X (2020) Oral administration of microencapsulated egg yolk immunoglobulin (IgY) in turbot (Scophthalmus maximus) to combat against Edwardsiella tarda 2CDM001 infections. Fish Shellfish Immunol 106:609–620

Xu J, Ren C, Wang S, Liu D, Cao L, Tao J (2013) Protection efficacy of multivalent egg yolk immunoglobulin against Eimeria tenella infection in chickens. Iran J Parasitol 8(3):449–458

Xu J, Tan X, Zhang X, Xia X, Sun H (2015) The diversities of staphylococcal species, virulence and antibiotic resistance genes in the subclinical mastitis milk from a single Chinese cow herd. Microb Pathog 88:29–38

Xu L, Xu Y, He L, Zhang M, Wang L, Li Z, Li X (2019) Immunomodulatory effects of chicken egg yolk antibodies (IgY) against experimental Shewanella marisflavi AP629 infections in sea cucumbers (Apostichopus japonicus). Fish Shellfish Immunol 84:108–119

Yokoyama H, Peralta RC, Diaz R, Sendo S, Ikemori Y, Kodama Y (1992) Passive protective effect of chicken egg yolk immunoglobulins against experimental enterotoxigenic Escherichia coli infection in neonatal piglets. Infect Immun 60(3):998–1007

Zhang X, Calvert RA, Sutton BJ, Doré KA (2017) IgY: a key isotype in antibody evolution. Biol Rev Camb Philos Soc 92(4):2144–2156

Zhang X, Diraviyam T, Li X, Yao G, Michael A (2016) Preparation of chicken IgY against recombinant E2 protein of bovine viral diarrhea virus (BVDV) and development of ELISA and ICA for BVDV detection. Biosci Biotech Bioch 80(12):2467–2472

Zhang JJ, Kang TY, Kwon T, Koh H, Chandimali N, Wang XZ, Kim N, Jeong DK (2019) Specific chicken egg yolk antibody improves the protective response against Gallibacterium anatis infection. Infection Immunity 87(3):e00618–e00619

Zhen YH, Jin LJ, Guo J, Li XY, Lu YN, Chen J, Xu YP (2008) Characterization of specific egg yolk immunoglobulin (IgY) against mastitis-causing Escherichia coli. Vet Microbiol 130 (1–2):126–133

Zhen YH, Jin LJ, Li XY, Guo J, Li Z, Zhang BJ, Fang R, Xu YP (2009) Efficacy of specific egg yolk immunoglobulin (IgY) to bovine mastitis caused by Staphylococcus aureus. Vet Microbiol 133(4):317–322

Zhong F, Li N, Huang X, Guo Y, Chen H, Wang X, Shi C, Zhang X (2011) Genetic typing and epidemiologic observation of bovine viral diarrhea virus in Western China. Virus Genes 42 (2):204–207

Zuo Y, Fan J, Fan H, Li T, Zhang X (2009) Prophylactic and therapeutic effects of egg yolk immunoglobulin against porcine transmissible gastroenteritis virus in piglets. Front Agric China 3(1):104–108

Applications of IgY in Human Medicine 16

Ricardo S. Vieira-Pires, Patricia M. Morgan, Ankit Choraria, Chinnappan Gandhimathi, Antonysamy Michael, Tiago Ôchoa-Pires, Jafar Salimian, Sentila Rajan, and Meenatchisundaram Subramani

Abstract

In this chapter we will focus on the potential applications of IgY in human medicine, summarizing some of in vivo and in vitro studies that support them. Respiratory infections caused by viruses and bacteria have a major impact in human health, and are associated with other major underlying conditions, such as cardiopulmonary disorders. Influenza and SARS-CoV-2 are good examples of

R. S. Vieira-Pires (✉) · T. Ôchoa-Pires
CNC – Center for Neuroscience and Cell Biology, University of Coimbra, Coimbra, Portugal
e-mail: ricardo.pires@biocant.pt

P. M. Morgan
Emerita Lecturer, School of Natural Sciences, National University of Ireland Galway, Galway, Ireland
e-mail: pat.morgan@nuigalway.ie

A. Choraria · A. Michael
PSG College of Arts and Science, Coimbatore, Tamil Nadu, India
e-mail: michael@psgcas.ac.in

C. Gandhimathi
Sri Ramakrishna Dental College and Hospital, Coimbatore, Tamil Nadu, India

J. Salimian
Applied Microbiology Research Center, Chemical Injuries Research Center, Systems Biology and Poisoning Institute, Baqiyatallah University of Medical Sciences, Tehran, Iran
e-mail: jafar.salimian@bmsu.ac.ir

S. Rajan
Bharathiar University, Coimbatore, Tamil Nadu, India

M. Subramani
Nehru Arts and Science College, Coimbatore, Tamil Nadu, India

X.-Y. Zhang et al. (eds.), *IgY-Technology: Production and Application of Egg Yolk Antibodies*, https://doi.org/10.1007/978-3-030-72688-1_16

how the effective airborne transmission via droplets and aerosols enables rapid viral spread, that ultimately lead to seasonal epidemic or pandemic outbreaks. These together with a number of emerging viral disease urges the development of vaccine alternatives where IgY immnunotherapies could be included. Moreover oral passive neutralization of bacteria causing dental caries or even chronic gastric inflamation leading to cancer has been sucessfully shown with IgY and will be presented herein. Finally, the robustness of IgY to neutralize lethal toxins and venoms will also be covered.

Keywords

Infectious diseases · Virulence factors · Antivirals · Antimicrobials · IgY immunotherapy

16.1 Introduction

The advent of the 2020 SARS-CoV-2 pandemic revealed the importance of continuous development of alternative and complementary strategies to prevent and treat infection diseases. Vaccines play a central role in prevention and therapy but face both economic and technical limitations. Indeed, their cost and turnover of production limit their accessibility, especially in developing countries; also, vaccines rely on healthy and competent immune systems of individuals, being less effective in immunocompromised or aged subjects and additionally can lose efficiency due to pathogen adaptation.

Vaccine design relies on the fundamental understanding of the molecular mechanism of virulence of a given pathogen. Antibodies are one the most useful biological research tools to help dissect the precise structural and functional features of complex mechanisms of pathogenicity. For example, antibodies developed for the detection of specific viral strains can ultimately present robust anti-viral properties, opening novel possibilities not initially considered. In this line, antibody customization for characterization of virulence determinants, as for example surface-exposed spike glycoproteins on enveloped viruses that mediate host cell infection, has been important for fundamental and applied knowledge. For this reason, both vaccine and antibody-based therapies are typically co-developed, and advances like monoclonal antibody technology (Chap. 13) have led to refined therapeutic targeting using pathogen-specific antibodies.

On the other hand, the versatility and accessibility of polyclonal antibody therapeutic formulations for human applications, including serum-based therapies using human plasma of convalescent patients, or plasma from hyperimmune animals, such as horses and pigs, are still many times a last resource that saves lives even today (Hifumi et al. 2017). There are common post exposure alternatives for treatment of deadly venoms of snakes and scorpions, or toxins from microbial pathogens if supplies could meet demand.

As described in Chap. 1, IgY comes as a natural alternative for affordable and cost-effective passive immunoprophylactic or immunotherapeutic applications for humans. Similarly to what we describe for veterinary applications (Chap. 15), a large part of these are passive therapies against bacterial, viral and fungal pathogens infecting humans.

In this chapter we cover a number of diseases affecting humans and present findings on how IgY antibodies contribute to their prevention or treatment. We detail the specific virulent determinants being targeted for each case and combine in vitro and in vivo results that support therapeutic outcomes or the potential for its development in the future. We focus on human respiratory viral diseases, that are highly transmissible diseases and very often the cause of epidemic and pandemic outbreaks. We then describe the impact of IgY in the prevention of dental caries and the promotion of good general oral health status. Though a non-communicable disease, dental caries is particularly relevant given the current implications with other major conditions such as endocarditis and cardiovascular diseases (Sanchez et al. 2017). Also central in the prevention of gastric ulcers and ultimately gastric cancer is the therapeutic potential of IgY antibodies against *Helicobacter pylori*, that we also review. Finally, we summarize the studies on anti-toxin and anti-venom approaches using IgY. Indeed, envenoming by snakes and scorpions for example are a serious health threat in tropical and subtropical countries for which there is high demand on customized antivenom formulations.

16.2 Human Respiratory Infections

16.2.1 Severe Acute Respiratory Syndrome Coronaviruses (SARS-CoV)

Severe Acute Respiratory Syndrome Coronaviruses (SARS-CoV) are enveloped non-segmented positive sense RNA viruses from the family Coronaviridae and the order Nidovirales (Kuljić-Kapulica and Budisin 1992). Coronaviruses are capable of crossing the species barrier to cause respiratory infections in humans that can range from mild symptoms to deadly pneumonia. To date three highly pathogenic strains have been responsible for major outbreaks and pandemics: SARS-CoV (Drosten et al. 2003; Zhu et al. 2020), Middle-East respiratory syndrome coronavirus (MERS-CoV) (Zaki et al. 2012) and the most recent SARS-CoV-2 (Huang et al. 2020; Zhu et al. 2020). All these belong to the betacoronavirus genus that include four other mildly pathogenic strains.

In late 2019, SARS-CoV-2 emerged in Wuhan, China, as a new type of coronavirus and became a global health emergency that led to a devastating humanitarian and socio-economic crisis requiring the development and implementation of effective diagnostic, preventive and therapeutic solutions to fight SARS-CoV-2 (Kotta et al. 2020). Currently, (December first, 2020), it has spread to 220 countries, infecting over 62 million people, of which 1.46 million (2.3%) have succumbed to the disease, as described by the World Health Organization.

The current search for a vaccine for human SARS-CoV-2, runs side-by-side with the development of novel therapeutic antibody alternatives. For both strategies one needs to elicit cross-reactive Neutralizing Antibodies (Nabs) ideally capable of broad protection against different variant isolates. For this it is fundamental to understand the function and structure of SARS-CoV-2 main virulent factors, namely those susceptible to immune targeting strategies. The main structural proteins of SARS-CoV-2 include the Spike (S), membrane (M), envelope (E) and nucleic capsid (N) proteins (Zhou and Zhao 2020). The S protein is a large type I transmembrane glycoprotein, that forms homotrimers protruding from the viral envelope and is responsible for receptor binding and membrane fusion, mediating host-cell entry (Zhou and Zhao 2020) (Figure 16.1) and is a potential drug target (He et al. 2004; Sui et al. 2004; Walls et al. 2020).

The protein itself comprises two functional subunits: the distal S1 subunit, that harbours the Receptor-Binding Domain (RBD) and is responsible for stabilizing a prefusion state and the membrane spanning S2 subunit containing the fusion machinery (Walls et al. 2020; Zhou and Zhao 2020). Structural and functional analysis of the SARS-CoV-2 shows that the SARS-CoV-2 S protein binds the Angiotensin-converting enzyme 2 (ACE2) receptor on human alveolar epithelial cells (Zhou and Zhao 2020). This suggests that SARS-CoV-2 uses the same receptor, ACE2, as SARS-CoV (Zhou and Zhao 2020). However, the SARS-CoV-2 S protein appears to bind to ACE2 with a higher affinity than SARS-CoV S. The high affinity of the S protein for human ACE2 may be responsible for the greater human-to-human transmission of SARS-CoV-2 (Zhou and Zhao 2020). Due to the structure and role of the S protein in SARS-CoV-2 infection, it is the main target for antibody mediated neutralization (Zhou and Zhao 2020) and most of SARS-CoV-2 NAbs generated so far are directed against the Receptor Binding Domain (RBD) and impair ACE2 binding.

The potential of chicken IgY for COVID-19 treatment has been explored by several labs (Lu et al. 2020; Pérez de la Lastra et al. 2020; Somasundaram et al. 2020) (Table 16.1). Fu et al. used the SARS virus strain BJ01 as an antigen to develop chicken anti-SARS IgY antibodies and successfully demonstrated their ability to neutralize the virus in VERO E6 cells infection assays (Fu et al. 2006). IgY was stable after lyophilization and showed no significant loss of SARS binding activity in the pH range 2-7 thus supporting the claim of a robust antibody suitable for mass production (Fu et al. 2006). In 2007, Lee et al. developed chicken-derived scFv libraries against SARS-CoV and two scFv fragments with robust reactivity against SARS-CoV-infected Vero cells were selected and shown to target an equivalent region of the SARS-CoV spike protein, within the segment 750-1000 amino acids (Lee et al. 2007a, b). Later multiple recombinant versions of SARS-CoV proteins were used as antigens in individual chicken immunization experiments and resulted in specific and reactive IgY antibodies being produced. Five scFv clones were characterized with one specific clone showing high-binding to the segment 450-650 amino acids of the spike protein and additional strong binding to SARS-CoV-infected Vero E6 cells (Lee et al. 2007a, b). This pointed for the existence of a domain epitope within the region 450-650 amino acids. These

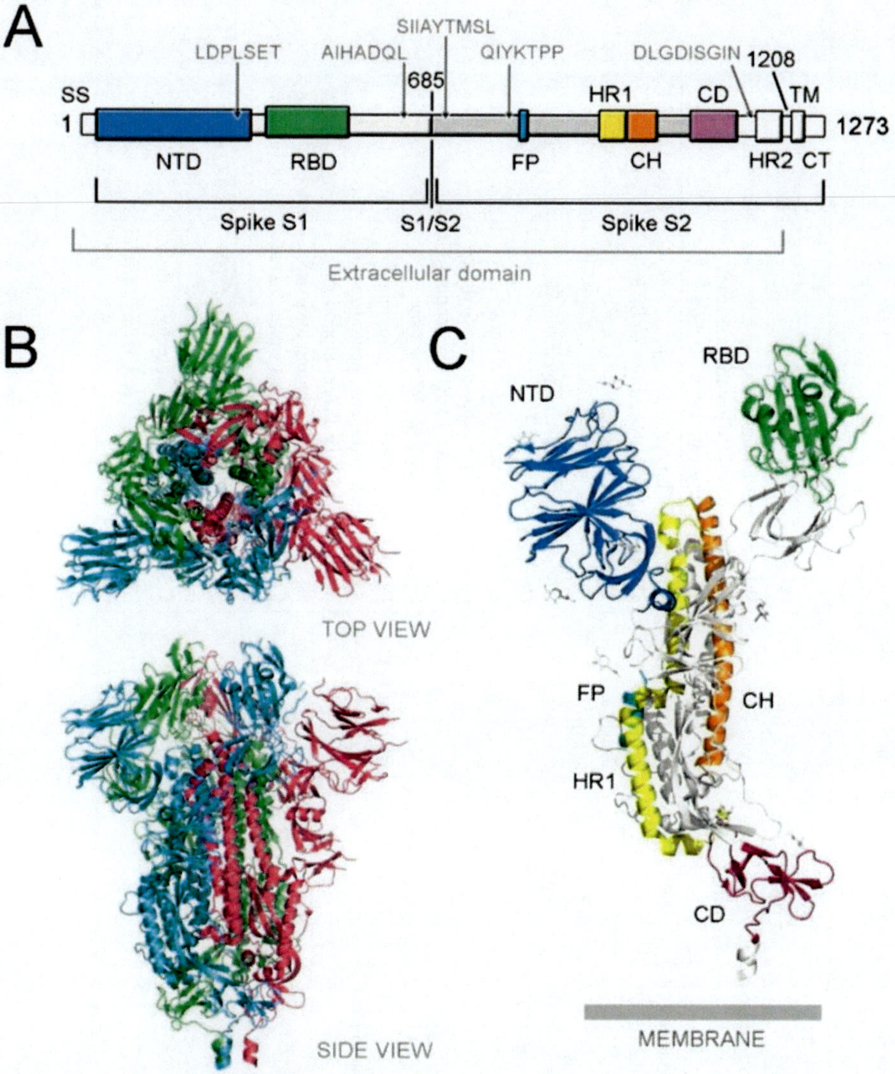

Fig. 16.1 Structural domain architecture of SARS-CoV-2. (**a**) Schematic organization of SARS-CoV-2 structural domains according to (Wrapp et al. 2020): NTD, N-terminal domain; RBD, receptor binding domain; FP, fusion peptide; HR1, heptad repeat 1; CH, central helix; CD, connector domain; HR2, heptad repeat 2; TM, transmembrane region; CT, C-terminal region; Spike S1, S2 and extracellular domain limits are shown; S1/S2, cleavage region. The sequences and locations of five IgY-binding epitopes mapped by microarray analysis by (Lu et al. 2020) are present by arrows. (**b**) Top and side view of the crystal structure of SARS-CoV-2 S trimer in ribbon representation (PDB: 6VXX)(Walls et al. 2020); monomers are presented in green/cyan/magenta. (**c**) Structural domains mapped on a single monomer with the same colour scheme as in A. Structure representations were generated using PYMOL (PYMOL molecular graphics system version 1.2r2; DeLano Scientific, LLC)

Table 16.1 IgY immunoprophylatic or immunotherapeutic potential in human respiratory infections and other human viral infections

Human Respiratory Infections					
Coronavirus					
Pathogen	**Immunogen**	**Model**	**IgY format**	**Description**	**Refs**
SARS-CoV	Whole inactivated SARS-CoV coronavirus	Vero E6 cells	Anti-SARS-CoV IgY	WB and neutralization assays show high reactivity, (titre–1:640);no significant loss of activity with lyophilization	(Fu et al. 2006)
SARS-CoV	Recombinant fragments of SARS-CoV[a] Spike (mix) *(E.coli)*	Vero E6 cells	Chicken scFv phage library & scFvs	Strong scFv binder to S region 456-650aa; strongly reactive in SARS-CoV-infected Vero E6 cells	(Lee et al. 2007a, b)
SARS-CoV-2	Recombinant SARS-CoV-2[2] Spike (S) protein *(BAC)*	*in vitro*	Chicken IgY	High immunoreactivity against SARS-CoV-2 S; epitope mapping reveals: 5 linear epitopes, 2 cross-reactive with SARS-CoV and 1 (SIIAYTMS) overlapping S1/S2 SARS-CoV-2 cleavage site	(Lu et al. 2020)
SARS-CoV-2	Recombinant SARS-CoV-2 Spike (S1) protein *(BAC)*	Hela cells	Chicken IgY	PsVNA using Hela cells expressing ACE2 reveals significant neutralization potency of αS1; ELISA competitions assays confirms αS1 binding to SARS-CoV-2 and SARS-CoV as well as to six SARS-CoV-2 S1 mutants	(Wei et al. 2020)
Avian Infectious Bronchitis Virus (IBV)	Whole inactivated M41 strain of IBV	Chick kidney cells	Ostrich IgY	Strong neutralizing activity against IBV infection in cultured primary chick kidney cells; lower IBV infectivity in chicks IM-treated with IgY prior to viral challenge	(Tsukamoto et al. 2018)

Influenza Virus

Pathogen	Immunogen	Model	IgY format	Description	Refs
Avian Influenza A (H5N1)	HA mix human influenza[4] or recombinantH5 protein	MDCK cells	Ostrich IgY	Strong inhibited cytopathic effects in MDCK cells; strong neutralization activity against H5N1; recombinant H5 suitable for IgY scalability	(Adachi et al. 2008)
Avian Influenza A (HPAIV)	H5N1 virus vaccines for poultry (Vietnam)	Mice	H5N1-specific IgY From supermarket eggs (Vietnam)	IgY cross-protects against infections with HPAIV H5N1 and related H5N2 strains in mice; intranasal route prevents and reduces infection	(Nguyen et al. 2010, 2011a, b)
Influenza A (IBA)[3]	Recombinant HA protein (H5)	MDCK cells	Chimeric chicken/ human Fab	Selection of highly specific Fab to H5N1; inhibition of H5N1 virus infection in MDCK cells	(Pitaksajjakul et al. 2010)
Influenza A A/H1N1 2009	Seasonal influenza vaccine 2009 (Japan)	MDCK cells; Erythrocytes	Ostrich IgY	Strong cross-reactivity to pandemic A/H1N1, seasonal A/H1N1, A/H3N2 and B viruses; inhibition of infection in MDCK cells; impair erythrocytes hemaggregation	(Adachi et al. 2011)
Influenza A/H1N1 2009	Swine influenza virus A/H1N1	MDCK cells; Erythrocytes	Ostrich IgY	ELISA and ICC reveal strong cross-reactivity; strong inhibited cytopathic effects in MDCK cells; impair erythrocytes hemaggregation	(M. Tsukamoto et al. 2011)
Influenza A (H1N1[5], H3N2[6], H5N1[7])	Whole inactivated viruses	Mice	Chicken IgY	H1N1, H3N2, and H5N1 cross-protection studies; H5N1 cross-protected against H1N1 (A/Puerto Rico/8/34); intranasal administration of anti-H5N1 protected 100% of mice in lethal challenges;	(Wallach et al. 2011)

(continued)

Table 16.1 (continued)

Human Respiratory Infections					
Influenza B (IBV)	Whole inactivated IBV	MDCK cells	Chicken IgY	IBV-specific IgY inhibited infection in MDCK cells; reduced IBV replication in mice lungs	(Wen et al. 2012)
Hantavirus					
Pathogen	**Immunogen**	**Model**	**IgY format**	**Description**	**Refs**
Andes virus (ANDV)	ANDV DNA vaccine encoding viral GP	Hamster	Duck IgY	αANDV IgY/IgYΔFc neutralized ANDV in vitro; hamster model of lethal HPS reveals protective effect of IgY/IgYΔFc up to 8 d or 5 d after IN or IM challenge respectively; HPS human survivor plasma also shows protection	(Brocato et al. 2012)
Andes virus (ANDV)	ANDV DNA vaccine encoding viral GP	Hamster	Goose IgY	PsVNA shows high titre (–1:100,000) of αANDV IgY and long term titre (1y); epitope mapping reveals target sequences of GP; hamster model of lethal HPS reveals protective effect of IgY/IgYΔFc	(Haese et al. 2015)
Pseudomonas aeruginosa					
Pathogen	**Immunogen**	**Model**	**IgY format**	**Description**	**Refs**
P. aeruginosa	*P.a.* whole cells	Human epithelial cells	Anti-*P.a.* IgY aqueous solution	Decreased adhesion of *P.a.* to human epithelial cells; but no bacterial growth inhibition	(D Carlander et al. 1999)
P. aeruginosa	*P.a.* whole cells of 6 strains	*in vitro*	Anti-*P.a.* IgY	MALDI-TOF-MS analysis reveals *P.a.* flagellin as major IgY antigen	(Nilsson et al. 2007a, b)
P. aeruginosa	*P.a.* whole cells	Human cystic fibrosis patients	Anti-*P.a.* IgY aqueous solution	Long-term (–10 to 12 y) treatment diminishes the number of positive P.a. cultures; postpones the onset of colonization; no adverse effects reported	(Kollberg et al. 2003; Nilsson et al. 2007a, b, 2008)

P. aeruginosa	*P.a.* whole cells	PMN cells	Anti-*P.a.* IgY	Bacterial immobilization by aggregation, enhanced bacterial killing by PMN-mediated phagocytosis	(Thomsen et al. 2015)
P. aeruginosa	*P.a.* whole cells	Mice (SPF BALB/c)	Anti-*P.a.* IgY	Significant reduction of bacterial burden and symptoms of lung inflammation	(Thomsen et al. 2016)
Other Human Viral Infections					
Rabies Virus					
Pathogen	**Immunogen**	**Model**	**IgY format**	**Description**	**Refs**
Rabies virus (RABV)	Recombinant G protein fragment (G-F2, 201-300aa) of rabies virus *(E.coli)*	Mice (BALB/c)	Chicken IgY	αG-F2 IgY showed effective virion binding in vitro; reduce mortality of challenged mice	(Motoi et al. 2005a, b)
Zika Virus					
Pathogen	**Immunogen**	**Model**	**IgY format**	**Description**	**Refs**
Zika virus	Whole inactivated ZIKV PRVABC59	Mice; K562 human cells	Goose IgY	IgY is able to neutralize the virus in vitro and in IFNAR−/− mice. ZIKV-specific IgY does not induce ADE; protection by IP injection of IgY to mice 24 h following a lethal ZIKV infection.	(O'Donnell et al. 2019)
Zika virus	ZIKV envelope (E) protein	In vitro	Chicken scFv phage library & scFvs	Selection of 4 ZIKV scFv with specific binding to E protein; diagnostic and therapeutic potential	(Mwale et al. 2020)
Dengue Virus					
Pathogen	**Immunogen**	**Model**	**IgY format**	**Description**	**Refs**
Dengue Virus	Recombinant Non-Structural-1 (NS1)	In vitro	Goose IgY	Anti-NS1 IgY revealed equivalent in vitro neutralization capacity as DENV2 IgY; anti-NS3 IgY did not neutralize DENV	(O'Donnell et al. 2017)

(continued)

Table 16.1 (continued)

Human Respiratory Infections					
Dengue Virus (DENV2)	Inactivated DENV2 Dengue Type 2 Antigen	Mice (AG129); U937 DC-SIGN cells	Goose IgY	Anti-DENV2 IgY neutralized DENV2 with no ADE observed in vitro; protection by IP injection of anti-DENV2 IgY to mice 24 hours following a lethal DENV2 infection; identification of DENV2 unique epitopes recognized by IgY	(Fink et al. 2017; O'Donnell et al. 2017, 2018)
Ebola Virus					
Pathogen	**Immunogen**	**Model**	**IgY format**	**Description**	**Refs**
Ebola virus (EBOV)	EBOV immunogens multiple platforms[8]	Mice (BALB/c)	Chicken IgY	αEBOV IgY shows high thermostability; VSVΔG/EBOVGP induces high αEBOV IgY titres; αEBOV show protective effect in lethal dose challenge assay in mice, with 2 h or 24 h p.i. dosage	(Zhang et al. 2020)
Norovirus					
Pathogen	**Immunogen**	**Model**	**IgY format**	**Description**	**Refs**
Norovirus (NoV)	NoV P particles	In vitro	Chicken IgY	αNoV P IgY strongly reacted with P in ELISA, WB and blocked NoV VLP and P binding to HBGA receptors (BT50 1:800); IgY stability assays showed activity retained after 70 °C and pH (4-9) treatments	(Dai et al. 2012)

[a]**SARS-CoV**, GenBank Accession NC_004718; ***E.coli***, *Escherichia coli* expression system. [2]**SARS-CoV-2,** GenBank No. YP_009724390; **BAC**, baculovirus-insect cell expression system; **ACE2,** Angiotensin-converting enzyme 2. **HA**, Hemagglutinin protein; [4]**HA mix**, mixture of HA antigens from vaccine strains of the human influenza virus ([5]**H1N1**/NewCaledonia/20/99, [6]**H3N2**/Hiroshima/ 52/2005 and B/Malaysia); [7]**H5N1**/A/Vietnam/1194/04). **MDCK**, Madin–Darby canine kidney cells; **Influenza A Virus** (A/Vietnam/1203/04); **GP**, viral glycoprotein. **HPS**, Hantavirus pulmonary syndrome. **IN,** intranasal; **IM**, intramuscular; **PsVNA**, pseudovirion neutralization assay; **P.a.,** *Pseudomonas aeruginosa*; **MALDI-TOF-MS**, Matrix-Assisted Laser Desorption/Ionisation Time-Of-Flight Mass Spectrometry; **PMN**, polymorphonuclear neutrophils; **SPF,** Specific Pathogen-Free; **ZIKV**, Zika virus; **ADE**, antibody-dependent enhancement; **IP**, intraperitoneally; [8]**EBOV immunogens**, DNA vaccine (pCAGGS/EBOVGP), recombinant protein (rEBOVGP) or virus-like particle (EBOV-117 VLP) subunit vaccines, and two viral vector vaccines (VSVΔG/EBOVGP, Ad5/EBOVGP). [3] **p.i.**, post-infection. **HBGA**, histo-blood group antigen

observations support the robustness of the avian immune repertoire for screening of novel SARS-CoV binders as well as epitope mapping studies.

Lu et al. have used the full-length extracellular domain of SARS-CoV-2 S protein (1-1213 amino acids) and Wei et al. used the SARS-CoV-2 Spike S1 domain in baculovirus-insect cell (BAC) expression system (Lu et al. 2020; Wei et al. 2020). The BAC system enables the expression of a native-like protein fold potentially enabling the development of structure-specific antibodies, critical for functional targeting of the spike protein. Lu et al. generated anti-Spike IgY antibodies with high reactivity to the SARS-CoV-2 S protein, and identified 5 particular epitopes by peptide microarray analysis, that were notably different from other epitopes previously reported, using convalescent sera of COVID-19 patients (Barnes et al. 2020; Brouwer et al. 2020; Chi et al. 2020; Poh et al. 2020): LDPLSET (C-terminus of N-terminal domain), SIIAYTMSL (S1/S2 cleavage region), QIYKTPP (between S1/S2 cleavage region and S2′ cleavage region), AIHADQL (within subdomain 2), DLGDISGIN (between connector domain and heptad repeat 2) (Lu et al. 2020) (Fig. 16.1). Cross reactivity with at least 2 of these epitopes was demonstrated also in a parallel SARS-CoV microarray. Notably, one of the selected conserved epitopes (SIIAYTMSL), overlaps with the S1/S2 cleavage region, whose proteolytic processing is a determinant in pre-membrane fusion stages of the virus. This suggests that the antibody could functionally block the enzyme's proteolytic activity thus halting viral invasion.

The study of Wei et al. using only the SARS-CoV-2 Spike S1 domain as immunogen, also reports the development of IgY antibodies capable of binding both SARS-CoV-2 and SARS-CoV. IgY neutralization assays were performed with lentiviral-derived pseudovirus harbouring S proteins on the envelope and a luciferase reporter gene. Quantifying the expression levels of luciferase reporter gene on ACE2-expressing Hela cells upon pseudovirus infection, demonstrated that anti-SARS-CoV-2 S1 IgY antibodies were effectively neutralizing the pseudovirus. Moreover, ELISA competition assays using serial dilutions of anti-S1 IgY competing with an Fc-tagged version of ACE2 for binding different S1 domain mutants, showed robust competition for all tested targets (WT SARS-CoV2, SARS-CoV, 6 SARSCoV-2 S1 mutants) (Wei et al. 2020).

16.2.2 Influenza

Influenza viruses are members of the family *Orthomyxoviridae*, a group of enveloped viruses containing a segmented negative-sense single-stranded RNA genome. There are three main types of influenza viruses: A, B, and C. Influenza A virus strains infect both humans and animals (e.g., ducks, chicken, pigs, horses, cats) and evolve rapidly giving rise to potential deadly variants responsible for major outbreaks and pandemics. The 1918 influenza A (H1N1) pandemic (Spanish flu) was the deadliest pandemic in recorded history, infecting an estimated 500 million people with at least 50 million deaths worldwide (CDC, Centre for Disease Control and Prevention; https://www.cdc.gov/flu/pandemic-resources). This was followed by other major influenza A pandemics, including A/H2N2 (Asian Flu, 1957-8), A/H3N2 (Hong Kong Flu, 1968) and A/H1N1 (Swine Flu, 2009). Interestingly, A/H3N2 (1968) continues to circulate worldwide, undergoing regular antigenic drifts and has become the leading cause of seasonal influenza illness and death over the last 50 years (Jester et al. 2020). Healthy patients can often deal with the viral infections and recover fully in few days, but complications in certain high-risk groups may lead to pneumonia and ultimately death.

The Influenza viral particle has its genome packed in a lipoprotein envelope lined from the inside with a matrix protein and with haemagglutinin (HA), neuraminidase (NA), and M2 proteins exposed on the outer surface. While HA mediates the attachment of the viral particle to the host cell receptors and triggers membrane fusion and viral penetration (Wu and Wilson 2020), NA is an enzyme that cleaves sialic and neuraminic acid linkages and is determinant during viral particle budding from the host cell and for viral motility (McAuley et al. 2019). HA and NA are therefore central virulent determinants and have been the major targets for vaccine and antibody development to fight Influenza.

Influenza vaccines are not 100% effective but have critically contributed in the last decades to control in particular seasonal epidemics and reduce mortality rates. Different strategies across the years have been used for vaccine optimization (Sautto et al. 2018), nevertheless alternative prophylactic and therapeutic approaches are constantly needed to complement the existing arsenal to fight influenza infections.

IgY antibodies against both Influenza A and B have been developed and characterized. Studies on avian influenza A (A/H5N1) preceded works in human variants (Table 16.1). Adachi et al. immunized ostriches with two sets of antigens: (i) H5 recombinant proteins from avian A/H5N1 and (ii) a mixture of HA antigens from vaccine strains of human influenza. Anti-H5 IgY showed strong inhibitory activities on H5N1, reducing cytopathic effects in MDCK cells and preventing the death of embryonated chicks after a viral inoculation. In contrast, anti-HA IgY generated from the human vaccine HA antigens presented weak immunoreactivity in MDCK cells assays (Adachi et al. 2008). Curiously, studies with H5N1-specific IgY recovered from commercial eggs of chickens subjected to mandatory vaccination programs against A/H5N1 in Vietnam, revealed protective effects against both A/H5N1 and A/H5N2 in a mouse lethal challenge model (Nguyen et al. 2010; 2011a, b). Pitaksajjakul et al. further refined the H5N1 IgY targeting approach by developing chimeric chicken/human Fab fragments specific against H5N1 and further demonstrating their inhibitory potential in MDCK cell infection model (Pitaksajjakul et al. 2010).

Several studies show successful and consistent results of IgY antibodies targeting human influenza variants, namely A/H1N1 but also A/H3N2. Adachi et al. generated antibodies in ostrich using a mixture of HA antigens of vaccine strains of seasonal influenza virus (Table 16.1) and reported strong cross reactivity to pandemic A/H1N1, seasonal A/H1N1, A/H3N2 and B viral strains. Also hemaggregation activities of all these strains was strongly inhibited by the ostrich IgY and neutralization assays in MDCK cell for pandemic A/H1N1 revealed effective viral infection inhibition (Adachi et al. 2011). Similar results were obtained by Tsukamoto et al., using a swine influenza A/H1N1 strain to produce ostrich IgY with strong cross reactivity with pandemic A/H1N1 showing robust inhibition of cytopathic effects by both pandemic A/H1N1 as well as swine influenza; observations further supported by hemaggregation assays (Tsukamoto et al. 2011). Wallach et al. used whole inactivated viral strains of A/H1N1, A/H3N2 and A/H5N1 to produce different chicken IgY antibodies and reported an effective cross-protection namely between A/H5N1 and A/H1N1 both in vitro and in mouse lethal challenge model (Wallach et al. 2011). In addition to influenza A strains, studies on influenza B have also been reported (Wen et al. 2012).

To complement this section, the reader is directed to Chap. 17 where we present examples of IgY-based products designed against influenza infections as well as further details on the real word company leading their development.

16.2.3 Hantavirus Pulmonary Syndrome

Hantavirus pulmonary syndrome (HPS) or Hantavirus cardiopulmonary syndrome (HCPS) and also haemorrhagic fever with renal syndrome (HFRS) are zoonotic diseases caused by Hantaviruses and are normally acquired by inhalation of aerosols contaminated with viral-containing rodents' faeces. Hantaviruses are enveloped virus with a negative sense, tri-segmented RNA genome belonging to the family *Bunyaviridae.* The virus infects platelets, endothelial cells and macrophages and leads to exacerbated inflammation, vascular permeability and accumulation of fluid in lungs and other body cavities causing respiratory and circulatory deficits and ultimate failure (Akram et al. 2020; Dheerasekara et al. 2020). There are currently no effective vaccines, prophylactics, or therapeutics to prevent or treat this highly pathogenic disease.

A work by Brocato et al. showed the successful generation of duck IgY antibodies and their use as a post exposure prophylactic treatment against Andes virus (ANDV), a species in the genus Hantavirus (Table 16.1). The study reports that both αANDV IgY as well as the truncated version IgYΔFc were capable of neutralizing ANDV in in vitro assays. Additionally, a hamster model of lethal HPS revealed the robust protective effect of both types of antibodies if administered as late as 8 days or 5 days after intranasal or intramuscular viral exposure, respectively (Brocato et al. 2012). Consistently, another work by Haese et al. also developed IgY antibodies using a DNA vaccine approach (as in the previous study) to elicit the generation of αANDV IgY antibodies in goose hosts (Haese et al. 2015). Interestingly, the work reports a long term monitoring of the αANDV IgY titres in goose and reveals the maintenance of very high titres for one year even without further immunization. Booster vaccine dosages resulted in even higher titres in the same birds (–1:100,000). Epitope mapping using IgY revealed reactivity against 5 epitopes in regions of ANDV glycoproteins Gn and Gc, different from those previously identified with serum of human ANDV HPS patients or rodent serum (Tischler et al. 2005). The hamster model of lethal HPS was used once again and similarly to the previous study a protective effect of both goose αANDV IgY and IgYΔFc was confirmed.

16.2.4 *Pseudomonas aeruginosa*

Pseudomonas aeruginosa is a gram-negative opportunistic bacterium and the primary cause of chronic respiratory infections in patients with cystic fibrosis (CF), bronchiectasis or chronic obstructive pulmonary disease (COPD), and also of acute infections in immunocompromised individuals (Sainz-Mejías et al. 2020). *P. aeruginosa* infections are responsible for a substantial economic burden for hospitals (Kaier et al. 2019), but notably give the wide arsenal of virulence factors that contribute to its complex pathogenicity mechanisms, there is no licensed vaccine to date, despite more than 50 year of research in the field.

A detailed case study on research developed on the immunotherapeutic use IgY antibodies against *P. aeruginosa* in cystic fibrosis patients is presented in Chap. 17. The reader is invited to review the case study of the company Immunsystem and the research, development and human clinical trials in cystic fibrosis patients promoted by the research teams associated with the company. Further references on *P. aeruginosa* IgY research are presented in Table 16.1.

16.3 Other Human Viral Infections

16.3.1 Rabies Virus

Rabies virus (RABV) is a negative-sense RNA enveloped virus and one of the 16 described viruses in the *Lyssavirus* genus belonging to the Rhabdoviridae family. Rabies is generally transmitted through saliva during bites of infected hosts, such as dogs or bats, and is still responsible for an estimated 55,000 human deaths per year. Infection symptoms include fever, headache, malaise, and upper respiratory and gastrointestinal tract disorders. Severe neurological symptoms can also be triggered, including encephalitis or paralysis. Rabies vaccine is rather safe and effective but requires multiple doses making it costly and thus less accessible, especially for developing countries (Ertl 2019). Vaccines are often given only after being bitten by an infected host, and post-exposure prophylaxis (PEP) typically include simultaneous administration of rabies immunoglobulin (RIG) preparation of either human (HRIG) or equine origin (ERIG), in the site of the wound.

RIG preparations are also costly to produce and IgY-based alternatives have been explored and supported by a number of studies (Inoue et al. 2015). Motoi et al. for example, developed IgY antibodies against a recombinant fragment (G-F2, 201-300 amino acids) of the surface-exposed rabies glycoprotein spike, critical for viral infectivity. The work showed effective virion binding of anti-G-F2 IgY and validated its protective effect in a BALB/c mice model, where high survival rates were observed upon viral challenge. The results support therefore the use of anti-G-F2 IgY as alternative to current RIGs. In other work, the same team, has focused on targeting rabies nucleoprotein (N) and phosphoprotein (P) for immunodiagnostic purposes. They successfully generated IgY against recombinant versions of both proteins (rN and rP) and reported IgY effective binding to CVS-11 strain of rabies virus by Western blotting, immunofluorescence assay and immunohistochemistry (Motoi et al. 2005a, b). In case of rabies N and P proteins, their reduced accessibility does limit translation towards an immunotherapeutic strategy.

16.3.2 Emerging Viral Threats

A number of other relevant viral threats affecting humans are the so called emerging infectious diseases, with a significant impact worldwide and which have been or have the potential to be responsible for deadly epidemic or pandemic outbreaks. This include diseases caused by Zika virus, Dengue virus, Ebola virus or even Norovirus, for which IgY antibodies have been successfully characterized (Table 16.1).

Dengue virus (DENG) is a mosquito-borne, single positive-stranded enveloped RNA virus of the family Flaviviridae. It is responsible for Dengue haemorrhagic fever (DHF) and dengue shock syndrome (DSS), which are severe disease manifestations with no licensed therapeutic currently available (Deng et al. 2020). Successful studies on Dengue virus neutralization with goose IgY antibodies have been reported using mice models (Fink et al. 2017) (Table 16.1); however to date there are no reports on IgY treatments of human patients. Zika virus (ZIKV) also belongs to the family Flaviviridae, and is responsible for mild fever, rash and muscle pain. In 2015, a Zika outbreak in Brazil coincided with a rise of birth defects, turning Zika into a growing public health concern (Heymann et al. 2016; Wen et al. 2017). Works on both chicken and goose IgY antibodies developed against Zika virus have reported the capacity for effective viral reactivity and neutralization in vivo and in vitro (Table 16.1). (Mwale et al. 2020; O'Donnell et al. 2019). IgY against DENG and ZIKV have the potential to integrate preventive and vector control interventions, critical to reduce and control the transmission of these two viral threats.

Ebola virus (EBOV) is an enveloped, single-stranded, negative sense RNA virus belonging to the Filoviridae family, and one of the deadliest viruses known to man. EBOV infections also result in severe haemorrhagic fevers and effective viral blocking can be life-saving. IgY anti-EBOV was shown to protect mice in lethal dose challenge assays (Table 16.1) (Y. Zhang et al. 2020). Norovirus are non-enveloped, single-stranded, positive sense RNA viruses, belonging to the Caliciviridae family and infect different mammalian species (Atmar 2010). In humans, norovirus is a major cause of acute gastroenteritis and studies with IgY antibodies for viral neutralization purposes have also been performed (Table 16.1) (Dai et al. 2012).

16.4 *Helicobacter pylori* and Gastric Pathogenesis

Helicobacter pylori is a human pathogen infecting over 50% of the population worldwide and with infection levels reaching over 70% in developing countries (Bravo et al. 2018). It has been described as the main aetiologic agent of gastric pathogenesis and cancer. *H. pylori* establishes a persistent infection that begins asymptomatically, but can develop to peptic ulcer, chronic gastritis, lymphoma of the gastric mucosa and eventually adenocarcinoma, thus causing a considerable rate of mortality and morbidity worldwide (De Brito et al. 2019). Antibiotic treatment can effectively eradicate *H. pylori* infections, yet in 10-20% of the cases multidrug resistance dictates the persistence of the infection and thus exploring alternative therapies is urgently needed.

One such alternative is passive immunization by oral administration of IgY developed against *H. pylori* (IgY-*Hp*). To date a number of studies has focused on the development IgY-*Hp* both using *H. pylori* whole-cell lysate (*Hp*-lysate) as antigen or rather specific virulence factors in recombinant formats and evaluated their antibacterial effects both in vitro and in vivo. Indeed, IgY antibodies produced against whole-cell lysates of different *H. pylori* strains showed high cross-strain inhibitory effects and this could potential translate into a multi-strain IgY therapy (Solhi et al. 2017). On the other hand refinement of virulence factor-specific IgY approaches has also revealed very encouraging results, with gains in targeting specificity and antibacterial efficiency and unveiled potential new *H. pylori* antigens, supporting the design a multivalent IgY (Sun et al. 2018) and overall development of novel *H. pylori* vaccines (Sutton and Boag 2019).

Prominent in vivo effects of anti-*H. pylori* IgY have been extensively evaluated in human volunteers (Horie et al. 2004; Shimamoto et al. 2002; Suzuki et al. 2004) and rodent models (e.g., mice, Mongolian gerbils) (Attallah et al. 2009; Malekshahi et al. 2011; Nomura et al. 2005; Wang et al. 2014; Yang et al. 2012) and most often supported by parallel in vitro studies for evaluation of the impact of IgY on host tissues and cells or on bacterial growth and virulence factors' activities. Shimamoto et al. showed that anti-*H. pylori* IgY effectively inhibited *H. pylori* growth and promoted increased bacterial agglutination in vitro. Moreover, the work demonstrated that anti-*H. pylori* IgY strongly inhibited *H. pylori*-associated urease activity and ammonia production as well as the cytopathic effect of this bacteria on cultured cells (Shimamoto et al. 2002).

Extensive studies have contributed so far to identify key *H. pylori* virulence factors, including the bacterial oncoprotein CagA (Tohidpour 2016), vacuolating cytotoxin A (VacA) (Foegeding et al. 2016), outer inflammatory protein A (OipA) (Yamaoka et al. 2002), neutrophil-activating protein A (NapA) (Satin et al. 2000), adhesin A (HpaA) (Mao et al. 2003), major flagellin FlaA, and minor flagellin FlaB (Pulić et al. 2014) and urease (Dunn and Phadnis 1998), as well as to elucidate their mechanisms of pathogenesis (Waskito et al. 2018; Xu et al. 2020; Yamaoka 2010). A study by Shin et al. used liquid chromatography–mass spectrometry (LC/MS) analysis to identify immunodominant proteins of *H. pylori*, reacting efficiently with IgY-*Hp* (against whole-cell lysate antigen) (Shin et al. 2003). Five immunodominant proteins were identified: urease α-subunit (26 kDa), urease β-subunit (62 kDa), heat-shock protein 60 (60 kDa), putative peroxiredoxin (22 kDa) and putative thiol peroxidase (18 kDa).

H. pylori urease enzyme (urea amidohydrolase) is considered a main factor for the colonization of host gastroduodenal mucosa and urease activity is essential to survival and pathogenesis of the bacterium (Dunn and Phadnis 1998). This multi-subunit enzyme is composed of two subunits, α (26–31 kDa) and β (61–66 kDa), assembling a dodecameric complex of repeating α-β subunits (Ha et al. 2001). The enzyme is produced in large amounts by the bacterium (6–10% of total protein) (Dunn and Phadnis 1998); is released by genetically programmed autolysis and adsorbs onto the surface of neighbouring intact bacteria. Once at the surface, urease catalyses the hydrolysis of urea to ammonia and carbamate, leading to a local increase in pH ensuring a neutral microenvironment surrounding *H. pylori.* Interfering with *H. pylori* urease activity has been successfully explored as therapeutic approach to reduce gastric mucosa colonization. In this line, also IgY antibodies targeting urease have been the subject of multiple works supporting an IgY-based control of *H. pylori* infections.

Horie et al. tested the efficacy of a commercial yogurt containing *Lactobacillus acidophilus* and *Bifidobacterium spp.* supplemented with 1% egg yolk IgY against urease to suppress *H. pylori* infections in humans (Horie et al. 2004). A clinical study conducted with forty-two (42) volunteers testing positive for *H. pylori* by 13C-urea breath test (UBT), demonstrated that regular consumption of 150 mL of yogurt, 3 times per day, for 4 weeks, considerably reduced urea breath test values. The work also showed that anti-urease IgY is stable up to 65 °C and declines only to 85% during a 3-weeksstorage period; both features thus support standard food industrialization procedures. Indeed, the development of functional foods and beverages supplemented with anti-urease IgY is becoming rather popular; for example, another study by Aiba et al. using mice models elucidated that the synergistic activity resulting from combined administration of *Lactobacillus johnsonii* No.1088 (LJ88) and IgY antibodies developed against *H. pylori* urease was more effective then each monotherapy (Aiba et al. 2019). Indeed, daily administration of living or heat-killed LJ88 promoted *H. pylori* co-aggregation and elimination from the gastric tract, and this activity was shown to be further enhanced when combined with anti-urease IgY, resulting in significant reduction of *H. pylori* infections in both germ-free mice infection models and human gut microbiota associated mice model.

Refinement of target specificity and efficacy against *H. pylori* urease has also been studied and achieved by peptide mapping. This is a prominent approach for revealing critical regions, also termed epitopes, typically at the surface of target antigens and is routinely used in peptide-based vaccine development, without exception for *H. pylori* vaccines (Takeshita et al. 2019). Shin et al. studied five synthetic peptides of *H. pylori* urease subunit β (UreB, more recently relabelled to UreC) designed by hydropathy plot-based analysis and identified a 15 amino acid peptide (UB-3; residues 396-410, DNDNFRIKRYLSKYT) showing high reactivity against whole cell IgY-Hp antibody (Shin et al. 2004). They further demonstrated that UB-3 peptide could be used successfully to immunize hens and to develop highly specific anti-urease IgY, with an effective inhibitory capacity against *H. pylori* urease activity *in vitro.* Consistently a study by Takeshita et al. also demonstrated that a peptide-based vaccine using another urease synthetic peptide could be effectively used to immunize BALB/c mice, inducing urease-neutralizing antibodies more robustly than those obtained by immunization with purified urease (Takeshita et al. 2019).

Antibodies against other *H. pylori* antigens such as CagA, VacA, NapA, HpaA, FlaA, and FlaB, have been detected in the sera of infected patients, reflecting their relevant immunogenic properties and making them interesting targets for IgY-based studies (Yan et al. 2005). Borhani et al. for example, showed that IgY antibodies developed against recombinant NapA promoted inhibitory effects on *H. pylori* attachment to AGS cell line (Borhani et al. 2016). Hong et al. also demonstrated that oral administration of anti-VacA IgY, produced against a mix of two recombinant Vacuolating cytotoxin A (VacA) forms, is correlated with a protective effect against *H. pylori* colonization and induced histological changes in gastric tissues of infected Female C57BL/6 mice (Hong et al. 2018). Sun et al. on the other hand developed a multivalent IgY antibody by immunizing hens with a cocktail of *H. pylori* virulence factors in recombinant format, namely antigen-binding adhesin A2 (BabA2), subunit β of urease (UreB) and minor flagellin (FlaB) (Sun et al. 2018). A bacteriostasis assay showed that the multivalent IgY had a significant inhibitory effect on the growth of *H. pylori* at 5 mg/mL, and this could be further combined with amoxicillin with improved growth reduction outputs.

Indeed, while a number of *H. pylori* virulence factors has been successfully targeted by IgY, it is definitely exciting to realize that these are only a few examples out of the dozens of possible targets involved in *H. pylori* mechanisms of pH homeostasis, host-cell adherence, motility and pathogenicity (Salillas and Sancho 2020), whose functions can be modulated using specific IgYs. Moreover, synergistic effects of IgY-Hp with either antibiotics (Sun et al. 2018) or antiacids (Suzuki et al. 2004; Yang et al. 2012) has also been successfully demonstrated supporting the use of combined IgY-drug therapy to control *H. pylori* infections, whose eradication currently relies on standard triple therapy (STT), consisting of a proton pump inhibitor (e.g. pantoprazol) and two antibiotics (clarithromycin and amoxicillin) (Gisbert et al. 2000; Molina-Infante and Gisbert 2014). Importantly synergistic effects of IgY-Hp with standard antibiotics may contribute to sustainable strategies of antibiotic dose reduction and thus are likely to be further explored in the future.

16.5 Mutans Streptococci and Dental Caries

16.5.1 Introduction

Oral diseases affect about 3.5 billion people worldwide and represent a major health burden in many countries. These are non-communicable oral health conditions including: dental caries (tooth decay), periodontal diseases, oral/pharyngeal cancers, oral lesions of HIV/AIDS origin, oro-dental trauma, cleft lip and palate, noma (severe orofacial gangrene) and Acute Necrotizing Ulcerative Gingivitis (ANUG) and affect people throughout their lifespan, leading to pain, discomfort, disfigurement and even death. Some are largely preventable and can be treated at early stages. However treatments are generally expensive, reaching e.g., 5% of total health expenditure in high-income countries and additionally are not part of universal health coverage (Dye 2017). Therefore, low cost and accessible solutions to prevent and treat oral health diseases are urgently needed. Dental caries is the major oral health disease affecting all age groups from paediatric to geriatric populations. According to WHO, it is estimated that 2.3 billion people suffer from caries of permanent teeth and more than 530 million children suffer from caries of primary teeth.

Understanding the different stages of dental plaque formation is critical to define preventive or therapeutic strategies. In the first stage of plaque formation, large hydrophobic molecules such as salivary glycoproteins disperse onto the surface of teeth and form a pellicle (within minutes) favouring bacterial adhesion. Subsequently, a bacteria seeding stage takes place, with attachment of initial colonizers (e.g., *Streptococcus mutans, S. mitis, S. godonii, S. sanguinis* and *Actinomyces*) in a reversible phase, followed by an irreversible phase mediated by extracellular polymers, fimbriae, and adhesins. Bacterial proliferation of adherent populations as well as the recruitment of secondary colonizers (e.g. Gram-negative cocci and rods, filamentous organisms, fusobacteria and spirochaetes) takes place next. Over time, deposition of extracellular material leads to an increase of biofilm complexity and thickness, until an equilibrium state is reached (Laurence 2009).

In 1924 Killian Clarke described deep dentinal caries lesions in *in vitro* teeth experiments using a streptococcus strain with unique acid production characteristics, that he named *Streptococcus mutans* (Clarke 1924). *Streptococcus mutans* is a facultatively anaerobic, gram-positive coccus. Its cariogenic potential results essentially from three characteristics: (1) highly effective synthesis of extracellular polymers of glucan from sucrose favouring local matrix formation and thus microbial colonization; (2) transport and processing of carbohydrates into organic acids (acidogenicity) and (3) high resistance to low pH (aciduricity) allowing in to thrive in such conditions (María Alejandra and Mariano Daniel 2020). In addition, the unique environment conditions created favours other acidogenic and aciduric species also to thrive and enhance cariogenic activities (Lemos et al. 2019).

Measures like the broad use of fluorides in oral products, or other chemoprophylactic agents such as chlorhexidine (CHX) (Bowden 1996; Emilson and Fornell 1976; Svante Twetman 1999) or Triclosan (Riley and Lamont 2013) are commonly administered to prevent or reduce *Streptococci mutans* colonization.

16.5.2 Vaccination Against Dental Caries

The concept of vaccination against *Streptococcus mutans* was introduced by Paul Keyes in the 60s (Keyes 1968) and his studies together with and other contemporary works supported the hypothesis of active immunization against dental caries pathogens (Fitzgerald and Keyes 1962; Wagner 1966). The reader is directed to a recent review on the development of vaccines which is outside the scope of this chapter (Patel 2020). More recently, experiments in a variety of animal models comprising rodents and primates have demonstrated the induction of salivary S-IgA and circulating IgG antibodies to mutans streptococcal antigens by oral or intranasal immunization with AgI/II, glucosyltransferase (GTF) or glucan-binding proteins (Childers et al. 2002; Koga et al. 2002; Russell et al. 1999, 2004; Smith 2002). Upon subsequent oral challenge with virulent *Streptococci mutans* and the institution of a high-sucrose diet, reductions in colonization and diminished development of dental caries lesions was shown.

16.5.3 Passive Immunization

Passive immunization against *Streptococcus mutans* using IgY antibodies (Table 16.3) has been extensively reported. However, herein we start by summarize results with other sources of preformed antibodies, namely murine monoclonal antibodies and bovine whey-derived antibodies that are relevant to consider (Table 16.2). Loimarantha et al. (Loimaranta et al. 1999) showed that specific bovine colostrum whey-derived antibodies could be produced against *Streptococcus mutans* and *Streptococcus sobrinus* and had significant inhibitory effects over *in vitro* glucan synthesis by *S. mutans* and supported the activation of human phagocytes against such pathogens. Mouth rinsing with this bovine immune whey by human subjects also decreased the relative number of *Streptococcus mutans*. Consistently a study showed that rats receiving a diet containing immune whey had lower plaque scores (Michalek et al. 1987). Also Mitoma et al. showed that bovine milk containing antibodies against PAcA-GB, a recombinant fusion of the saliva-binding alanine-rich region (PAcA) of the cell surface protein antigen (PAc) of *S. mutants* with the glucan-binding (GB) domain of glucosyltransferase I (GTF-I), when fed to rats, was able to significantly reduced caries development in comparison with control animals (Mitoma et al. 2002).

Notably, local passive immunization with murine monoclonal antibodies against streptococcal antigens (SA) I/II (Ag I/II) of *S. mutans* has also been successfully demonstrated in rats (van Raamsdonk et al. 1993), monkeys (Lehner et al. 1985) and humans (Ma et al. 1987, 1989). Lehner et al. showed that topical applications of mAb to deciduous teeth of Rhesus monkeys prevented significant colonization of smooth surfaces and fissures of teeth (Lehner et al. 1985). Similarly, Ma et al. reported that application of monoclonal antibodies on the teeth of human volunteers also prevented colonization by *S. mutans* and subsequent development of caries for a year. (Ma et al. 1987, 1989). Curiously, even an anti-Ag I/II monoclonal antibody produced in a transgenic plant system revealed consistent anti-cariogenic effects in comparison with its parent murine IgG antibody (Ma et al. 1998).

Chicken egg yolk IgY antibodies have also been successfully used in passive immunization studies against *Streptococcus mutans* (Table 16.3). Indeed, a study conducted by Carlander et al. was one of the initial works revealing IgY persistence in the oral cavity of humans for a period of hours, supporting the use of IgY as described for non-IgY antibodies in dental caries prevention (Carlander et al. 2002). The study used anti-*Pseudomonas aeruginosa* IgY antibodies and revealed that after a mouth rinse with an IgY solution in the evening, 18 out of 19 subjects still had active antibodies in the saliva the next morning, even though the antibody could not be detected 24 h after administration in most of the subjects. Moreover the duration

Table 16.2 Effects of *in vivo* passive immunization against *Streptococcus mutans* using mammalian and plant antibodies

Host	Ab origin	Specificity	Route	Efficacy	Refs
Rhesus Monkey	Murine Monoclonal antibodies	Ag I/II	Topical application	Decreased colonization of *S. mutans*	Lehner et al. (1985)
Rats (SPF)	Murine Monoclonal antibodies	SpaA (Antigen B)	Topical application	Reduced colonization of *S. sobrinus*	van Raamsdonk et al. (1993)
Human	Murine Monoclonal antibodies	Ag I/II	Topical application	Decreased colonization by exogenous strain of *S. mutans*	Ma et al. (1987)
Human	Murine Monoclonal antibodies	Ag I/II	Topical application	Inhibition of re-colonization by indigenous *S. mutans*. Protection up to 1 year after application	Ma et al. (1989)
Rats (gnotobiotic)	Bovine milk	*S. mutans* (serotype a–g)	Diet	Inhibition of colonization by *S. mutans*; inhibition of caries development	Michalek et al. (1987)
Rats	Murine Monoclonal antibodies	PAg *S. sobrinus*	Diet	Significantly lower levels of colonization by *S. sobrinus* on rats' teeth; lower caries scores on treated rats vs control	Zhang et al. (2000)
Rats (SPF)	Bovine milk	PAcA-GB	Diet	Significantly less caries development	Mitoma et al. (2002)
Humans	Bovine milk (Colostrum)	*S. mutans/ S. sobrinus*	Mouth rinse	Reduction in relative number of *S. mutans* in plaque	Loimaranta et al. 1999
Humans	Bovine milk	(PAcA-GB)	Mouth rinse	Inhibition of recolonization by indigenous *S. mutans*	Slowinski and Lawson (2002)
Humans	PlantS.Ig A/G	Ag I/II	Topical application	Specific protection of recolonization by indigenous *S. mutans* for at least 4 months	Ma et al. (1998)

PAcA-GB, recombinant fusion of the saliva-binding alanine-rich region (PAcA) of cell surface protein antigen (PAc) of *S. mutants* with the glucan-binding (GB) domain of glucosyltransferase I (GTF-I); **PAg**, same as streptococcal antigens (SA) I/II, also termed Antigen I/II or Antigen B or Pac or even P1

Table 16.3 Effects of *in vivo* passive immunization against *Streptococcus mutans* with chicken IgY

Host	Ab origin	Specificity	Route	Efficacy	Refs
Humans	IgY	*S. mutans*	Mouth rinse	Reduction in ratio of *S. mutans* per total streptococci in saliva	Hatta et al. (1997)
Humans	IgY	*S. mutans*	Spray	*Streptococcus mutans* in dental plaque largely suppressed with 3-week treatment; low levels persisted for 5 weeks	Zhou et al. (2003)
Humans	IgY	*S. mutans*	Spray	Significant decrease in caries activity after 2-months of use	Zhang et al. (2004)
Humans	IgY	*S. mutans*	Tooth paste	Significant prevention of caries on deciduous teeth in children	Wang and Wang (2008)
Humans	IgY	CA-GTF	Lozenges	Decrease in *S. mutans* count from a mean of 7 to a mean of 1	Nguyen et al. (2011a, b)
Rats SPF	IgY	CA-GTF, CF-GTF and whole cells of *S. mutans*	Diet	Significant reduction of dental plaque accumulation with anti-CA-GTF IgY; not observed with IgY to whole cell and CF-GTF	Hamada et al. (1991)
Rats SPF	IgY	*S. mutans*	Diet	60% inhibition of total caries development	Otake et al. (1991)
Rats	IgY	GBP-B protein	Diet and drinking water	Inhibition of *S. mutans* colonization and dental caries; no need for continuous administration	Smith et al. (2001)
Rats	IgY	*S. mutans*	Diet; Egg yolk powder	Significant lower mean of caries scores in treated groups	Fan et al. (2003)
Rats	IgY	CA-GTF	Diet; Egg yolk powder	Significant reduction of smooth and sulcal surface carious lesions	Krüger et al. (2004)
Rats	scFv expressed on *Lactobacillus*	*S. mutants* SA I/II adhesin antigen	Diet	Significant reduction in caries using anti-SA I/II single-chain antibody (scFv) expressed on the surface of *Lactobacillus*	Krüger et al. (2005)
Rats	IgY	–	Gel	Anti-*S. mutans* IgY gel reduced the quantity of *S. mutans* on the tooth surface	Bachtiar et al. (2016)

GBP-B, Glucan binding protein B; **CA-GFT**, Cell-Associated Glucosyltransferase; **CF-GFT**, Cell-Free Glucosyltransferase; **SA I/II**, streptococcal antigens (SA) I/II, also termed Antigen I/II or Antigen B or Pac or even P1

of the rinse seemed to be well correlated with antibody titre, with a 2-min mouth rinse resulting in a higher ELISA signal than a 1-min rinse (David Carlander et al. 2002).

The efficacy of oral IgY anti-*S. mutans* rinses in human volunteers, revealed a tendency for reduction of *S. mutans* ratio per total streptococci in the plaque of treated individuals (Table 16.3) (Hatta et al. 1997). The effect of an anti-*S. mutans* IgY spray in adult volunteers showed a significant decrease in *S. mutans* colonies in the test group after three weeks of IgY application (Zhou et al. 2003). Similar results were seen in human volunteers after 2 months of use of an anti-*S. mutans* IgY spray (Zhang et al. 2004). Moreover, anti-*S. mutans* IgY incorporated into toothpaste was found to be effective in reducing caries in deciduous teeth in human volunteers (Wang and Wang 2008) and the use of lozenges containing IgY against the cell-associated glucosyltransferase (CA-GTF) of *S. mutans* significantly decreased in *S. mutans* count in the human test groups (Nguyen et al. 2011a, b).

Parallel studies in rodent models also reveal the robustness of IgY anti-cariogenic effects (Table 16.3). Otake et al. reported that passive oral administration of crude IgY produced against *S. mutans* whole cells exhibited protection against *S. mutans*-induced dental caries in rats (Otake et al. 1991). The immunological specificities of IgY raised against whole-cells or developed against *S. mutans* cell-associated glucosyltransferase (CA-GFT) or cell-free glucosyltransferase (CF-GFT) was examined (Hamada et al. 1991). Notably, the work demonstrated a significant reduction of dental plaque accumulation in rats fed only with anti-CA-GTF IgY (Hamada et al. 1991). Smith et al. reported that the administration of IgY anti-*S. mutans* glucan binding protein B (GBP-B) via the diet and drinking water of experimentally infected rats caused a significant decrease in *S. mutans* aggregation on dental biofilms and reduction in the incidence of dental caries (Smith et al. 2001). Furthermore, the decrease in the *S. mutans* infection rate did not require continuous IgY administration.

Altogether, these data support the use of IgY-based formulations as a low-cost preventive measure to reduce *S. mutans* colonization in the oral cavity and help reducing the incidence of dental caries.

16.6 Antitoxin IgY Therapies and Approaches

16.6.1 Introduction

Toxins are harmful substances of different chemical nature (e.g., cytotoxic proteins, polypeptides, alkaloids) produced within living cells and organisms. A number of microbial strains produce toxins which cause diseases in humans. To counteract their effects antitoxins are used. Antitoxins are usually prepared by hyper-immunization of horses or pigs and plasma-derived polyclonal antibodies and are administered to patients requiring the antitoxin therapy. At present, antitoxins are prescribed by physicians in clinics providing treatment for diphtheria, tetanus, botulism diseases, and toxic shock syndrome (Alouf et al. 2015; Schmitt et al. 1999). Adverse reactions such as serum sickness, anaphylaxis and risk of transmission of mammalian blood-borne pathogens to humans poses some limitations for antitoxin therapy. Chickens have shown different resistance and reactivity to toxin immunizations and IgY-based antitoxin formulations can be effectively produced and help overcoming some of the limitations of mammalian derived antitoxins.

Herein are presented several examples of antitoxin IgY antibodies targeting different microbial pathogens and associated virulence factors. The in vitro and in vivo studies presented support the potential translation to antitoxin therapies in humans (Table 16.4).

16.6.2 *Escherichia coli*

Enterohaemorrhagic *Escherichia coli* (EHEC) is an important bacterial pathogen causing several human diseases including haemorrhagic colitis, diarrhoea, renal failure and haemolytic uremic syndrome (HUS)(Paton and Paton 1998). It has been shown that EHEC produces shiga-like toxins (Stxs) as the main virulence factor associated with the pathogenicity of EHEC infections. This pathogen produces two types of Stxs, grouped as Stx-1 and Stx-2.Typically Stxs enter into the intestinal lumen, get released into the systemic circulation and bind to specific receptors on cells, where they cause cell death. Stxs belong to the family of AB5 toxins, comprised of one catalytic A subunit (32 kDa) and a pentameric B subunit, itself formed by identical B subunits (7.7 kDa each), involved in target specificity of the toxin. The A subunit binds to the pentamer of B subunits through non-covalent interactions. The B pentamer binds to recipient cell-surface glycolipid receptors including galabiosylceramide (Gb_2-Cer) and globotriaosylceramide (Gb_3-Cer). After binding, the toxin is internalized in the cells, and the A subunit is cleaved into two subunits A1 (28 kDa) and A2 (4 kDa). The A1 fragment has a RNA *N*-glycosidase activity, which leads to inhibition of protein synthesis via inactivation of 28S rRNA (Saxena et al. 1989). The management of EHEC infection with antibiotics is questionable because of the elevated risk of haemolytic uremic syndrome (HUS) due to the increase in Stx stimulation/release (Wong et al. 2000). Multiple lines of evidence have revealed that some antibiotics significantly improve Stx-2 production from EHEC via stimulation of Stx-encoding bacteriophages and release Stx-1 that is maintained in the periplasm of EHEC in high quantities (Zhang et al. 2000). These results suggest that an effective platform for treatment and prevention of Stx-mediated diseases is required worldwide. A number of independent studies support the use of IgY-based alternatives for treatment of EHEC infections as detailed in Table 16.4. IgY antibodies were raised against either of the two toxins (Stx 1 and 2) and were initially shown to block the cytotoxicity in Vero or HeLa cells. Further studies showed that IgY protected mice from various levels of the toxins (Table 16.4). These findings support the idea that anti-Stx IgY antibodies could be used as preventive agents for diseases associated with Stx in EHEC infections. The findings confirmed that anti-shiga toxin IgY antibodies are cost-

Table 16.4 Studies on Anti-toxin IgY antibodies

Bacterial strain	Toxin/ Protein	IgY applications and experimental observations	Model	Ref
Enterohemorrhagic *Escherichia coli* (EHEC)	Stx2B	4.3μg of anti-Shiga toxin2B IgY blocked cytotoxicity of STX on Vero cells 8.75μg of IgY protected mice against 5 LD50 of STX	Vero cells Mice	(Parma et al. 2011)
	Stx1 and Stx2	Anti-Stx-1 IgY and anti-Stx-2 IgY inhibited the cytotoxicity of Stx-1 and Stx-2 to HeLa 229 Anti-Stx-1 IgY and anti-Stx-2 IgY protected mice against injection of STX-1 and STX-2	HeLa cells Mice	(Neri et al. 2011)
	STXB	17.8μmol/L IgY neutralized cytotoxicity of 1 IC50 (0.428μmol/L) Stx-1 on HeLa 229 225μg IgY protected mice against 5 LD50 of STX	HeLa cells Mice	(Wang et al. 2010)
	Stx2e	Anti STX2e IgY neutralized 5 LD50 STX toxin when simultaneously administrated to mice	Mice	(Arimitsu et al. 2014)
Enterotoxigenic *Escherichia coli (ETEC)*	LTB-STa-STb	Anti STa-STb IgY neutralize STa and STb	Mice	(You et al. 2014b)
Clostridium tetani	Tetanus toxin	Anti-TeNT IgY (1320 lf-eq) protected mice against 2 MLD of *C. Tetani* Anti-TeNT IgY (4300 lf-eq) protected donkeys against 1 MLD of *C. Tetani*	Mice Donkey	(Selim et al. 2015)
Clostridium botulinum	BoNT/A	Use of IgY against substrate peptide of BoNT in detection system Limit of detection 0.1 LD50 (0.04 pg/ml)	Mice	(Rossetto et al. 2011)
	BoNT/B (BHc)	50 ng/mouse of specific IgY protected mice against 4 LD50 toxin	Mice	(You et al. 2014a)

(continued)

Table 16.4 (continued)

Bacterial strain	Toxin/ Protein	IgY applications and experimental observations	Model	Ref
	BoNT/A, B Ricin	Anti-BoNT IgY neutralized ten-fold LD50 of toxin in mice 3.7-7.4μg specific IgY inhibited cell death	Birds Vero cells	(Pauly et al. 2009)
	BoNT/A, B and E	Specific IgY for detection of human anti-BoNT in ELISA-ELCA system	Human	(Doellgast et al. 1997)
Clostridium difficile	FliD protein	Anti FliD IgY inhibited *C. difficle* adherence	Hamster	(Mulvey et al. 2011)
	TcdA	Anti-TcdA IgY neutralized toxin in rabbit intestinal loop	Rabbit	23
Clostridium perfringens	Toxin A (CPA)	IgY against recombinant CPA neutralized CPA	Birds	(Cooper et al. 2009)
	Toxin A (CPA) Enterotoxin (CPE)	Anti-CPA- CPE IgY in immune capture PCR ELISA detected 1 pg of toxin	PBS and artificially contaminated samples	(Das et al. 2019)
Staphylococcus aureus	SEB	IgY used in passive immunotherapy protected mice against SEB and thesis mommy against aerosolized SEB	Mice and monkey	(LeClaire et al. 2002)

TeNT: Tetanus Neurotoxin, **BoNT**: Botulinum neurotoxin, **LTB**: heat-labile enterotoxin B subunit, **STa**: heat-stable enterotoxin a, **STb**: heat-stable enterotoxin b, **Stx1**: Shiga toxin 1, **Stx2**: Shiga toxin 2, **Stx2e**: shiga toxin 2e, **AFB1**: AflatoxinB1, **SE**: Staphylococcal enterotoxin, **SEA**: Staphylococcal enterotoxin A, **SEB**: staphylococcal enterotoxin B, **TcdA**: *Clostridium difficle* toxin A, **Pir-A** and **Pir-B**: *Photorhabdus* insect-related (Pir) toxins, **AHPND**: Acute hepatopancreatic necrosis disease, **GtxA**: *Gallibacterium* toxin A

effective agents to be considered for prophylactic and/or detection kits for oedema diseases.

Enterotoxigenic *Escherichia coli* (ETEC) is the major cause of diarrhoea in children younger than 5 years of age living in developing countries. Heat labile (LT) and heat stable (ST) enterotoxins are major virulent factors of ETEC. LT contains one A subunit (LTA) of 27 kDa that harbours the toxic ADP ribosyl transferase activity, and a 55 kDa homopentamer of 11.6 kDa B subunit peptides (LTB) which binds to the GM-1 ganglioside on the surface of enterocytes. Due to its large molecular weight LTB is highly immunogenic while also free of toxicity (Field 1979). ST, which is a small peptide, binds to the guanylate cyclase C receptor and cause secretion of fluids and electrolytes. Moreover, ST is very poorly immunogenic given its size (W. Zhang et al. 2006). To overcome this fact, You et al. produced a recombinant fusion protein LTB-STa-STb (or Bab) in *E. coli*, consisting of LBT fused with the two ST subtypes STa and STb, and developed a specific IgY against this form. This fusion protein elicited a significant titre of IgY in hens which means STa and STb maintain their immunogenicity in fusion form (Table 16.4). A suckling mouse assay showed that anti-Bab IgY could neutralize the natural toxicity of STa and STb (You et al. 2014b).

16.6.3 *Clostridium tetani*

Tetanus neurotoxin (TeNT), the product of *Clostridium tetani*, is the causative agent of the fatal disease tetanus. It is estimated 58,000 neonates and an unknown number of mothers die every year from tetanus (Thwaites et al. 2015). The tetanus bacteria typically enter the body through a cut or puncture wound and release the toxin in increasing amounts. When the neurotoxin reaches the nervous system it triggers increased rigidity of voluntary muscles, mainly those of the face, body, legs, neck, and tail (in animals). The steady and prolonged rigidity of the affected muscles ultimately leads to spasms and death. The neurotoxin is composed by two chains, one of 100 kDa (heavy chain) and another of 50 kDa (light chain), interacting through an interchain disulphide bond. The heavy chain is responsible for binding to polysialogangliosides and nidogen of nerve cell membranes and the light chain has a catalytic function, cleaving VAMP/synaptobrevin and blocking inhibitory synaptic vesicle release. Finally, an imbalance occurs between the inhibitory and excitatory synaptic vesicles on the motor neurons, leading to interruptions in muscle contraction and spastic paralysis (Surana et al. 2018). In a study, Selim et al., investigated the application of IgY both prophylactically and therapeutically for tetanus treatment (Selim et al. 2015) (Table 16.4). Results showed that all mice in the therapeutic groups as well as a prophylactic groups survived after a challenge with 2 minimum lethal dose (MLD) of *C. tetani*; similar results were obtained in donkeys. These data confirmed that the IgY approach was as effective as the equine IgG approach in tetanus therapy (Selim et al. 2015).

16.6.4 *Clostridium botulinum*

Botulinum neurotoxin (BoNT) is produced by *Clostridium botulinum* under anaerobic conditions and is known as one of the most poisonous substances in the world (Peck 2009). To date a total of seven different BoNT toxin types are described (A to G). Botulism usually occurs as a food poisoning caused by botulinum neurotoxin produced by *C. botulinum*. BoNT consists of a heavy chain (100 KDa) and a light chain (50 KDa) that interact via a disulphide bond. Just like TeNT, the heavy chain binds to polysialogangliosides on the nerve plasma membrane, in particular to G1b gangliosides, with high specificity and affinity and whole toxin is internalized to the cytoplasm via endocytic vesicles. The light chain has an endopeptidase activity and cleaves the SNARE proteins VAMP/synaptobrevin 1-3 and syntaxin. This phenomena inhibits neurotransmitter (acetylcholine) vesicle release which results in muscle fibre paralysis (Rossetto et al. 2011). In a number of studies, BoNT/A or BoNTB or a combination were used to generate IgY for treatment of mice or birds (Li et al. 2013; Pauly et al. 2009; You et al. 2014a) (Table 16.4). In all cases IgY showed a protective effect.

Another important fact to significantly reduce botulism mortality, is the need for a fast diagnosis of BoNT toxin. Indeed a rapid and accurate test for the botulinum neurotoxin is essential for BoNT prevention and therapy. IgY was developed against either a linear peptide substrate (SNAP25) (Li et al. 2013) or BoNT A/B and D (Doellgast et al. 1997) and both showed robust results for toxin detection applications, namely in food or clinical conditions. Notably, Doellgast et al., develop an enzyme linked immunosorbent assay and an enzyme linked coagulation assay (ELISA-ELCA) for high sensitivity detection of anti-neurotoxin in human sera (Doellgast et al. 1997).

16.6.5 *Clostridium difficile*

Among toxin producing bacteria, *Clostridium difficile* is a major cause of nosocomial diarrhoea in the developed and developing countries, particularly among patients over 65 years of age (Curcio et al. 2019). The disruption of the normal gastrointestinal microbiota and/or antibiotic treatment can lead to colon colonization by *C. difficile*. Subsequently, the bacterium generates a high level of the important virulence agents including toxin A (TcdA) and toxin B (TcdB) with molecular weights of 380 KDa and 270 KDa, respectively. TcdA and TcdB have two subunits, A and B. The B subunit functions by delivering the enzymatic A subunit into the target cell cytosol. The A subunit is a glucosyl transferase enzyme which glucosylates members of the Rho family of small GTPases, inactivating them and leading to disruption of cell functions. It seems these toxins together are responsible for disease features such as enterotoxicity, cytotoxicity and pro-inflammatory effects in the human colon (Chandrasekaran and Lacy 2017). The gold standard therapy for *C. difficile* infection (CDI) is the utilization of antibiotics such as fidaxomicin, vancomycin, and metronidazole. These treatments however show a 12-30% re-occurrence rate. Moreover, the use of recombinant fragments or toxoids as vaccines are associated with different and limited outcomes in human and animal models. It has been shown that injection of flagellin D (FliD)-specific IgY could be associated with protective effects against death in a CDI model of hamster (Mulvey et al. 2011) (Table 16.4). The development of microbeads (Zhang et al. 2015) or chitosan-Ca pectinated microbeads (Xing et al. 2017) as coating for IgY were two strategies employed to overcome gastrointestinal denaturation with the latter increasing drug loading and being more effective (Table 16.4).

16.6.6 *Clostridium perfringens*

Clostridium perfringens type A is the most common bacterial infection associated with necrotic enteritis (NE). The main toxin produced by *Clostridium perfringens* type A is alpha toxin (CPA), which can have critical roles in pathogenesis of NE. This toxin is able to induce mucosal damage in chicken intestinal loops and has been associated with NE lesions in germ-free chickens. Anti-CPA serum applications have been shown to effectively neutralize the effects of the toxin (Doellgast et al. 1997). Specific IgY has been shown to be effective in protecting birds lowering the presentation of NE lesions (Table 16.4).

16.6.7 *Staphylococcus aureus*

Staphylococcus aureus is an important human pathogen that produces a wide range of exotoxins. *Staphylococcus aureus* enterotoxins (SEs) are heat stable toxins causing common food poisoning in humans. These toxins are small proteins (20-30 kDa), belonging to the superantigen family, known to be highly resistant to protease degradation and heat denaturation. Five types of enterotoxins are the usual causes of food poisoning; these include SEA, SEB, SEC, SED and SED toxins. A study from LeClaire and colleagues used anti-SEB IgY as passive immunotherapy against SEB and showed protection in mice and rhesus monkeys against aerosolized SEB (LeClaire et al. 2002).

16.7 Antivenoms

16.7.1 Snake Venom

The venom of the snake is considered to be highly modified saliva containing zootoxins which facilitates the immobilization and digestion of prey, and defends against threats. Venoms contain more than 20 different compounds, mostly proteins and polypeptides. Some of the proteins in snake venom have very specific effects on various biological functions including blood coagulation, blood pressure regulation, transmission of the nervous or muscular impulse and have been developed for use as pharmacological or diagnostic tools and even useful drugs (Slowinski and Lawson 2002).

Envenomation from snakes and scorpions are considered to be a serious health hazard in tropical and subtropical countries (Bawaskar 2004). The incidence is relatively higher in tropical countries than in developed countries (Warrell 1996). In case of snake bite, it has been estimated that more than six million cases occur per year with estimate of fatalities ranging from 50,000–100,000 (Pillay 2003; Warrell 1996). The effective treatment for the envenomation is immunotherapy which is based on immunization of animals and purification of the immunoglobulins, mostly the serum. In the year 2009, WHO included snake bite in the list of neglected disease. WHO has estimated that annually ten million antivenom vials are needed globally.

16.7.2 Clinical Manifestations of Envenoming

Envenoming with viperid snake venoms result in severe pathological outcomes, such as haemorrhage, dermonecrosis, blistering, myonecrosis and oedema, typically associated with pain (Gutiérrez et al. 2009; Mara 2012; Warrell 2004). These are local manifestations that occur at the site of bite and may cause permanent tissue damage and loss (Dart et al. 1992; Otero et al. 2002). The impact may be more drastic, depending on the dose and its systemic distribution resulting in coagulopathies, bleeding, renal alterations and hemodynamic manifestations leading to cardiovascular shock and multisystem organ failure (Gutiérrez et al. 2009; Warrell 2004). In addition, intravascular haemolysis can also be observed, occasionally associated with microthrombi formation (Warrell 1996).

Interestingly there are exceptions to the common symptoms and effects described above and general trends of envenoming are not always observed. For example, envenoming by the South American and some populations of North American rattlesnakes, and also by some viperids in the Old World, are characterized by complex neurotoxic outcomes (Azevedo-Marques et al. 2009; Ferquel et al. 2007). Also venoms of the Caribbean viperid species *Bothrops caribbaeus* and *Bothrops lanceolatus* induce unique thrombotic effects (Thomas et al. 1996), and envenoming by *Daboia russelli* may lead to pituitary apoplexy, a rare condition resulting from acute haemorrhagic infarction of the pituitary gland (Tun-Pe et al. 1987). Indeed despite the general trends, these uncommon clinical manifestations reflect the complexity of snakebite envenoming urging continuous research for their fundamental understanding and treatment.

16.7.3 IgY Antivenoms: the Key Therapy for Snakebite Envenoming

The majority of manufacturers use horses for immunization, although a few use sheep and donkeys (Gutiérrez et al. 2011). In most cases, plasma fractionation involves the digestion of proteins with pepsin or, by few producers, with papain, followed by the purification of antibody fragments by salting-out with ammonium salts or caprylic acid fractionation and, in some cases, with chromatographic procedure (dos Santos et al. 1989). The ability of antivenoms to neutralize venom toxins is based on the capacity of antivenom antibodies, or antibody fragments, to bind and neutralize the most relevant toxins in venom.

In a work by Meenatchisundaram et al. (Meenatchisundaram et al. 2008) chickens were immunized Cobra, Krait, Russell's viper and Saw-scaled viper venoms and IgY extracted and purified from eggs as described in Chap. 11 and further characterized for antivenon activity, as described below in the pharmacology section. Antibodies were detected in egg yolk after a week and reached an average yield of 80 mg per egg yolk at 180th day of immunization period and the highest titre of 1:10000 was observed during 135th day. These results were comparable to the work done by Almeida et al. (1998) who showed that IgY antibodies against snake venom components started to appear in serum two weeks after immunization began and reached high titre at 45th day and remained stable till 168th day observation (Almeida et al. 1998). Anti-venom egg yolk antibodies are present in the egg for up to 100 days after the immunization as reported by Devi et al. (Devi et al. 2002). Rungsiwongse and Ratanabanangkoon (1991) work showed a significant correlation between their ELISA and the antivenom potency tested *in vivo* against *Naja naja* venom (Rungsiwongse and Ratanabanangkoon 1991). Consistently Heneine et al. (1998) have shown a good correlation between ELISA and the *in vivo* potency of Bothropic antivenoms when crude Bothrops venom was fractionated and the purified fractions were used as the antigen for ELISA (Heneine et al. 1998).

16.7.4 Pharmacological Characterization and Neutralization of Venoms

Some of the pharmacological effects of venom have also been studied. The lethal toxicity for 18 g of mice was found to be 10μg for Cobra, 3μg for Krait, 8μg for Russell's viper and 12μg for Saw-scaled viper venoms (Meenatchisundaram et al. 2008). When venom and antivenom were given as independent injections the neutralization responses were found to be less effective. The median effective dose (ED_{50}) for $2LD_{50}$ of venom was found to be 1.24 mg for Cobra, 0.83 mg for Krait, 0.96 mg for Russell's viper and 1.27 mg for Saw-scaled viper venoms. Control animals injected with 2LD50 of venom alone did not show any survival. I another study by Carroll et al. using mouse protection assays, purified IgY antibodies against *Crotalus atrox* and *T. flavoviridis* antivenoms were reported to be 6.3 and 2.0 times as potent as equine antivenoms in neutralizing the lethal activities. (Carroll et al. 1992). The results demonstrate that the chicken egg yolk antibodies are effective in neutralizing the lethality and several pharmacological effects of the venom. Laying hens respond well immunologically to the wide range of different toxic compounds present in venoms and avian antivenoms are significantly more potent than existing mammalian-derived antivenoms.

References

Adachi K, Handharyani E, Sari DK, Takama K, Fukuda K, Endo I, Yamamoto R, Sawa M, Tanaka M, Konishi I, Tsukamoto Y (2008) Development of neutralization antibodies against highly pathogenic H5N1 avian influenza virus using ostrich (Struthio camelus) yolk. Mol Med Rep 1(2):203–209

Adachi K, Takama K, Tsukamoto M, Inai M, Handharyani E, Hiroi S, Tsukamoto Y (2011) Ostrich produce cross-reactive neutralization antibodies against pandemic influenza virus A/H1N1 following immunization with a seasonal influenza vaccine. Exp Ther Med 2(1):41–45. https://doi.org/10.3892/etm.2010.180

Aiba Y, Umeda K, Rahman S, Nguyen SV, Komatsu Y (2019) Synergistic effect of anti-*Helicobacter pylori* urease immunoglobulin Y from egg yolk of immunized hens and *Lactobacillus johnsonii* No.1088 to inhibit the growth of *Helicobacter pylori* in vitro and in vivo. Vaccine 37(23):3106–3112. https://doi.org/10.1016/j.vaccine.2019.04.045

Akram SM, Mangat R, Huang B (2020) Hantavirus Cardiopulmonary Syndrome. StatPearls. http://www.ncbi.nlm.nih.gov/pubmed/29083610

Almeida CMC, Kanashiro MM, Rangel Filho FB, Mata MFR, Kipnis TL, Dias Da Silva W (1998) Development of snake antivenom antibodies in chickens and their purification from yolk. Vet Rec 143(21):579–584. https://doi.org/10.1136/vr.143.21.579

Alouf J, Ladant D, Popoff MR (2015) The Comprehensive Sourcebook of Bacterial Protein Toxins. https://doi.org/10.1016/C2013-0-14258-4

Arimitsu H, Sasaki K, Kohda T, Shimizu T, Tsuji T (2014) Evaluation of Shiga toxin 2e-specific chicken egg yolk immunoglobulin: Production and neutralization activity. Microbiol Immunol 58(11):643–648. https://doi.org/10.1111/1348-0421.12197

Atmar RL (2010) Noroviruses: State of the Art. Food and Environ Virol 2(3):117–126. https://doi.org/10.1007/s12560-010-9038-1

Attallah AM, Abbas AT, Ismail H, Abdel-Raouf M, El-Dosoky I (2009) Efficacy of Passive Immunization with IgY Antibodies to a 58-kDa *H. pylori* Antigen on Severe Gastritis in BALB/c Mouse Model. J Immunoass Immunochem 30(4):359–377. https://doi.org/10.1080/15321810903187922

Azevedo-Marques M, Hering S, Cupo P (2009) Acidente crotálico. In: Wen FH, Málaque CMS, Haddad JV (eds) Animais Peconhentos no Brasil: Biologia Clinica e Terapeutica dos Acidentes. São Paulo, Sarvier Editora, pp 91–98

Bachtiar EW, Afdhal A, Meidyawati R, Soejoedono RD, Poerwaningsih E (2016) Effect of topical anti-*Streptococcus mutans* IgY gel on quantity of *S. mutans* on rats' tooth surface. Acta Microbiol Immunol Hung 63(2):159–169. https://doi.org/10.1556/030.63.2016.2.2

Barnes CO, West AP, Huey-Tubman KE, Hoffmann MAG, Sharaf NG, Hoffman PR, Sharaf NG, Hoffman PR, Koranda N, Gristick HB, Gaebler C, Muecksch F, Lorenzi JCC, Finkin S, Hägglöf T, Hurley A, Millard K, Weisblum Y, Schmidt F, Hatziioannou T, Bieniasz PD, Caskey M, Robbiani DF, Nussenzweig MC, Bjorkman PJ (2020) Structures of Human Antibodies Bound to SARS-CoV-2 Spike Reveal Common Epitopes and Recurrent Features of Antibodies. Cell 182(4):828–842.e16. https://doi.org/10.1016/j.cell.2020.06.025

Bawaskar HS (2004) Snake venoms and antivenoms: Critical supply issues. J Assoc Phy Ind 52 (JAN):11–13

Borhani K, Mohabati Mobarez A, Khabiri AR, Behmanesh M, Khoramabadi N (2016) Inhibitory effects of rHP-NAP IgY against *Helicobacter pylori* attachment to AGS cell line. Microb Pathog 97:231–235. https://doi.org/10.1016/j.micpath.2016.06.004

Bowden GH (1996) Mutans streptococci caries and chlorhexidine. J Can Dent Assoc 62(9)., 700:703–707

Bravo D, Hoare A, Soto C, Valenzuela MA, Quest AFG (2018) *Helicobacter pylori* in human health and disease: Mechanisms for local gastric and systemic effects. World J Gastroenterol 24 (28):3071–3089. https://doi.org/10.3748/wjg.v24.i28.3071

Brocato R, Josleyn M, Ballantyne J, Vial P, Hooper JW (2012) DNA vaccine-generated duck polyclonal antibodies as a postexposure prophylactic to prevent hantavirus pulmonary syndrome (HPS). PLoS One 7(4). https://doi.org/10.1371/journal.pone.0035996

Brouwer PJM, Caniels TG, van der Straten K, Snitselaar JL, Aldon Y, Bangaru S, Torres, Jonathan L, Okba NMA, Claireaux M, Kerster G, Bentlage AEH, van Haaren MM, Guerra D, Burger JA, Schermer EE, Verheul KD, van der Velde N, van der Kooi A, van Schooten J, van Breemen MJ, Bijl TPL, Sliepen K, Aartse A, Derking R, Bontjer I, Kootstra NA, Wiersinga WJ, Vidarsson G, Haagmans BL, Ward AB, de Bree GJ, Sanders RW, van Gils MJ (2020) Potent neutralizing antibodies from COVID-19 patients define multiple targets of vulnerability. Science 369(6504):643–650. https://doi.org/10.1126/science.abc5902

Carlander D, Kollberg H, Larsson A (2002) Retention of specific yolk IgY in the human oral cavity. BioDrugs 16(6):433–437. https://doi.org/10.2165/00063030-200216060-00004

Carlander D, Sundstrom J, Berglund A, Larsson A, Wretlind B, Kollberg H (1999) Immunoglobulin Y (IgY) A new tool for the prophylaxis against *Pseudomonas aeruginosa* in cystic fibrosis patients. Pediatric Pulmonol Suppl 19:240

Carroll SB, Thalley BS, Theakston RDG, Laing G (1992) Comparison of the purity and efficacy of affinity purified avian antivenoms with commercial equine crotalid antivenoms. Toxicon 30 (9):1017–1025. https://doi.org/10.1016/0041-0101(92)90046-8

Chandrasekaran R, Lacy DB (2017) The role of toxins in *Clostridium difficile* infection. FEMS Microbiol Rev 41(6):723–750. https://doi.org/10.1093/femsre/fux048

Chi X, Yan R, Zhang J, Zhang G, Zhang Y, Hao M, Zhang Z, Fan P, Dong Y, Yang Y, Chen Z, Guo Y, Zhang J, Li Y, Song X, Chen Y, Xia L, Fu L, Hou L, Xu J, Yu C, Li J, Zhou Q, Chen W (2020) A neutralizing human antibody binds to the N-terminal domain of the Spike protein of SARS-CoV-2. Science 369(6504):650–655. https://doi.org/10.1126/science.abc6952

Childers NK, Tong G, Li F, Dasanayake AP, Kirk K, Michalek SM (2002) Humans Immunized with *Streptococcus mutans* Antigens by Mucosal Routes. J Dent Res 81(1):48–52. https://doi.org/10.1177/002203450208100111

Clarke JK (1924) On the Bacterial Factor in the Ætiology of Dental Caries. Br J Exp Pathol 5 (3):141–147

Cooper KK, Trinh HT, Songer JG (2009) Immunization with recombinant alpha toxin partially protects broiler chicks against experimental challenge with *Clostridium perfringens*. Vet Microbiol 133(1–2):92–97. https://doi.org/10.1016/j.vetmic.2008.06.001

Curcio D, Cané A, Fernández FA, Correa J (2019) *Clostridium difficile*-associated Diarrhea in Developing Countries: A Systematic Review and Meta-Analysis. Infect Dis Ther 8(1):87–103. https://doi.org/10.1007/s40121-019-0231-8

Dai YC, Wang YY, Zhang XF, Tan M, Xia M, Wu XB, Jiang X, Nie J (2012) Evaluation of anti-norovirus IgY from egg yolk of chickens immunized with norovirus P particles. J Virol Methods 186(1–2):126–131. https://doi.org/10.1016/j.jviromet.2012.07.002

Dart RC, McNally J, Spaite D, Gustafson R (1992) The sequelae of pitviper poisoning in the United States. In: Campbell J, Brodie E (eds) Biology of the Pitvipers, 1st edn. Selva Publishing, pp 395–404

Das S, Majumder S, Nag M, Kingston JJ (2019) A sandwich duplex immuno PCR for rapid and sensitive identification of *Clostridium perfringens* alpha and enterotoxin. Anaerobe 57:63–74. https://doi.org/10.1016/j.anaerobe.2019.03.015

De Brito BB, Antônio F, Soares AS, Pereira VA, Santos MLC, Sampaio MM, Neves PHM, Melo FFD (2019) Pathogenesis and clinical management of *Helicobacter pylori* gastric infection. World J Gastroenterol 25(37):5578–5589. https://doi.org/10.3748/wjg.v25.i37.5578

Deng S-Q, Yang X, Wei Y, Chen J-T, Wang X-J, Peng H-J (2020) A Review on Dengue Vaccine Development. Vaccine 8(1):63. https://doi.org/10.3390/vaccines8010063

Devi CM, Bai MV, Krishnan LK (2002) Development of viper-venom antibodies in chicken egg yolk and assay of their antigen binding capacity. Toxicon 40(7):857–861. https://doi.org/10.1016/S0041-0101(01)00258-6

Dheerasekara K, Sumathipala S, Muthugala R (2020) Hantavirus Infections—Treatment and Prevention. Curr Treat Options Infect Dis. https://doi.org/10.1007/s40506-020-00236-3

Doellgast GJ, Brown JE, Koufman JA, Hatheway CL (1997) Sensitive assay for measurement of antibodies to *Clostridium botulinum* neurotoxins A, B, and E: use of hapten-labeled-antibody elution to isolate specific complexes. J Clin Microbiol 35(3):578–583. https://doi.org/10.1128/jcm.35.3.578-583.1997

dos Santos MC, D'Império Lima MR, Furtado GC, Colletto GMDD, Kipnis TL, Dias da Silva W (1989) Purification of F(ab′)2 anti-snake venom by caprylic acid: A fast method for obtaining IgG fragments with high neutralization activity, purity and yield. Toxicon 27(3):297–303. https://doi.org/10.1016/0041-0101(89)90177-3

Drosten C, Günther S, Preiser W, van der Werf S, Brodt HR, Becker S, Rabenau H, Panning M, Kolesnikova L, Fouchier RAM, Berger A, Burguière AM, Cinatl J, Eickmann M, Escriou N, Grywna K, Kramme S, Manuguerra JC, Müller S, Rickerts V, Stürmer M, Vieth S, Klenk HD, Osterhaus ADME, Schmitz H, Doerr HW (2003) Identification of a novel coronavirus in patients with severe acute respiratory syndrome. N Engl J Med 348(20):1967–1976. https://doi.org/10.1056/NEJMoa030747

Dunn BE, Phadnis SH (1998) Structure, function and localization of *Helicobacter pylori* urease. Yale J Biol Med 71(2):63–73

Dye BA (2017) The Global Burden of Oral Disease: Research and Public Health Significance. J Dent Res 96(4):361–363. https://doi.org/10.1177/0022034517693567

Emilson CG, Fornell J (1976) Effect of toothbrushing with chlorhexidine gel on salivary microflora, oral hygiene, and caries. Eur J Oral Sci 84(5):308–319. https://doi.org/10.1111/j.1600-0722.1976.tb00495.x

Ertl HCJ (2019) New Rabies Vaccines for Use in Humans. Vaccine 7(2):54. https://doi.org/10.3390/vaccines7020054

Fan M, Jiang Q, Bian Z (2003) The study on specific anti-*Streptococcus mutans* IgY against dental caries in rats. West China J Stomatol 21(5):339–341. http://www.ncbi.nlm.nih.gov/pubmed/14650981

Ferquel E, de Haro L, Jan V, Guillemin I, Jourdain S, Teynié A, D'Alayer J, Choumet V (2007) Reappraisal of Vipera aspis Venom neurotoxicity. PLoS One 2(11):e1194. https://doi.org/10.1371/journal.pone.0001194

Field M (1979) Modes of Action of Enterotoxins from *Vibrio cholerae* and *Escherichia coli*. Clin Infect Dis 1(6):918–926. https://doi.org/10.1093/clinids/1.6.918

Fink AL, Williams KL, Harris E, Alvine TD, Henderson T, Schiltz J, Nilles ML, Bradley DS (2017) Dengue virus specific IgY provides protection following lethal dengue virus challenge and is neutralizing in the absence of inducing antibody dependent enhancement. PLoS Negl Trop Dis 11(7):1–17. https://doi.org/10.1371/journal.pntd.0005721

Fitzgerald R, Keyes PH (1962) Attempted immunization of albino hamsters against induced dental caries. Int Assn Dental Research Programs and Abstracts

Foegeding NJ, Caston RR, McClain MS, Ohi MD, Cover TL (2016) An overview of *Helicobacter pylori* VacA toxin biology. Toxins 8(6):1–21. https://doi.org/10.3390/toxins8060173

Fu CY, Huang H, Wang XM, Liu YG, Wang ZG, Cui SJ, Gao HL, Li Z, Li JP, Kong XG (2006) Preparation and evaluation of anti-SARS coronavirus IgY from yolks of immunized SPF chickens. J Virol Methods 133(1):112–115. https://doi.org/10.1016/j.jviromet.2005.10.027

Gisbert JP, González L, Calvet X, García N, López T, Roqué M, Gabriel R, Pajares JM (2000) Proton pump inhibitor, clarithromycin and either amoxycillin or nitroimidazole: A meta-analysis of eradication of *Helicobacter pylori*. Aliment Pharmacol Ther 14(10):1319–1328. https://doi.org/10.1046/j.1365-2036.2000.00844.x

Gutiérrez JM, León G, Lomonte B, Angulo Y (2011) Antivenoms for snakebite envenomings. Inflammation and Allergy–Drug Targets 10(5):369–380. https://doi.org/10.2174/187152811797200669

Gutiérrez JM, Lomonte B, León G, Alape-Girón A, Flores-Díaz M, Sanz L, Angulo Y, Calvete JJ (2009) Snake venomics and antivenomics: Proteomic tools in the design and control of antivenoms for the treatment of snakebite envenoming. J Proteome 72(2):165–182. https://doi.org/10.1016/j.jprot.2009.01.008

Ha NC, Oh ST, Sung JY, Cha KA, Lee MH, Oh BH (2001) Supramolecular assembly and acid resistance of *Helicobacter pylori* urease. Nat Struct Biol 8(6):505–509. https://doi.org/10.1038/88563

Haese N, Brocato RL, Henderson T, Nilles ML, Kwilas SA, Josleyn MD, Hammerbeck CD, Schiltz J, Royals M, Ballantyne J, Hooper JW, Bradley DS (2015) Antiviral biologic produced in DNA vaccine/ goose platform protects hamsters against hantavirus pulmonary syndrome when administered post-exposure. PLoS Negl Trop Dis 9(6):1–19. https://doi.org/10.1371/journal.pntd.0003803

Hamada S, Horikoshi T, Minami T, Kawabata S, Hiraoka J, Fujiwara T, Ooshima T (1991) Oral passive immunization against dental caries in rats by use of hen egg yolk antibodies specific for cell-associated glucosyltransferase of *Streptococcus mutans*. Infect Immun 59(11):4161–4167. https://doi.org/10.1128/iai.59.11.4161-4167.1991

Hatta H, Tsuda K, Ozeki M, Kim M, Yamamoto T, Otake S, Hirasawa M, Katz J, Childers NK, Michalek SM (1997) Passive Immunization against Dental Plaque Formation in Humans: Effect of a Mouth Rinse Containing Egg Yolk Antibodies (IgY) Specific to *Streptococcus mutans*. Caries Research 31(4):268–274. https://doi.org/10.1159/000262410

He Y, Zhou Y, Wu H, Luo B, Chen J, Li W, Jiang S (2004) Identification of immunodominant sites on the spike protein of severe acute respiratory syndrome (SARS) coronavirus: implication for developing SARS diagnostics and vaccines. J Immunol 173(6):4050–4057. https://doi.org/10.4049/jimmunol.173.6.4050

Heneine LGD, Carvalho ÁDD, Barbosa CDF, Dos Santos MRA (1998) Development of an ELISA to assess the potency of horse therapeutic polyvalent antibothropic antivenom. Toxicon 36 (10):1363–1370. https://doi.org/10.1016/S0041-0101(98)00014-2

Heymann DL, Hodgson A, Sall AA, Freedman DO, Staples JE, Althabe F, Baruah K, Mahmud G, Kandun N, Vasconcelos PFC, Bino S, Menon KU (2016) Zika virus and microcephaly: why is this situation a PHEIC? The Lancet 387(10020):719–721. https://doi.org/10.1016/S0140-6736(16)00320-2

Hifumi T, Yamamoto A, Ato M, Sawabe K, Morokuma K, Morine N, Kondo Y, Noda E, Sakai A, Takahashi J, Umezawa K (2017) Clinical Serum Therapy: Benefits, Cautions, and Potential Applications. Keio J Med 66(4):57–64. https://doi.org/10.2302/kjm.2016-0017-IR

Hong KS, Ki MR, Ullah HMA, Lee EJ, Kim YD, Chung MJ, Elfadl AK, Park JK, Jeong KS (2018) Preventive effect of anti-VacA egg yolk immunoglobulin (IgY) on *Helicobacter pylori*-infected mice. Vaccine 36(3). https://doi.org/10.1016/j.vaccine.2017.11.082

Horie K, Horie N, Abdou AM, Yang JO, Yun SS, Chun HN, Park K, Kim M, Hatta H (2004) Suppressive effect of functional drinking yogurt containing specific egg yolk immunoglobulin on *Helicobacter pylori* in humans. J Dairy Sci 87(12):4073–4079. https://doi.org/10.3168/jds.S0022-0302(04)73549-3

Huang C, Wang Y, Li X, Ren L, Zhao J, Hu Y, Zhang L, Fan G, Xu J, Gu X, Cheng Z, Yu T, Xia J, Wei Y, Wu W, Xie X, Yin W, Li H, Liu M, Xiao Y, Gao H, Guo L, Xie J, Wang G, Jiang R, Gao Z, Jin Q, Wang J, Cao B (2020) Clinical features of patients infected with 2019 novel coronavirus in Wuhan, China. The Lancet 395(10223):497–506. https://doi.org/10.1016/S0140-6736(20)30183-5

Inoue S, Park CH, Hatta H (2015) Production of polyclonal rabies virus antibodies in birds. In: Current laboratory techniques in rabies diagnosis, research and prevention, Volume 2. Cambridge, MA, USA, Academic Press, pp 305–316

Jester BJ, Uyeki TM, Jernigan DB (2020) Fifty years of influenza A(H3N2) following the pandemic of 1968. Am J Public Health 110(5):669–676. https://doi.org/10.2105/AJPH.2019.305557

Kaier K, Heister T, Götting T, Wolkewitz M, Mutters NT (2019) Measuring the in-hospital costs of *Pseudomonas aeruginosa* pneumonia: methodology and results from a German teaching hospital. BMC Infect Dis 19(1):1028. https://doi.org/10.1186/s12879-019-4660-5

Keyes PH (1968) Research in dental caries. J Am Dent Assoc 76(6):1357–1373. https://doi.org/10.14219/jada.archive.1968.0186

Koga T, Oho T, Shimazaki Y, Nakano Y (2002) Immunization against dental caries. Vaccine 20 (16):2027–2044. https://doi.org/10.1016/S0264-410X(02)00047-6

Kollberg H, Carlander D, Olesen H, Wejåker PE, Johannesson M, Larsson A (2003) Oral administration of specific yolk antibodies (IgY) may prevent *Pseudomonas aeruginosa* infections in patients with cystic fibrosis: a phase I feasibility study. Pediatr Pulmonol 35(6):433–440. https://doi.org/10.1002/ppul.10290

Kotta S, Aldawsari HM, Badr-Eldin SM, Alhakamy NA, Md S, Nair AB, Deb PK (2020) Combating the pandemic COVID-19: clinical trials, therapies and perspectives. Front Mol Biosci 7. https://doi.org/10.3389/fmolb.2020.606393

Krüger C, Pearson SK, Kodama Y, Vacca Smith A, Bowen WH, Hammarström L (2004) The effects of egg-derived antibodies to glucosyltransferases on dental caries in rats. Caries Res 38 (1):9–14. https://doi.org/10.1159/000073914

Krüger C, Hultberg A, Van Dollenweerd C, Marcotte H, Hammarström L (2005) Passive immunization by lactobacilli expressing single-chain antibodies against *Streptococcus mutans*. Mol Biotechnol 31(3):221–231. https://doi.org/10.1385/MB:31:3:221

Kuljić-Kapulica N, Budisin A (1992) Coronaviruses: An RNA proofreading machine regulates replication fidelity and diversity. RNA Biol 120(7–8):215–218. https://doi.org/10.4161/rna.8.2.15013

Laurence JW (2009) Microbiology (6th Ed.). The Royal Australasian College of Dental Surgeons Incorporated

LeClaire RD, Hunt RE, Bavari S (2002) Protection against bacterial superantigen staphylococcal enterotoxin B by passive vaccination. Infect Immun 70(5):2278–2281. https://doi.org/10.1128/IAI.70.5.2278-2281.2002

Lee YC, Leu SJC, Hung HC, Wu HH, Huang IJ, Hsieh WS, Chiu WT, Hsieh MS, Cheng TF, Yang YY (2007a) A dominant antigenic epitope on SARS-CoV spike protein identified by an avian single-chain variable fragment (scFv)-expressing phage. Vet Immunol Immunopathol 117 (1–2):75–85. https://doi.org/10.1016/j.vetimm.2007.02.001

Lee YC, Leu SJC, Hu CJ, Shih NY, Huang IJ, Wu HH, Hsieh WS, Chiang BL, Chiu WT, Yang YY (2007b) Chicken single-chain variable fragments against the SARS-CoV spike protein. J Virol Methods 146(1–2):104–111. https://doi.org/10.1016/j.jviromet.2007.06.010

Lehner T, Caldwell J, Smith R (1985) Local passive immunization by monoclonal antibodies against streptococcal antigen I/II in the prevention of dental caries. Infect Immun 50 (3):796–799. https://doi.org/10.1128/iai.50.3.796-799.1985

Lemos JA, Palmer SR, Zeng L, Wen ZT, Kajfasz JK, Freires IA, Abranches J, Brady LJ (2019) The biology of *Streptococcus mutans*. Microbiol Spectrum 7(1). https://doi.org/10.1128/microbiolspec.GPP3-0051-2018

Li T, Liu H, Cai K, Tian M, Wang Q, Shi J, Gao X, Wang H (2013) Hypersensitive detection and quantitation of BoNT/A by IgY antibody against substrate linear-peptide. PLoS One 8(3): e58908. https://doi.org/10.1371/journal.pone.0058908

Loimaranta V, Nuutila J, Marnila P, Tenovuo J, Korhonen H, Lilius E-M (1999) Colostral proteins from cows immunised with *Streptococcus mutans/S. sobrinus* support the phagocytosis and killing of mutans streptococci by human leucocytes. J Med Microbiol 48(10):917–926. https://doi.org/10.1099/00222615-48-10-917

Lu Y, Wang Y, Zhang Z, Huang J, Yao M, Huang G, Ge Y, Zhang P, Huang H, Wang Y, Li H, Wang W (2020) Generation of chicken IgY against SARS-COV-2 spike protein and epitope mapping. J Immunol Res 2020. https://doi.org/10.1155/2020/9465398

Ma JK-C, Hikmat BY, Wycoff K, Vine ND, Chargelegue D, Yu L, Hein MB, Lehner T (1998) Characterization of a recombinant plant monoclonal secretory antibody and preventive immunotherapy in humans. Nat Med 4(5):601–606. https://doi.org/10.1038/nm0598-601

Ma JKC, Hunjan M, Smith R, Lehner T (1989) Specificity of monoclonal antibodies in local passive immunization against *Streptococcus mutans*. Clin Exp Immunol 77(3):331–337

Ma JKC, Smith R, Lehner T (1987) Use of monoclonal antibodies in local passive immunization to prevent colonization of human teeth by *Streptococcus mutans*. Infect Immun 55(5):1274–1278. https://doi.org/10.1128/iai.55.5.1274-1278.1987

Malekshahi ZV, Latif S, Gargari M, Rasooli I, Ebrahimizadeh W (2011) Treatment of *Helicobacter pylori* infection in mice with oral administration of egg yolk-driven anti-UreC immunoglobulin. Microb Pathog 51(5):366–372. https://doi.org/10.1016/j.micpath.2011.06.002

Mao Y-F, Yan J, Li L-W (2003) Cloning, expression and identification of hpaA gene from a clinical isolate of *Helicobacter pylori*. J Zhejiang Univ Med Sci 32(1):9–12. Retrieved from http://www.ncbi.nlm.nih.gov/pubmed/12640701

Mara J (2012) Snakebite envenoming: a public health perspective. In: Public health– methodology, environmental and systems issues. https://doi.org/10.5772/36076

María Alejandra B, Mariano Daniel O (2020) Virulence factors of *Streptococcus mutans* related to dental caries. In: Staphylococcus and Streptococcus. https://doi.org/10.5772/intechopen.85807

McAuley JL, Gilbertson BP, Trifkovic S, Brown LE, McKimm-Breschkin JL (2019) Influenza Virus Neuraminidase Structure and Functions. Front Microbiol 10. https://doi.org/10.3389/fmicb.2019.00039

Meenatchisundaram S, Parameswari G, Michael A, Ramalingam S (2008) Studies on pharmacological effects of Russell's viper and Saw-scaled viper venom and its neutralization by chicken egg yolk antibodies. Int Immunopharmacol 8(8):1067–1073. https://doi.org/10.1016/j.intimp.2008.03.017

Michalek SM, Gregory RL, Harmon CC, Katz J, Richardson GJ, Hilton T et al (1987) Protection of gnotobiotic rats against dental caries by passive immunization with bovine milk antibodies to *Streptocococcus mutans*. Infect Immun 55(10):2341–2347. https://doi.org/10.1128/iai.55.10.2341-2347.1987

Mitoma M, Oho T, Michibata N, Okano K, Nakano Y, Fukuyama M, Koga T (2002) Passive immunization with bovine milk containing antibodies to a cell surface protein antigen-glucosyltransferase fusion protein protects rats against dental caries. Infect Immun 70 (5):2721–2724. https://doi.org/10.1128/IAI.70.5.2721-2724.2002

Molina-Infante J, Gisbert JP (2014) Optimizing clarithromycin-containing therapy for *Helicobacter pylori* in the era of antibiotic resistance. World J Gastroenterol 20(30):10338–10347. https://doi.org/10.3748/wjg.v20.i30.10338

Motoi Y, Inoue S, Hatta H, Sato K, Morimoto K, Yamada A (2005a) Detection of rabies-specific antigens by egg yolk antibody (IgY) to the recombinant rabies virus proteins produced in *Escherichia coli*. Jpn J Infect Dis 58(2):115–118. http://www.ncbi.nlm.nih.gov/pubmed/15858294

Motoi Y, Sato K, Hatta H, Morimoto K, Inoue S, Yamada A (2005b) Production of rabies neutralizing antibody in hen's eggs using a part of the G protein expressed in *Escherichia coli*. Vaccine 23(23):3026–3032. https://doi.org/10.1016/j.vaccine.2004.11.071

Mulvey GL, Dingle TC, Fang L, Strecker J, Armstrong GD (2011) Therapeutic potential of egg yolk antibodies for treating *Clostridium difficile* infection. J Med Microbiol 60(8):1181–1187. https://doi.org/10.1099/jmm.0.029835-0

Mwale PF, Lee CH, Lin LT, Leu SJ, Huang YJ, Chiang LC, Mao YC, Yang YY (2020) Expression, purification, and characterization of anti-Zika virus envelope protein: Polyclonal and chicken-derived single chain variable fragment antibodies. Int J Mol Sci 21(2):1–21. https://doi.org/10.3390/ijms21020492

Neri P, Tokoro S, Kobayashi R, Sugiyama T, Umeda K, Shimizu T, Tsuji T, Kodama Y, Oguma K, Mori H (2011) Specific Egg Yolk Immunoglobulin as a New Preventive Approach for Shiga-Toxin-Mediated Diseases. PLoS One 6(10):e26526. https://doi.org/10.1371/journal.pone.0026526

Nguyen SV, Icatlo FC, Nakano T, Isogai E, Hirose K, Mizugai H, Kobayashi-Sakamoto M, Isogai H, Chiba I (2011a) Anti–cell-associated glucosyltransferase immunoglobulin Y suppression of salivary mutans streptococci in healthy young adults. J Am Dent Assoc 142(8):943–949. https://doi.org/10.14219/jada.archive.2011.0301

Nguyen HH, Tumpey TM, Park HJ, Han GY, Lee J, Byun YH, Song JM, Tran LD, Nguyen VD, Kilgore PE (2011b) Avian antibodies to combat potential H5N1 pandemic and seasonal influenza. Influenza Respi Viruses 5:233–236

Nguyen HH, Tumpey TM, Park HJ, Byun YH, Tran LD, Nguyen VD, Kilgore PE, Czerkinsky C, Katz JM, Seong BL, Song JM, Kim YB, Do HT, Nguyen T, Nguyen CV (2010) Prophylactic and therapeutic efficacy of avian antibodies against influenza virus H5N1 and H1N1 in mice. PLoS One 5(4):1–11. https://doi.org/10.1371/journal.pone.0010152

Nilsson E, Kollberg H, Johannesson M, Wejåker PE, Carlander D, Larsson A (2007a) More than 10 years' continuous oral treatment with specific immunoglobulin Y for the prevention of *Pseudomonas aeruginosa* infections: a case report. J Med Food 10(2):375–378. https://doi.org/10.1089/jmf.2006.214

Nilsson E, Amini A, Wretlind B, Larsson A (2007b) *Pseudomonas aeruginosa* infections are prevented in cystic fibrosis patients by avian antibodies binding *Pseudomonas aeruginosa* flagellin. J Chromatogr B Anal Technol Biomed Life Sci 856(1–2):75–80. https://doi.org/10.1016/j.jchromb.2007.05.029

Nilsson E, Larsson A, Olesen HV, Wejåker PE, Kollberg H (2008) Good effect of IgY against *Pseudomonas aeruginosa* infections in cystic fibrosis patients. Pediatr Pulmonol 43 (9):892–899. https://doi.org/10.1002/ppul.20875

Nomura S, Suzuki H, Masaoka T, Kurabayashi K, Ishii H, Kitajima M, Nomoto K, Hibi T (2005) Effect of dietary anti-urease immunoglobulin Y on *Helicobacter pylori* infection in Mongolian gerbils. Helicobacter 10(1):43–52. https://doi.org/10.1111/j.1523-5378.2005.00290.x

O'Donnell KL, Espinosa D, Nilles ML, Harris E, Bradley DS (2018) Avian IgY antibodies recognize novel dengue NS1 epitopes with the ability to neutralize infection and reduce vascular leakage without inducing antibody dependent enhancement. Am J Trop Med Hyg 99(412). Retrieved from https://www.ncbi.nlm.nih.gov/pmc/articles/PMC7565424/?report=classic

O'Donnell KL, Fink A, Nilles ML, Bradley DS (2017) Dengue NS1-specific IgY antibodies neutralizes dengue infection without inducing antibody dependent enhancement. J Immunol 198:225.3. Retrieved from https://www.jimmunol.org/content/198/1_Supplement/225.3.abstract

O'Donnell K, Meberg B, Schiltz J, Nilles M, Bradley D (2019) Zika virus-specific IgY results are therapeutic following a lethal zika virus challenge without inducing antibody-dependent enhancement. Viruses 11(3):301. https://doi.org/10.3390/v11030301

Otake S, Nishihara Y, Makimura M, Hatta H, Kim M, Yamamoto T, Hirasawa M (1991) Protection of rats against dental caries by passive immunization with hen-egg-yolk antibody (IgY). J Dent Res 70(3):162–166. https://doi.org/10.1177/00220345910700030101

Otero R, Gutiérrez J, Beatriz Mesa M, Duque E, Rodríguez O, Luis Arango J, Gómez F, Toro A, Cano F, María Rodríguez L, Caro E, Martínez J, Cornejo W, Mariano Gómez L, Luis Uribe F, Cárdenas S, Núñez V, Díaz A (2002) Complications of Bothrops, Porthidium, and Bothriechis snakebites in Colombia. A clinical and epidemiological study of 39 cases attended in a university hospital. Toxicon 40(8):1107–1114. https://doi.org/10.1016/S0041-0101(02)00104-6

Parma YR, Chacana PA, Rogé A, Kahl A, Cangelosi A, Geoghegan P, Lucchesi PMA, Fernández-Miyakawa ME (2011) Antibodies anti-Shiga toxin 2 B subunit from chicken egg yolk: Isolation, purification and neutralization efficacy. Toxicon 58(4):380–388. https://doi.org/10.1016/j.toxicon.2011.07.009

Patel M (2020) Dental caries vaccine: are we there yet? Lett Appl Microbiol 70(1):2–12. https://doi.org/10.1111/lam.13218

Paton JC, Paton AW (1998) Pathogenesis and diagnosis of Shiga toxin-producing *Escherichia coli* infections. Clin Microbiol Rev 11(3):450–479. https://doi.org/10.1128/cmr.11.3.450

Pauly D, Dorner M, Zhang X, Hlinak A, Dorner B, Schade R (2009) Monitoring of laying capacity, immunoglobulin Y concentration, and antibody titer development in chickens immunized with ricin and botulinum toxins over a two-year period. Poult Sci 88(2):281–290. https://doi.org/10.3382/ps.2008-00323

Peck MW (2009) Biology and genomic analysis of *Clostridium botulinum*. Adv Microb Physiol. 55:183–265, 320. https://doi.org/10.1016/S0065-2911(09)05503-9

Pérez de la Lastra JM, Baca-González V, Asensio-Calavia P, González-Acosta S, Morales-Delanuez A (2020) Can Immunization of Hens Provide Oral-Based Therapeutics against COVID-19? Vaccine 8(3):1–17. https://doi.org/10.3390/vaccines8030486

Pillay VV (2003) Comprehensive Medical Toxicology, 1st Ed. Paras Pub

Pitaksajjakul P, Lekcharoensuk P, Upragarin N, Barbas CF, Ibrahim MS, Ikuta K, Ramasoota P (2010) Fab MAbs specific to HA of influenza virus with H5N1 neutralizing activity selected from immunized chicken phage library. Biochem Biophys Res Commun 395(4):496–501. https://doi.org/10.1016/j.bbrc.2010.04.040

Poh CM, Carissimo G, Wang B, Amrun SN, Lee CY-P, Chee RS-L et al (2020) Two linear epitopes on the SARS-CoV-2 spike protein that elicit neutralising antibodies in COVID-19 patients. Nat Commun 11(1):2806. https://doi.org/10.1038/s41467-020-16638-2

Pulić I, Loconte V, Zanotti G (2014) Structural characterization at the atomic level of a molecular nano-machine: the state of the art of *Helicobacter pylori* flagellum organization. Am J Biochem Biotechnol 10(3):143–161. https://doi.org/10.3844/ajbbsp.2014.143.161

Riley P, Lamont T (2013) Triclosan/copolymer containing toothpastes for oral health. In: Riley P (ed) Cochrane database of systematic reviews. https://doi.org/10.1002/14651858.CD010514.pub2

Rossetto O, Pirazzini M, Bolognese P, Rigoni M, Montecucco C (2011) An update on the mechanism of action of tetanus and botulinum neurotoxins. Acta Chim Slov 58(4):702–707

Rungsiwongse J, Ratanabanangkoon K (1991) Development of an ELISA to assess the potency of horse therapeutic antivenom against Thai cobra venom. J Immunol Methods 136(1):37–43. https://doi.org/10.1016/0022-1759(91)90247-D

Russell MW, Childers NK, Michalek SM, Smith DJ, Taubman MA (2004) A caries vaccine? The state of the science of immunization against dental caries. Caries Res 38(3):230–235. https://doi.org/10.1159/000077759

Russell MW, Hajishengallis G, Childers NK, Michalek SM (1999) Secretory immunity in defense against cariogenic mutans streptococci. Caries Res 33(1):4–15. https://doi.org/10.1159/000016490

Sainz-Mejías M, Jurado-Martín I, McClean S (2020) Understanding *Pseudomonas aeruginosa*–host interactions: the ongoing quest for an efficacious vaccine. Cell 9(12):2617. https://doi.org/10.3390/cells9122617

Salillas S, Sancho J (2020) Flavodoxins as novel therapeutic targets against *Helicobacter pylori* and other gastric pathogens. Int J Mol Sci 21(5). https://doi.org/10.3390/ijms21051881

Sanchez P, Everett B, Salamonson Y, Ajwani S, Bhole S, Bishop J, Lintern K, Nolan S, Rajaratnam R, Redfern J, Sheehan M, Skarligos F, Spencer L, Srinivas R, George S (2017) Oral health and cardiovascular care: perceptions of people with cardiovascular disease. PLoS One 12(7):e0181189. https://doi.org/10.1371/journal.pone.0181189

Satin B, Del Giudice G, Della Bianca V, Dusi S, Laudanna C, Tonello F, Kelleher D, Rappuoli R, Montecucco C, Rossi F (2000) The neutrophil-activating protein (HP-NAP) of *Helicobacter pylori* is a protective antigen and a major virulence factor. J Exp Med 191(9):1467–1476. https://doi.org/10.1084/jem.191.9.1467

Sautto GA, Kirchenbaum GA, Ross TM (2018) Towards a universal influenza vaccine: different approaches for one goal. Virol J 15(1):17. https://doi.org/10.1186/s12985-017-0918-y

Saxena SK, O'Brien AD, Ackerman EJ (1989) Shiga toxin, Shiga-like toxin II variant, and ricin are all single-site RNA N-glycosidases of 28 S RNA when microinjected into *Xenopus* oocytes. J Biol Chem 264(1):596–601. https://doi.org/10.1016/S0021-9258(17)31302-9

Schmitt CK, Meysick KC, O'Brien AD (1999) Bacterial toxins: friends or foes? Emerg Infect Dis 5 (2):224–234. https://doi.org/10.3201/eid0502.990206

Selim AM, Ibrahim EM, El M, Hamouda FK (2015) Development of IgY antibodies for control of tetanus. Biotechnol Anim Husband 31(1):109–122. https://doi.org/10.2298/BAH1501109S

Shimamoto C, Tokioka S, Hirata I, Tani H, Ohishi H, Katsu K (2002) Inhibition of *Helicobacter pylori* infection by orally administered yolk-derived anti-*Helicobacter pylori* antibody. Hepato-Gastroenterology 49(45):709–714. Retrieved from http://www.ncbi.nlm.nih.gov/pubmed/12063975

Shin JH, Nam SW, Kim JT, Yoon JB, Bang WG, Roe IH (2003) Identification of immunodominant *Helicobacter pylori* proteins with reactivity to *H. pylori*-specific egg-yolk immunoglobulin. J Med Microbiol 52(3):217–222. https://doi.org/10.1099/jmm.0.04978-0

Shin JH, Roe IH, Kim HG (2004) Production of anti-*Helicobacter pylori* urease-specific immunoglobulin in egg yolk using an antigenic epitope of *H. pylori* urease. J Med Microbiol 53 (1):31–34. https://doi.org/10.1099/jmm.0.05327-0

Slowinski JB, Lawson R (2002) Snake phylogeny: evidence from nuclear and mitochondrial genes. Mol Phylogenet Evol 24(2):194–202. https://doi.org/10.1016/S1055-7903(02)00239-7

Smith DJ (2002) Dental caries vaccines: prospects and concerns. Critical Rev Oral Biol Med 13 (4):335–349. https://doi.org/10.1177/154411130201300404

Smith DJ, King WF, Godiska R (2001) Passive transfer of immunoglobulin Y antibody to *Streptococcus mutans* glucan binding protein B can confer protection against experimental dental caries. Infect Immun 69(5):3135–3142. https://doi.org/10.1128/IAI.69.5.3135-3142.2001

Solhi R, Alebouyeh M, Khafri A, Rezaeifard M, Aminian M (2017) In vitro evaluation of cross-strain inhibitory effects of IgY polyclonal antibody against *H. pylori*. Microb Pathog:110. https://doi.org/10.1016/j.micpath.2017.03.025

Somasundaram R, Choraria A, Antonysamy M (2020) An approach towards development of monoclonal IgY antibodies against SARS CoV-2 spike protein (S) using phage display method: A review. International Immunopharmacology, (January). https://doi.org/10.1016/j.intimp.2020.106654

Sui J, Li W, Murakami A, Tamin A, Matthews LJ, Wong SK, Moore MJ, Tallarico ASC, Olurinde M, Choe H, Anderson LJ, Bellini WJ, Farzan M, Marasco WA (2004) Potent neutralization of severe acute respiratory syndrome (SARS) coronavirus by a human mAb to S1 protein that blocks receptor association. Proc Natl Acad Sci U S A 101(8):2536–2541. https://doi.org/10.1073/pnas.0307140101

Sun L, Zhang Z, Zeng B, Chen Y, Li H, Zhou H, Zhang Z, Li Z (2018) Preparation of multivalent egg yolk IgY antibodies against *Helicobacter pylori*. Chinese J Cell Molecular Immunol 34 (6):555–560. Retrieved from http://www.ncbi.nlm.nih.gov/pubmed/30236210

Surana S, Tosolini AP, Meyer IFG, Fellows AD, Novoselov SS, Schiavo G (2018) The travel diaries of tetanus and botulinum neurotoxins. Toxicon 147:58–67. https://doi.org/10.1016/j.toxicon.2017.10.008

Sutton P, Boag JM (2019) Status of vaccine research and development for *Helicobacter pylori*. Vaccine 37(50):7295–7299. https://doi.org/10.1016/j.vaccine.2018.01.001

Suzuki H, Nomura S, Masaoka T, Goshima H, Kamata N, Kodama Y, Ishii H, Kitajima M, Nomoto K, Hibi T (2004) Effect of dietary anti-Helicobacter pylori-urease immunoglobulin Y on Helicobacter pylori infection. Alimentary Pharmacol Therapeut, Suppl 20(1):185–192. https://doi.org/10.1111/j.1365-2036.2004.02027.x

Svante Twetman LGP (1999) Interdental caries incidence and progression in relation to mutans streptococci suppression after chlorhexidine-thymol varnish treatments in schoolchildren. Acta Odontol Scand 57(3):144–148. https://doi.org/10.1080/000163599428878

Takeshita H, Watanabe E, Norose Y, Ito Y, Takahashi H (2019) Neutralizing antibodies for *Helicobacter pylori* urease inhibit bacterial colonization in the murine stomach in vivo. Biomed Res (Japan) 40(2):87–95. https://doi.org/10.2220/biomedres.40.87

Thomas L, Tyburn B, Lang J, Ketterle J (1996) Early infusion of a purified monospecific F(ab') 2 antivenom serum for *Bothrops lanceolatus* bites in Martinique. Lancet 347(8998):406. https://doi.org/10.1016/s0140-6736(96)90590-5

Thomsen K, Christophersen L, Bjarnsholt T, Jensen P, Moser C, Høiby N (2015) Anti-*Pseudomonas aeruginosa* IgY antibodies induce specific bacterial aggregation and internalization in human polymorphonuclear neutrophils. Infect Immun 83(7):2686–2693. https://doi.org/10.1128/IAI.02970-14

Thomsen K, Christophersen L, Bjarnsholt T, Jensen P, Moser C, Høiby N (2016) Anti-*Pseudomonas aeruginosa* IgY antibodies augment bacterial clearance in a murine pneumonia model. J Cyst Fibros 15(2):171–178. https://doi.org/10.1016/j.jcf.2015.08.002

Thwaites CL, Beeching NJ, Newton CR (2015) Maternal and neonatal tetanus. The Lancet 385 (9965):362–370. https://doi.org/10.1016/S0140-6736(14)60236-1

Tischler ND, Galeno H, Rosemblatt M, Valenzuela PDT (2005) Human and rodent humoral immune responses to Andes virus structural proteins. Virology 334(2):319–326. https://doi.org/10.1016/j.virol.2005.01.031

Tohidpour A (2016) CagA-mediated pathogenesis of *Helicobacter pylori*. Microb Pathog 93:44–55. https://doi.org/10.1016/j.micpath.2016.01.005

Tsukamoto M, Hiroi S, Adachi K, Kato H, Inai M, Konishi I, Tanaka M, Yamamoto R, Sawa M, Handharyani E, Tsukamoto Y (2011) Antibodies against swine influenza virus neutralize the pandemic influenza virus A/H1N1. Mol Med Rep 4(2):209–214. https://doi.org/10.3892/mmr.2011.410

Tsukamoto Y, Nakano Y, Adachi K (2018) Protection against infectious bronchitis virus, a corona virus infection, using ostrich antibodies. Health 10(10):1294–1308. https://doi.org/10.4236/health.2018.1010100

Tun-Pe, Warrell DA, Tin-Nu-Swe, Phillips RE, Moore RA, Myint-Lwin, Burke CW (1987) Acute and chronic pituitary failure resembling sheehan's syndrome following bites by russell's viper in burma. Lancet 330(8562):763–767. https://doi.org/10.1016/S0140-6736(87)92500-1

van Raamsdonk M, de Soet JJ, de Graaff J (1993) Effect of monoclonal antibodies on the colonization of rats by *Streptococcus sobrinus*. Caries Res 27(1):31–37. https://doi.org/10.1159/000261512

Wagner M (1966) A study of the effects of specific immunization on experimental dental caries in the gnotobiotic rat. Purdue University

Wallach MG, Webby RJ, Islam F, Walkden-Brown S, Emmoth E, Feinstein R, Gronvik KO (2011) Cross-protection of chicken immunoglobulin Y antibodies against H5N1 and H1N1 viruses passively administered in mice. Clin Vaccine Immunol 18(7):1083–1090. https://doi.org/10.1128/CVI.05075-11

Walls AC, Park YJ, Tortorici MA, Wall A, McGuire AT, Veesler D (2020) Structure, function, and antigenicity of the SARS-CoV-2 Spike Glycoprotein. Cell 180:1–12. https://doi.org/10.1016/j.cell.2020.02.058

Wang Q, Hou X, Cai K, Li T, Liu Y, Tu W, Xiao L, Bao S, Shi J, Gao X, Liu H, Tian R, Wang H (2010) Passive protection of purified yolk immunoglobulin administered against Shiga toxin 1 in mouse models. Can J Microbiol 56(12):1003–1010. https://doi.org/10.1139/W10-087

Wang Z, Wang Q (2008) Clinical study of anticarious IgY tooth paste in deciduous teeth caries prevention. Chin J Conserv Dent 18(6):334–337

Wang B, Yang J, Cao S, Wang H, Pan X, Zhu J, Zhou Y, Gao L, Li W, Li M (2014) Preparation of specific anti-*Helicobacter pylori* yolk antibodies and their antibacterial effects. Int J Clin Exp Pathol 7(10):6430–6437

Warrell DA (1996) Clinical features of envenoming from snake bites. In: Bon C, Goyffon M (eds) Envenomings and their treatments. Fondation Marcel Mérieux, Lyon, France, Editions, pp 63–76

Warrell DA (2004) Snakebites in Central and South America: epidemiology, clinical features and clinical management. In: Campbell JA, Lamar WW (eds) The Venomous reptiles of the Western Hemisphere. Ithaca: Cornell University Press; 2004. pp 709–761

Waskito LA, Salama NR, Yamaoka Y (2018) Pathogenesis of *Helicobacter pylori* infection. Helicobacter 23:1–6. https://doi.org/10.1111/hel.12516

Wei S, Duan S, Liu X, Wang H, Ding S, Chen Y, Xie J, Tian J, Yu N,Ge P, Zhang X, Chen X, Li Y, Meng Q (2020) Chicken Egg Yolk Antibodies (IgYs) block the binding of multiple SARS-CoV-2 spike protein variants to human ACE2. Int Immunopharmacol (September):107172. https://doi.org/10.1016/j.intimp.2020.107172

Wen Z, Song H, Ming G (2017) How does Zika virus cause microcephaly? Genes Dev 31 (9):849–861. https://doi.org/10.1101/gad.298216.117

Wen J, Zhao S, He D, Yang Y, Li Y, Zhu S (2012) Preparation and characterization of egg yolk immunoglobulin Y specific to influenza B virus. Antivir Res 93(1):154–159. https://doi.org/10.1016/j.antiviral.2011.11.005

Wong CS, Jelacic S, Habeeb RL, Watkins SL, Tarr PI (2000) The Risk of the Hemolytic–Uremic Syndrome after Antibiotic Treatment of *Escherichia coli* O157:H7 Infections. N Engl J Med 342(26):1930–1936. https://doi.org/10.1056/NEJM200006293422601

Wrapp D, Wang N, Corbett KS, Goldsmith JA, Hsieh CL, Abiona O, Graham BS, McLellan JS (2020) Cryo-EM structure of the 2019-nCoV spike in the prefusion conformation. Science 367 (6483):1260–1263. https://doi.org/10.1126/science.abb2507

Wu NC, Wilson IA (2020) Structural biology of influenza hemagglutinin: an amaranthine adventure. Viruses 12(9):1053. https://doi.org/10.3390/v12091053

Xing P, Shi Y, Dong C, Liu H, Cheng Y, Sun J, Li D, Li M, Sun K, Feng D (2017) Colon-targeted delivery of IgY against *Clostridium difficile* toxin A and B by encapsulation in Chitosan-Ca pectinate microbeads. AAPS PharmSciTech 18(4):1095–1103. https://doi.org/10.1208/s12249-016-0656-2

Xu C, Soyfoo DM, Wu Y, Xu S (2020) Virulence of *Helicobacter pylori* outer membrane proteins: an updated review. Eur J Clin Microbiol Infect Dis 39(10):1821–1830. https://doi.org/10.1007/s10096-020-03948-y

Yamaoka Y (2010) Mechanisms of disease: *Helicobacter pylori* virulence factors. Nat Rev Gastroenterol Hepatol 7(11):629–641. https://doi.org/10.1038/nrgastro.2010.154

Yamaoka Y, Kikuchi S, ElZimaity HMT, Gutierrez O, Osato MS, Graham DY (2002) Importance of *Helicobacter pylori oipA* in clinical presentation, gastric inflammation, and mucosal interleukin 8 production. Gastroenterology 123(2):414–424. https://doi.org/10.1053/gast.2002.34781

Yan J, Mao YF, Shao ZX (2005) Frequencies of the expression of main protein antigens from *Helicobacter pylori* isolates and production of specific serum antibodies in infected patients. World J Gastroenterol 11(3):421–425. https://doi.org/10.3748/wjg.v11.i3.421

Yang YH, Park D, Yang G, Lee SH, Bae DK, Kyung J, Kim D, Choi E, Son JC, Hwang SY, Kim YB (2012) Anti-*Helicobacter pylori* effects of IgY from egg yolk of immunized hens. Laboratory Animal Research 28(1):55. https://doi.org/10.5625/lar.2012.28.1.55

You Z, Yang H, Xin W, Kang L, Gao S, Wang J, Zhang T, Wang J (2014a) Preparation of egg yolk antibodies against BoNT/B and their passive protection in mouse models. Hum Vaccin Immunother 10(8):2321–2327. https://doi.org/10.4161/hv.29433

You J, Xu Y, Li H, Wang L, Wu F, Xu F, Jin L, Li S, Li X (2014b) Chicken egg yolk immunoglobulin (IgY) developed against fusion protein LTB-STa-STb neutralizes the toxicity of *Escherichia coli* heat-stable enterotoxins. J Appl Microbiol 117(2):320–328. https://doi.org/10.1111/jam.12525

Zaki AM, Van Boheemen S, Bestebroer TM, Osterhaus ADME, a.M. Fouchier, R. (2012) Isolation of a novel coronavirus from a man with pneumonia in Saudi Arabia. N Engl J Med 367 (19):1814–1820. https://doi.org/10.1056/NEJMoa1211721

Zhang W, Berberov EM, Freeling J, He D, Moxley RA, Francis DH (2006) Significance of heat-stable and heat-labile enterotoxins in porcine colibacillosis in an additive model for pathogenicity studies. Infect Immun 74(6):3107–3114. https://doi.org/10.1128/IAI.01338-05

Zhang X, McDaniel AD, Wolf LE, Keusch GT, Waldor MK, Acheson DWK (2000) Quinolone antibiotics induce shiga toxin–encoding bacteriophages, toxin production, and death in mice. J Infect Dis 181(2):664–670. https://doi.org/10.1086/315239

Zhang Y, Wei Y, Li Y, Wang X, Liu Y, Tian D, Jia X, Gong R, Liu W, Yang L (2020) IgY antibodies against Ebola virus possess post-exposure protection and excellent thermostability. BioRxiv:1–28. https://doi.org/10.1101/2020.05.21.108159

Zhang S, Xing P, Guo G, Liu H, Lin D, Dong C, Li M, Feng D (2015) Development of microbeads of chicken yolk antibodies against *Clostridium difficile* toxin A for colonic-specific delivery. Drug Deliv:1–8. https://doi.org/10.3109/10717544.2015.1022836

Zhang M, Xu X, Tsutomu S (2004) Influence of Olitin Tooth Protecting Spray (Chicken Egg Yolk Antibodies-IgY) on Caries Activity. Stomatology 5

Zhou G, Zhao Q (2020) Perspectives on therapeutic neutralizing antibodies against the Novel Coronavirus SARS-CoV-2. Int J Biol Sci 16(10):1718–1723. https://doi.org/10.7150/ijbs.45123

Zhou Z, Zhou R, Tang Z (2003) Effects of topical application of immunoglobulin yolk on mutans streptococci in dental plaque. West China J Stomatol 21(4):295–297. Retrieved from http://www.ncbi.nlm.nih.gov/pubmed/14513588

Zhu N, Zhang D, Wang W, Li X, Yang B, Song J et al (2020) A novel coronavirus from patients with pneumonia in China, 2019. N Engl J Med 382(8):727–733. https://doi.org/10.1056/NEJMoa2001017

IgY Industries and Markets

17

Ricardo S. Vieira-Pires, Hyeong Chul Ahn, Marina Bok, Catherine D. Caulfield, Pablo Chacana, Fazle Elahi, Anders O. Larsson, Carlos Leónidas Leiva, Leonard M. Moran, Patricia M. Morgan, Viviana Parreño, Celina G. Vega, and Andrés Wigdorovitz

R. S. Vieira-Pires (✉)
CNC – Center for Neuroscience and Cell Biology, University of Coimbra, Coimbra, Portugal
e-mail: ricardo.pires@biocant.pt

H. C. Ahn · F. Elahi
ADBioTech Co., Chuncheon, South Korea
e-mail: adex_adbio@naver.com; elahidr@adbiotech.com

M. Bok
INCUINTA, CICVyA-INTA, Buenos Aires, Argentina
e-mail: bok.marina@inta.gob.ar

C. D. Caulfield · L. M. Moran
Ovagen Group Ltd, Mayo, Ireland
e-mail: Catherine.Caulfield@ovagen.com; leonard.moran@ovagen.com

P. Chacana · C. L. Leiva
Instituto de Patobiología - UEDD INTA-CONICET IPVet, Instituto Nacional de Tecnología Agropecuaria, Buenos Aires, Argentina
e-mail: chacana.pablo@inta.gob.ar

A. O. Larsson
Department of Medical Sciences, Uppsala University, Uppsala, Sweden
e-mail: anders.larsson@akademiska.se

P. M. Morgan
Emerita Lecturer, School of Natural Sciences, National University of Ireland Galway, Galway, Ireland
e-mail: pat.morgan@nuigalway.ie

V. Parreño
INCUINTA, CICVyA-INTA, CONICET, Buenos Aires, Argentina
e-mail: parreno.viviana@inta.gob.ar

C. G. Vega · A. Wigdorovitz
Bioinnovo; INCUINTA, CICVyA-INTA, CONICET, Buenos Aires, Argentina
e-mail: vega.celina@inta.gob.ar

X.-Y. Zhang et al. (eds.), *IgY-Technology: Production and Application of Egg Yolk Antibodies*, https://doi.org/10.1007/978-3-030-72688-1_17

Abstract

Applications of the knowledge of avian IgY biology from establishing proof-of-concept studies to the formation of new industries and markets has been very poorly documented. It is particularly relevant to clearly present to a broader audience the real-world companies and industries that are translating the unique capabilities of birds and their antibodies into new products and services with worldwide impact. Herein we showcase a selection of what we termed IgY companies and elucidate how these are exploiting opportunities in areas such as biopharmaceuticals, functional foods or cosmetics, animal nutrition, large scale IgY manufacture or even vaccine production. A brief overview of the current IgY-technology intellectual property landscape and its peculiarities is also presented.

Keywords

Biopharmaceuticals · Immunoprophylaxis and immunodiagnosis · Human and animal health · Large-scale IgY manufacture · Vaccines · Intellectual property

17.1 Introduction

In this chapter we present a selection of real-world companies and industries developing a broad range of solutions based on IgY antibodies, IgY Technology and chicken hosts. The chapter is divided in the form of extended company and/or product case-studies, or rather brief company profiles, aiming to convey a clear picture of how the potential of IgY antibodies and related technological advances is being exploited in industrial and commercial endeavours.

Some of the major players in the global market of IgY-based products and services are listed in Table 17.1. From research, development (R&D) and innovation of antibodies, to human and veterinary medicine, nutraceuticals, foods, cosmeceuticals and even scientific education, the spectrum of activities of these IgY companies is quite broad. There is also a large number of companies retailing products and services based on IgY-derived technologies for research purposes, including R&D antibodies, antibody-based solutions and diagnostic kits that have not been exhaustively listed herein. For reference, these include companies such as Abcore Inc. (US), Agrisera AB (Sweden), Covalab (France), Creative Biolabs Inc. (US), Immuno Reagents Inc. (US), Innovagen AB (Sweden) and OriGene Technologies Inc. (US).

To date there is no systematic report on the IgY antibody markets, so one can only extrapolate the market value of particular IgY applications in specific market niches, and it is rather challenging and maybe misleading, to combine all of them in a single number reflecting the overall IgY businesses market value. For example in the Research Antibodies & Reagents Market, that is projected to reach USD 14.1 billion by 2025 from USD 10.1 billion in 2020, at a CAGR of 6.7% during the forecast

Table 17.1 Companies, Industries, Brands and Markets with activities based on IgY antibodies and IgY Technology

Name	Website	Activities	Country
AD Biotech Co.	http://adbiotech.com	1,3,4,5,6	South Korea
Aves Labs, Inc.	http://www.aveslab.com/	1,2	US
Avianax LLC	https://www.avianax.com/	1,3	US
Bioinnovo	http://bioinnovo.com.ar/	1,3	Argentina
Crystal Bioscience *	http://www.crystalbioscience.com/	1,5	US
DAN Biotech, Inc.	http://www.danbio.com	1,3,4,5,6	South Korea
Davids Biotechnologie GmbH	https://www.davids-bio.de/	1,2	Germany
Eggcellent Proteins	https://www.eggcellentproteins.com/	1,2,3,5	Scotland
EW Nutrition	https://ew-nutrition.com/	1,3,5,6	Germany
Gallus Immunotech, Inc. **	https://www.exalpha.com/	1,2	US
Gentian AS	https://www.gentian.com/	1,2	Norway
Good Biotech Corp.	http://www.good-biotech.com/	1,2,4	China
HenBiotech	http://www.henbiotech.com/	1,2	Portugal
IgY Immunologix	http://www.igylx.com/	1,2,3,5	India
IGY Life Sciences, Inc.	http://www.igylifesciences.com/	1,4,5,6	Canada
IgY Nutrition	https://www.igynutrition.com/	1,5,6	US
IgYTechnology.com	https://www.igytechnology.com/	7	Portugal
Immune Biosolutions	https://immunebiosolutions.com/	1,2,5	Canada
NABAS	http://www.nabas.no/	1,2,3	Norway
Ovagen Group Limited	http://www.ovagen.ie/	1,2,3,5	Ireland

(1) Research and Development (R&D); (2) Research Antibodies and antibody services; (3) Veterinary Medicine Products; (4) Cosmeceuticals and personal care; (5) Human medicine; (6) Nutraceutical and Food supplements; (7) Science Education; *Acquired by Ligand Pharmaceuticals Inc. in 2017 (CA, USA); **Acquired by Exalpha Biologicals, Inc. in 2018 (MA, USA)
Working webpages as of December 2020

period (MarketsandMarkets™Inc 2020), one can do a rough estimation of what chicken antibodies would represent in this market. According to a BenchSci platform (www.benchsci.com) search, that uses an AI-based algorithm to screen the literature for published antibodies, there are 4.6 M antibody product entries for rabbit hosts, –2 M for mice, –282.1 K for goats, –170.2 K for rats and –17.2 K for chickens. This means that chicken antibodies would represent a –0.24% share of the market (–USD 33.8 M in 2025), but certainly one should carefully consider such an extrapolation.

In the last section of this Chapter, we discuss the current status of the IgY-technology intellectual property (IP) landscape, namely describing how companies and inventors are fulfilling the patentability criteria of novelty, inventiveness and industrial applicability, for applications of egg yolk antibodies, that happen to be a product already existing in nature, thus presenting IP protection peculiarities.

17.2 Case-studies of IgY Companies

17.2.1 Gentian AS (Moss, Norway)

Anders O. Larsson

Introduction

Gentian AS is the first diagnostic company that has focused on avian antibody-based particle-enhanced turbidimetric immunoassays (PETIA) for high volume chemistry analysers (Fig. 17.1). Gentian develops and manufactures *in vitro* diagnostic products for the detection and quantification of specific markers for use on a wide range of clinical chemistry analysers. The company has its headquarters in Moss, near Oslo but it also works from Sweden (Stockholm and Gothenburg) and China (Beijing).

Some of the PETIA products are for the determination of human serum/plasma calprotectin, faecal calprotectin, neutrophil gelatinase-associated lipocalin (NGAL), cystatin C and canine C reactive protein. In the pipeline of products to be launched in the next few years are tests addressing cardiovascular diseases, viral infections and cancer diagnostics.

IgY Based Particle-Enhanced Turbidimetric Immunoassay (PETIA) for Determination of Cystatin C

Brief Overview of the Methodology

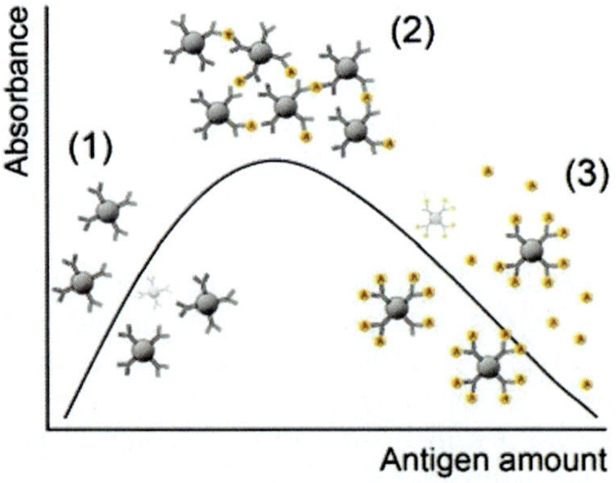

Fig. 17.1 The effects of increasing amounts of antigen on absorbance values. Initially the absorbance increases as the amount of antigen increases until the absorbance reaches a plateau; after the plateau the absorbance decreases again. (**1**) Nanoparticles (e.g. latex beads) coated with antigen specific antibodies are initially dispersed in solution, resulting in lower absorbances; (**2**) As the antigen amount increases, the nanoparticles tend to form immune complexes that are detected by an absorbance change (increase); (**3**) In the presence of an excess of antigen, the nanoparticles get their available binding sites saturated and no immune complexes are favoured (absorbance decrease)

The first antibody-based immunoassays were based on mammalian antibodies and the precipitation of antigen-antibody complexes in a gel or solution. They are still used in hospital laboratories in large volumes adapted for chemistry analysers. The concentration of proteins such as albumin, IgG, IgA and IgM are in the g/L range in plasma. It is impractical to analyse such high concentrations by ELISA as it would require very high dilutions. Instead, in the chemistry analyser the plasma samples are diluted with a buffer and the antibodies are added and the absorbance change caused by the immune complexes is monitored. The amount of protein is determined by comparing the absorbance change in the samples with the calibration curve. A normal assay time is approximately 10 min and can be used for proteins that are present in concentrations higher than approximately 0.1 g/L.

With increasing amounts of antigen in the solution larger immune complexes are formed, and thus higher absorbance, until a point is reached when the size of the immune complexes decreases again. This decrease is due to antigen excess (Fig. 17.1). The amount of the antigen in the excess situation is erroneously interpreted as being much lower. Thus, it is important to have a safety zone so that even samples with very high antigen concentrations do not reach antigen excess. A weakness with IgY, when it is used in free solution, is that the decrease in absorbance in the antigen excess situation is much more rapid than for rabbit antibodies. The situation is quite different when IgY is used in PETIA assays. The use of PETIA rather than free antibodies allows for a stronger signal and detection of lower amounts of antigen. The limit of detection is dependent on many factors, but a general guideline is that a PETIA can be used for proteins present at concentration of approximately 0.1 mg/L and higher. C-reactive protein is one of the proteins that often are determined using PETIA. Other examples are cystatin C and urine albumin.

The Cystatin C PETIA

It is now more than 10 years since Gentian developed their original cystatin C PETIA. It utilizes affinity-purified chicken antibodies to cystatin C bound to nanoparticles (Flodin et al. 2007). The reagent consists of two components, a buffer (Reagent 1) and the particles (Reagent 2) and calibrators. Today there are optimized protocols for all the major chemistry analysers.

The application on Architect c8000 can be seen as a typical example for the IgY based cystatin C PETIA. Cystatin C is a frequently requested analyte in our hospital, and we are performing approximately 35,000 to 60,000 routine cystatin C assays per year as markers of kidney function. The kinetics of samples with different concentrations of cystatin C are shown in Fig. 17.2.

Cystatin C concentration may be measured in the range of 0.30– 8.00 mg/L representing normal plasma or serum sample ranges. Values above 8 mg/L are only seen in patients with total loss of kidney function and are confirmed by dilution and re-running of the sample within an additional 10 minutes. The CV for duplicate samples in one study was 0.58%. In contrast to free IgY antibodies, the PETIA has a very slow decline in absorbance in the antigen excess region. Even a cystatin C value of 50 mg/L was higher than the highest standard point and was not misinterpreted. There is thus a very broad safety margin for the assay. We have also looked at the

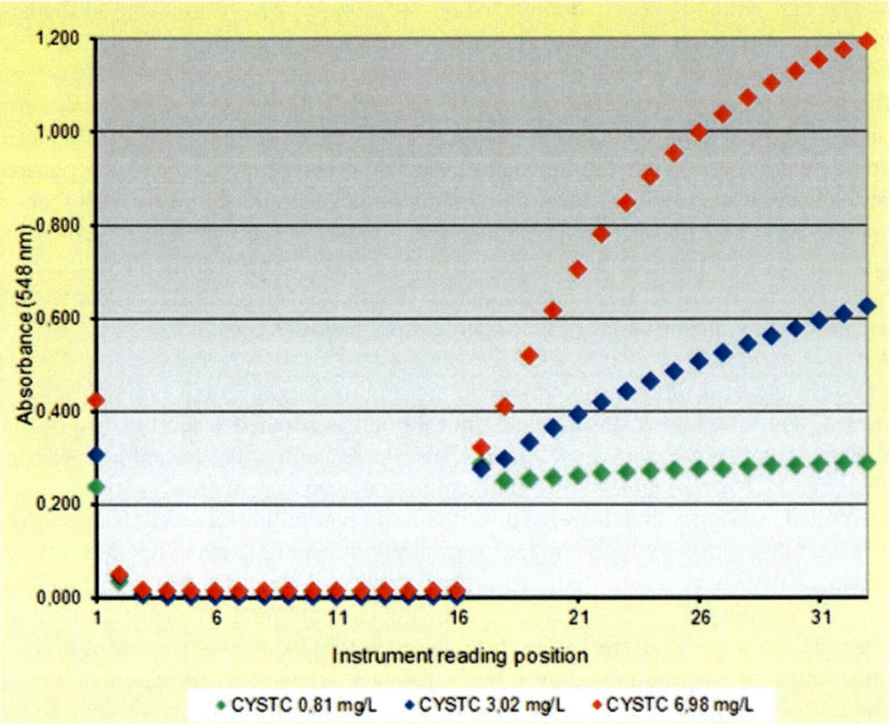

Fig. 17.2 Kinetics of the absorbance change for three plasma samples containing 0.81, 3.02 and 6.98 mg/L of cystatin C. The increase in absorbance just after read position 16 is due to the addition of the antibody-coated particles (Reagent 2). A blank reading is performed at position 18 and an endpoint reading is performed at position 33. The absorbance change is used to calculate the cystatin C concentration in the patient samples

reproducibility over time with data from a number of routine batches. Using ten patient pools stored at −70 °C we showed that all tested batches gave assay results within 3% of the originally assigned values for the patient pools.

When comparing the reaction kinetics of the Gentian IgY based cystatin C method with rabbit antibody-based cystatin C reagents, we found similar kinetics, but the IgY based immunoparticles resulted in a higher delta absorbance with the Architect c8000. This may be related to the phylogenetical differences between birds and mammals.

The cystatin C results are automatically converted from a cystatin C concentration to an estimated glomerular filtration rate (GFR). Estimated GFR is used as a marker for kidney function and is considered as the best test to monitor kidney function in clinical routine. GFR is used to monitor patients with a number of diseases that may affect the kidney function including cardiovascular diseases. The kidney function is also measured as eGFR for correct dosage of drugs. GFR estimates measured with creatinine or cystatin C is thus one of our most frequently requested laboratory tests.

17.2.2 Immunsystem AB (Uppsala, Sweden)

Anders O. Larsson

Introduction

Immunsystem AB is a small Swedish biotech company working with the production of avian antibodies. The company started in 1983 and initially focused on the production of avian antibodies for diagnostic purposes. The focus during the last 20 years have been on the use of avian antibodies for therapeutic applications.

Oral Immunotherapy with Anti-*Pseudomonas* IgY in Cystic Fibrosis Patients

Immunsystem AB is focussed on oral immunotherapy with yolk antibodies to prevent *Pseudomonas aeruginosa* infections in patients with Cystic Fibrosis (CF). CF is a hereditary disorder affecting the gene for the cystic fibrosis transmembrane conductance regulator. This protein acts as a transport channel for chloride ions. These ions are necessary for the production of freely flowing mucus, sweat, saliva and tears. CF patients suffer from repeated respiratory infections and the thick mucus reduces the ability to clear the bacteria from the respiratory tract. The patients usually become infected by *P. aeruginosa* bacteria sometime during their life. Initially, it is possible to eradicate these infections but over time the patients become chronically infected and it is then impossible to eradicate the infection. Chronic *P. aeruginosa* infections are major causes of morbidity and mortality in CF patients. Patients are regularly treated with high doses of antibiotics to reduce the bacterial loads, but it also causes antibiotic resistance.

This troublesome situation was the background to Immunsystem's focus on *P. aeruginosa* infections in CF. Oral administration of avian antibodies against *P. aeruginosa* could be an alternative for the prevention of *P. aeruginosa* infections in CF patients, and one that would not lead to antibiotic resistance. The idea was that the treatment would clear the oral cavity of *P. aeruginosa* bacteria thus reducing the risk of the bacteria reaching the lower respiratory tract. The antibody treatment is highly specific and would only target the *P. aeruginosa* bacteria thus leaving the normal bacterial flora intact.

In 1995 the first patient was enrolled in the study. The study was an open study without a double-blind control group. It was important that the treatment was simple for the patients and that the preparation did not involve any toxic substances. To avoid any toxic substances the antibodies were extracted with the water dilution method and stored frozen in 70 mL volumes. The patients thawed the bottle and gargled once daily (in the evening) and swallowed the solution. The gargling resulted in the presence of active antibodies against *P. aeruginosa* throughout the night i.e., for at least 8 hours. The initial antibody formulation was based on egg yolk preparations from hens immunized with two *P. aeruginosa* strains. A total of seventeen Swedish CF patients were enrolled in this study (Kollberg et al. 2003). A Danish group of CF patients was used as a control group. In the IgY treated group there were 2.5 *P. aeruginosa* positive sputum cultures and in the control group 13.7

which corresponds to a more than 80% reduction in *P. aeruginosa* infections. There were no signs of adverse events related to the IgY treatment. This was considered so promising that Immunsystem received an approval from the Swedish Medical Products Agency (MPA) to continue the treatment of the patient group with reimbursement from the Swedish government (Nilsson et al. 2008). Still, a non-randomized study could not be the basis for a registration of the treatment as a pharmaceutical. Immunsystem AB has over the years showed that it is possible to produce avian antibodies against *Pseudomonas* (=Anti-*Pseudomonas* IgY) in a reproducible manner that fulfil MPA requirement. Immunsystem has also received Orphan Drug Designation by the EMA (ref. no. EU13/08/56) for the prevention of *P. aeruginosa* infections in CF patients. To continue the development of this treatment strategy, Immunsystem together with partners received a research grant from EU of 5.35 million Euro to perform a placebo controlled double-blind multicentre study (Larsson 2016). The study was called IMPACTT and was officially started in November 2011. The study included 164 patients in nine European countries and a total of 47 sites. The patients were treated with either Anti-*Pseudomonas* IgY or with placebo solution, gargling every night; IgY activity against PA is present in saliva and oro-pharynx overnight and prevents PA from entering the lungs. The Anti-*Pseudomonas* IgY was a similar egg yolk formulation as in the original Swedish open study, but the number of strains used for immunization had been increased from two *P. aeruginosa* strains to six. The placebo is made by the exact same procedure as the active product, the only difference being the starting material which is eggs from non-vaccinated hens. The treatment period was a maximum of two years. Well over 100,000 daily doses were produced for the study under MPA approved conditions. The last patient completed the clinical trial in 2018. The results so far indicate that the treatment is safe and that there have been no serious adverse events in the study that could be linked to the treatment. The preliminary results of the study showed no significant difference between the patients receiving active antibody formulation from the patients receiving the non-specific antibody formulation. The results in the active arm was as expected but the control group had less infections than expected. A possible explanation for this is that both groups received yolk formulations that contain, apart from the antibodies, also egg proteins that are antibacterial (e.g., lactoferrin and avidin). These antibacterial proteins may thus have protected against the bacterial infections in both groups reducing the treatment effects between the two groups. Oral administration of water extracted egg yolk proteins very much resembles the use of eggs as part of the diet. Thus, it seems reasonable that oral immunotherapy with yolk antibodies should be safe for the patient. The amount of egg proteins in a daily dose corresponds to an amount less than half an egg.

The IMPACTT study also included preclinical parts with the same type of antibodies as used for the patient treatment. A safety study in mice was performed showing that the anti-*Pseudomonas* antibodies did not cause damage to the gastrointestinal mucosa and did not alter the normal bacterial flora in the gastrointestinal tract. Mode-of-action studies were also performed and these showed that Anti-*Pseudomonas* IgY promoted bacterial opsonization and augmented the phagocytic activity of human polymorphonuclear neutrophils (Thomsen et al. 2015). The unspecific IgY preparation also had an antibacterial effect although clearly less than the specific IgY formulation. IgY do not interact with mammalian Fc-receptors so the phagocytotic activity was not Fc-receptor mediated but most likely due to bacterial aggregation that facilitated phagocytosis. The same antibodies could also facilitate clearance of *P. aeruginosa* bacteria from the lungs of mice (Thomsen et al. 2016). Administration of the antibodies into the lung reduced the bacterial burden 100-fold compared to controls. The effect on bacterial counts after 24 hours were also accompanied by reduced clinical symptom scores and diminished lung inflammation in the treated animals.

Thus, there are preclinical data that support the positive effect of the treatment. Considering the lack of superior effect of the specific formulation over the placebo group, Immunsystem has put the development of this formulation on hold and stopped the open Swedish study.

Detection of Protein A in Mammalian IgG Containing Solutions using the Immunsystem ABs ELISA

Protein A Determination Assays

Immunsystem AB also produce ISO certified kits for protein A determination. Monoclonal antibodies administered intravenously have to be highly purified to minimize reactions to impurities in the antibody preparations. The most widely used affinity chromatography ligand is *Staphylococcal* protein A that has a very good affinity for mammalian immunoglobulins (Lindmark et al. 1983). Companies that purify their antibodies on protein A matrixes have to document that no protein A has leached into the final preparation (less than 1 ppm) as protein A has several Ig-binding regions and will cause the formation of immune complexes *in vivo*. Immunological determination of protein A in an IgG containing solution offers considerable challenges especially when using mammalian detection antibodies as they will compete with the antibodies in the solution for the binding to protein A. One of the few antibodies that do not bind to protein A is IgY. Immunsystem AB thus developed an ELISA for determination of protein A based on affinity purified chicken antibodies (Larsson et al. 1992). The ELISA is a sandwich immunoassay with a chicken anti-protein A as capture and another chicken anti-protein A antibody as detector (Fig. 17.3). The assay can detect less than 1 ng/mL of protein A both in solutions free of IgG and solutions containing mg amounts of IgG. The production of this kit has recently been transferred to IgY Lab Systems AB (Sweden).

17.2.3 Bioinnovo (Castelar, Argentina)

Andrés Wigdorovitz, Celina G. Vega, Marina Bok, and Viviana Parreño

Introduction

Bioinnovo® is a biotechnology company founded in July 2014 dedicated to the research, design and development of IgY antibodies, recombinant vaccines and the

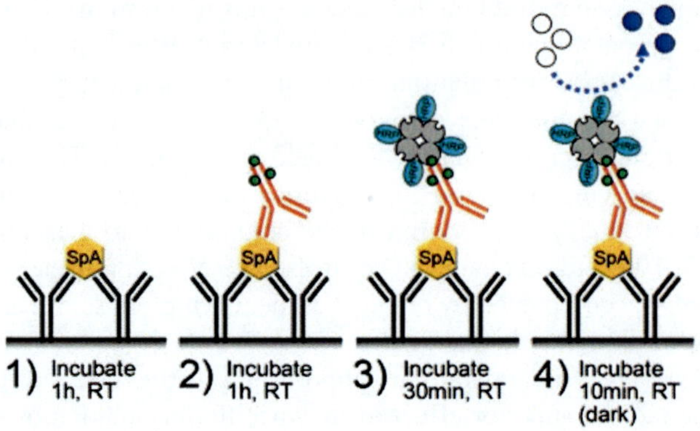

Fig. 17.3 Principle of Immunsystem AB ELISA assay for determination of protein A. (1) *Staphylococcal* protein A (SpA) from the sample is bound to wells precoated with the primary chicken anti-SpA IgY antibody (capture antibody; black). (2) Biotinylated chicken anti-SpA IgY antibody is added as secondary antibody (detector antibody; orange). (3) Streptavidin HRP (gray-cyan complex) is added and binds to the biotin (green circles) conjugate. (4) Substrate (white circles) is added and reacts with HRP to produce a coloured product (blue circles). Finally, stop solution is added to terminate the chromogenic reaction

provision of specialized animal health services. It was co-founded by Andrés Wigdorovitz, Celina Vega and Viviana Parreño. The products developed by Bioinnovo® are based on two biotechnological platforms: the IgY Technology and the patented APCH1 platform, derived from the research work of the founding group and the Algenex® company. In addition, Bioinnovo® provides specialized services for the evaluation of products for the poultry industry on its experimental farm. The headquarters of Bioinnovo are located at the Center for Research in Veterinary and Agricultural Sciences (CICVyA) of Instituto Nacional de Tecnología Agropecuaria (INTA) Castelar, in Argentina. It comprises four sectors: (1) a fully-equipped R&D laboratory for its two production platforms and quality control; (2) a layer production unit; (3) a plant for IgY immunoglobulin production (Fig. 17.4); and (4) an experimental poultry farm for the evaluation of own and third-party products. The Bioinnovo plant is independent, but as it is located inside the CICVyA. The facilities of the centre include the Incuinta laboratory, P2 biosafety cabinets, the Tissue Culture sector and the vivarium, and high-technology equipment. It is associated with the facilities of the international veterinary laboratory Vetanco S.A., where regulatory affairs, quality control and logistics, packaging and distribution are located.

Bioinnovo has developed two biological products: Bioinnovo IgY DNT and VEDEVAX block. Internationally, Bioinnovo has a unique strategic advantage due to its location in Argentina: its products are free from Foot-and-mouth disease and Avian Influenza viruses as well as from Bovine Spongiform Encephalopathy (BSE), meeting the requirements of the most demanding markets worldwide.

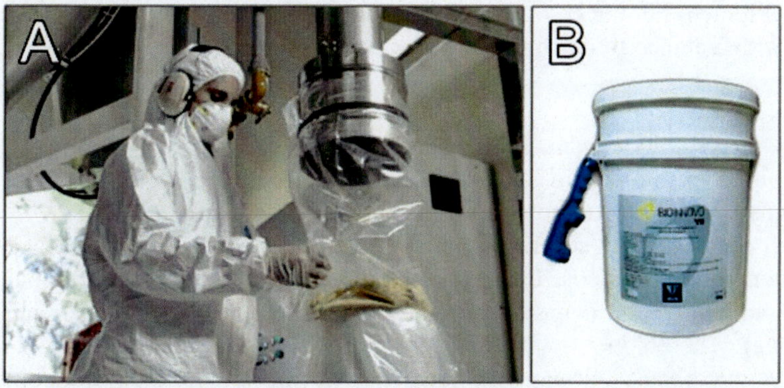

Fig. 17.4 Bioinnovo plant for IgY production. (**a**) Egg yolk spray-drier apparatus of the plant unit for IgY immunoglobulin production. (**b**) Example of a commercial unit (6 Kg) of non-nutritional additive Bioinnovo IgY (Pictures taken from http://bioinnovo.com.ar/ with permission)

Effective Health Management in Intensive Animal Production

Managing Neonatal Calf Diarrhoea (NCD)

Neonatal diarrhoea of artificially-reared calves is an example of challenges in intensive milk production systems (Chap. 15). These require standardized and effective health management measures to prevent diseases. Such practices must also be sustainable and have little or no impact on human populations and on the environment. Indeed, the current global trend is the limited and targeted use of antibiotics aiming at preserving their efficacy both in animals and humans, preventing the emergence of multi-resistant bacterial strains, as well as reducing treatment-associated costs. Climate, management practices and the calf's immune and nutritional status also influence the emergence, severity, and consequences of the disease.

Bioinnovo IgY DNT is the perfect ally for the control of NCD. It is based on specific avian IgY immunoglobulins against the main infectious agents that cause NCD (Vega et al. 2011, 2015). The administration of Bioinnovo IgY DNT reduces the severity and duration of scours, minimizing weight loss and improving the performance of calves. It is also highly palatable, easy to administer and complements vaccination and treatment schemes in cows and calves. Bioinnovo IgY DNT controls not only diarrhoea of viral origin, but also those caused by bacteria, reducing or even avoiding the use of antibiotics, thus benefitting herd health. In addition, by shortening the duration of diarrhoea episodes, the shedding of infectious agents in the environment, which is the main source of infection for susceptible animals, is reduced. Its inclusion in the feed, as a complementary strategy, improves animal health and significantly reduces losses due to disease and mortality of artificially-reared calves. The product is added directly to the calf's feed (milk or milk replacer) twice a day to maintain optimal antibody levels in the intestinal lumen. For the commercialisation of the Bioinnovo IgY DNT, Bioinnovo agreed with Vetanco to outsource the commercial strategy.

Conclusions

Bioinnovo IgY DNT advantages are summarized as follows:

- High efficacy, proven in field conditions against the pathogens for which it is designed (Rotavirus, Coronavirus, *Salmonella* and *Escherichia coli*)
- Lower incidence and severity of clinical symptoms and mortality due to NCD
- Reduction of gut mucosa colonization by pathogens, promoting intestinal health
- Easy administration and storage at room temperature (it does not require cold storage)
- Lower microbial environmental load by reducing the duration and severity of scours and as a result of reduced pathogen shedding in the faeces (main source of infection)
- High-quality nutritional supplement.

17.2.4 Crystal Biosciences (California, USA)

Ricardo S. Vieira-Pires

Crystal Biosciences Incorporated (CB) is a privately-held company based in Emeryville, CA, USA, that was founded in 2008 to develop a platform for therapeutic antibody discovery featuring wild-type and engineered chickens. CB became a leader company in the area of difficult-to-address-epitopes of antibody research, having developed the proprietary Gel Encapsulated Microenvironment (GEM) assay platform, as well as the HuMab technology, claimed to be the world's only chicken-based human antibody platform to successfully generate human antibodies (Ching et al. 2018). In 2017, CB was acquired by Ligand Pharmaceuticals Incorporated (Emeryville, CA, USA) a NASDAQ listed biopharmaceutical company focused on developing or acquiring technologies that help pharmaceutical companies to discover and develop new medicines (www.ligand.com). CB's HuMab technology was integrated as part of Ligand's OmniAb® platform (www.omniab.com) and rebranded as OmniChicken™. For simplicity reasons the name Crystal Biosciences (CB) will be kept in this brief profile addressing the company.

As described in Chap. 3, the immune response developed in mice, rats, rabbits and other mammals to human proteins is very poor, since these hosts present natural tolerance to mammalian conserved-epitopes. CB has exploited the potential of chickens as phylogenetic distant hosts to generate antibodies with high affinity and broad epitope coverage to conserved targets. The company developed two prominent technologies, the (1) Gel Encapsulated Microenvironment (GEM) assay platform and (2) OmniChicken™, that are promising and disruptive approaches to revisit and reboost chicken antibody repertoires in the field of antibody discovery.

The gel encapsulated microenvironment (GEM) assay platform (Izquierdo et al. 2016) has been described in Chap. 13.3.4. Major advantages of the GEM platform include: (i) screening capacity up to 100 million cells per single experiment,

(ii) direct recover of non-immortalized antibody-secreting B cells from tissue of any species, (iii) multiparameter screening of antibody features from a single B cell, (iv) direct identification of biologically active antibodies, (v) screening for antibodies binding with low nanomolar affinity or better, (vi) retention of native heavy and light chain pairs created by *in vivo* affinity maturation and (vii) fast selection of more than 100 clones after harvesting B cells from immunized hosts. By combining the GEM platform with chicken B cell repertoires, CB was able to successfully select unique antibodies against particularly relevant therapeutic targets, including membrane proteins channels and transporters (Bednenko et al. 2018) as well as membrane receptors, such G protein-coupled receptors (GPCRs) (Könitzer et al. 2017). These are intrinsically challenging targets since normally the antibody is required to bind to a limited surface-exposed portion of the membrane protein, most often through a structure-sensitive interaction and/or is required to additionally modulate the protein function (Dodd et al. 2018).

CB also developed the ground technology for the current OmniChicken™ platform. The technology enables generation of fully-human antibodies based on genetically engineered chickens expressing human immunoglobulin repertoires (Leighton et al. 2015; Ching et al. 2018). Indeed, human therapeutic monoclonal antibodies must be as similar to native human antibodies as possible in order to minimize adverse immunogenicity in patients (Dodd et al. 2018). Similarly to other in vivo systems for expression of human antibodies in transgenic animal platforms, including genetically modified mice, rats, rabbits, and cows (Brüggemann et al. 2015), OmniChicken™ core advantage relies on the fact that the antibodies are produced in the intact immune system of the animal and undergo highly rigorous native selection stages including (i) for target specificity, (ii) counter-selection to endogenous off-target proteins and (ii) high-level expression in plasma cells. Grounded on successful data in DT40 cell line, where CB performed a proof of concept showing that arrays of human-derived immunoglobulin gene sequences could be diversified by chicken B cells (Schusser et al. 2013; Leighton et al. 2015), the company moved forward to generate a transgenic chicken line—HuMab chicken—expressing chimeric antibodies with human V genes and chicken C genes for both heavy and light chains (Ching et al. 2018). Notably, the transgenic birds retained a phenotypically normal B cell development, with transgenes diversified as in native chickens (by gene conversion and somatic hypermutation), while delivering affinity-matured human-sequence antibodies with chicken-like immune recognition and responsiveness, namely to mammalian conserved targets.

17.2.5 Ovagen Group Ltd.

Catherine D. Caulfield and Leonard M. Moran

Ovagen Group Ltd. is an innovation-based biotechnology company, led by serial entrepreneurs and located in Mayo in the west of Ireland. It was founded in 2007. The company has previously extensive experience in maintenance of SPF hens

(Chap. 8). Currently, the company specialises in avian germ-free technology and avian transgenics to Good Manufacturing Practice standards.

Development of Germ-Free eggs

Following seven years of R&D the founders have developed the world's first germ-free eggs for supply to global vaccine manufacturers and for use in new biopharmaceuticals. This is a novel, patented, innovative and disruptive technology. Over 1 billion (Bn) eggs are used annually in vaccine manufacture servicing a vaccine market size estimated to be worth $104.87 Bn by 2017 with a CAGR of 10.7%. (Vaccines Market Size, Share, Growth Global Industry Report 2027). Today 82% of the world's vaccines are produced using egg-based technology and there is an enormous demand for vaccines for COVID-19 and seasonal influenza. Ovagen is committed to substantially improving the global production (volume & cost) of vaccines which are essential in the fight against disease and pandemics such as coronavirus or other zoonotic viruses associated with viral pandemics. Ovagen's germ free eggs will revolutionise the vaccine manufacturing process by: eliminating contaminating bacteria; increasing the yield of vaccines per egg; and increasing quality and security thus enabling the accelerated production of higher quality vaccines to support volume production and security of supply. When vaccines are made, the virus is grown in the developing embryo in the yolk of fertilised hens' eggs. Specific Pathogen Free (SPF) eggs, which are the best quality eggs available to vaccine manufacturers currently, are not always free of bacteria or viruses. Bacterial contamination of eggs is caused by both the unique anatomy of the avian species (Chap. 2) and the porosity of the shell. The pores of the egg shell are particularly vulnerable to bacterial ingress during laying and for approximately 5 minutes after the egg is laid whilst the outer shell cuticle dries. In chickens, the reproductive tract merges with the digestive tract with a single point of exit from the hen for eggs and faecal matter (Chap. 2). This results in the porous egg coming into contact with the chicken's faeces prior to being laid and consequently contamination of the fertilised egg within the shell.

Ovagen has developed a process to address this anatomical problem of egg contamination with a specialised procedure which combines the extensive experience of the technical team in germ-free technology with a wealth of veterinary, scientific and research expertise. A surgical procedure is required for creation of the first generation of germ-free eggs/birds; subsequent generations are produced from germ-free eggs laid naturally by germ-free birds housed in specialised germ-free isolators (Fig. 17.5). Vaccine yield per egg is expected to be significantly higher using germ-free eggs due to reduced interferon levels as the immune system is not challenged (Chap. 3). This is another game changer for vaccine producers.

Transgenic Chicken Technology at Ovagen Group Ltd

Ovagen Group Ltd. also have the capacity to produce transgenic and Germ-free chickens in this avian facility (Chap. 16). The company has developed and perfected avian Transgenic technology and plan to develop a number of biosimilar drugs in the

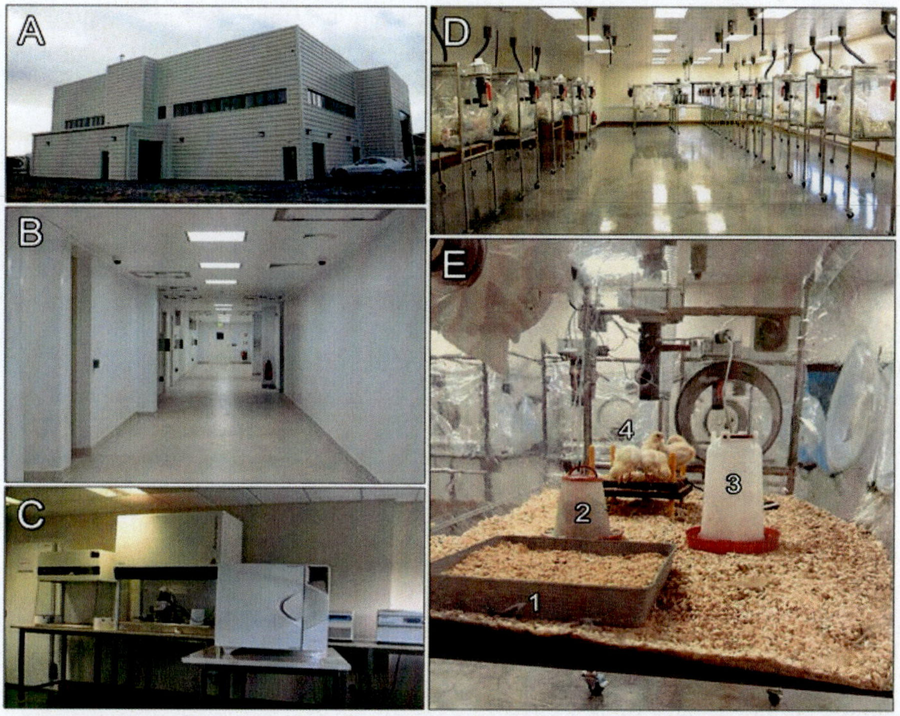

Fig. 17.5 Ovagen Group Ltd. state of the art avian facility. The main avian building is shown in (**a**). The central corridor of the avian SPF facility is shown in (**b**) which provides entry to five individual SPF rooms. A fully equipped and dedicated Transgenic laboratory with a high-definition imaging system, microinjection capabilities and dedicated incubation facilities is shown in (**c**). The avian isolators which are used to house the germ-free birds are shown in (**d**). Germ-free chickens are shown in panel (**e**) in an enriched isolator. A dust bath is shown in (1), a diet feeder (2), water drinker (3) and chickens on a heater panel (4). Photographs courtesy of Ovagen Group Ltd.

future. The methods for the production of transgenic proteins in chickens has been outlined in Chap. 14. This results in the targeted production of recombinant protein into the egg-white of all eggs laid by transgenic hens. Protein production by transgenic technology is now being realised as a viable alternative to current methodology, with therapeutic drugs, produced from transgenic chickens, receiving full FDA approval. Eggs from transgenic hens have been shown to contain approximately 300 mg of recombinant protein per egg. This is equivalent to 10 g/litre which is better than most cell-based expression systems. With the average chicken laying 300 eggs per year, a single chicken can produce 60 g of recombinant protein per year. The time-scale to produce a transgenic chicken line is approximately 18 months. Once a transgenic line of chicken has been produced, it is simply a case of standard avian husbandry to maintain the protein-producing flock. This is an area where Ovagen personnel have many years of experience in contract research. Stability of transgenes has been shown to be very high, with no variance observed in either DNA or protein in over 20 chicken generations. Chicken produced proteins have been shown to have a very good safety profile, as chickens have been the source of most vaccines produced for over 70 years.

17.2.6 ADBioTech (South Korea)

Hyeong Chul Ahn and Fazle Elahi

ADBiotech Co Ltd. was established in 2000, with the aim of transforming natural resources into therapeutic products. This company focuses on development of alternative therapeutic and healthy food for livestock, aquaculture and human sectors (Table 17.2). It holds the premier position in Korea for the IgY technology and employs 53 people.

Ig-Guard Calf Prevents Diarrhoea in New-Born Calves

Passive immunization with IgY is an effective and economical method to control calf diarrhoea (Vega et al. 2011) (Chap. 15). We have developed a product that contains antibodies against several pathogens such as *E. coli*, *Salmonella*, ***Clostridium,*** rotavirus, coronavirus, and *Cryptosporidium*. IgY-Guard Calf in animal passive protection studies prevented diarrhoea and enhanced the growth performance during 1 to 2 and 2 to 3 months of age on growing period in Hanwoo calves (Table 17.3). The product also has the capacity to enhance immunity in Hanwoo calves by stimulating IgA levels (Table 17.4).

Ig-Guard Shrimp

Recently, farmed shrimp production has decreased drastically due to Early Mortality Syndrome (EMS) (FAO 2013). Generally, this disease can be treated with antibiotics, but this has constrained the exporting of shrimp due to antibiotic resistance, public health and environmental issues. ADBioTech has developed Ig-Guard SH by using IgY technology to combat the EMS problem. We did a challenge experiment with *Vibrio spp*. and the results demonstrated that the product prevents the mortality of shrimp, increased body weight and decreased the Feed Conversion Ratio (FCR) rate. Moreover, this product had an immune enhancer function by increasing the immune parameters like total haemocytes number, phagocytosis, phenol oxidase activity, lysozyme activity and even bactericidal activity. As white spot syndrome virus (WSSV) is another big problem in shrimp farming (Chap. 15), we have already developed a IgY product against it in 2007.

Ig-Guard Helico

Helicobacter pylori is one of the most frequently occurring global infectious agents and infects around 50% of the world's population (Chap. 16). ADBioTech developed Ig-Guard Helico and tested the *in vitro* anti-bacterial effect at 16 mg/mL and higher concentrations; the product completely inhibited the growth of *H. pylori* following 3-day incubation of the bacteria (1×10^8 CFU/mL) with a serially diluted IgY (Table 17.5). In an *in vivo* test, *H. pylori* (1×10^8 CFU, 3 times) inoculated in

Table 17.2 Product Categories of ADBioTech

Products	Species	Application/ Use /Indication
Livestock		
Ig-Guard Calf	Calf	Calf diarrhoea
Ig-Guard Swine	Piglet	Piglet diarrhoea
Ig-Guard Duck	Duck	Duck viral hepatitis, Colibacillosis, Salmonellosis
Ig-Guard Poultry	Broiler and layer chicken	Colibacillosis, Salmonellosis
Ig-Guard Puppy	Dog	Distemper, Parvovirus, Bordetella, coronavirus
Aquaculture		
Ig-Guard (SH)	Shrimp	Early Mortality Syndrome (EMS) or Hepatopancreatic Acute Necrosis Syndrome (APHN)
Ig-Guard (SA)	Salmon	Salmonid Rickettsial Septicaemia (SRS)
Ig-Guard (AE)	Eel	*Edwardsiella spp* infection
Ig-Guard (AY)	Yellowtail	*Streptococcus spp* and *Nocardia spp*
Human Health		
Ig-Guard Helico	Human	Prevents *Helicobacter pylori* infection, gastric ulcer
Ig-Guard Cholesterol	Human	Decreases cholesterol
Ig-Guard Rota	Children	Prevents rotavirus infection
Ig-Guard Mutant	Human	Prevents dental caries
Ig-Guard Acne	Skin care for human	Protects and prevents acne

the stomach of to C57BL/6 mice demonstrated *H. pylori* infection. Mice groups orally treated with IgY twice a day for 18 days showed less infected mice in comparison with a control (vehicle) group (Table 17.5). By comparison, 25% (2/8) positive reaction was achieved following treatment with pantoprazole a medication used to treat stomach ulcers (30 mg/kg).

Gastric mucosal specimens were cultured after treatment with Ig-Guard Helico or vehicle (Table 17.6). Bacteria grew in all the samples from mice (100%) treated with vehicle. However, IgY treatment inhibited the bacterial growth in a dose-dependent manner (Table 17.6). Pantoprazole (30 mg/kg) also decreased bacterial growth to 25%. Histopathological results revealed that IgY also elevates the sloughing off of the gastric villi and the inflammation lesion returned to normal at 100–500 mg/kg

Table 17.3 Ig-Guard Calf enhances the growth performance in Hanwoo calves

Parameters	Calves					
	From 1 to 2 months of age			From 2 to 3 months of age		
	Control	TF[2]	P	Control	TF[2]	P
Initial body weight, kg	50.67 ± 4.3	48.61 ± 4.9	0.965	68.51 ± 4.5[a]	70.50 ± 4.5[b]	0.021
Off test body weight, kg	68.51 ± 4.5[a]	70.50 ± 6.8[b]	0.021	87.05 ± 8.2[a]	93.04 ± 8.9[b]	0.013
Average daily gain, kg	0.59 ± 0.08[a]	0.72 ± 0.06[b]	0.051	0.61 ± 0.06[a]	0.74 ± 0.07[b]	0.001
Feed and nutrient intakes	**Kg/day, 30 days**					
Concentrates	0.41	0.46	0.637	1.07	1.15	0.521
Roughages	0.07	0.07	0.998	0.2	0.2	0.536
DDMI[3]	0.48	0.53	0.512	1.15	1.21	0.249
DDMI/ADG[4]	0.81 ± 0.21[b]	0.74 ± 0.26[a]	0.042	1.81 ± 0.84[a]	1.64 ± 0.65[b]	0.035`

[a]Means with difference superscript in the same row are significantly different ($p < 0.05$)
[b]Number of Hanwoo calves in each group is 20,[2] Test feed starter. [3] Daily dry matter intake. [4] Daily dry matter intake per average daily body gain

Table 17.4 Ig-Guard Calf enhances IgA levels in Hanwoo calves

Immune parameter	Weeks	Treatments[1]		SEM	P
		Control	TF[2]		
Ig A		---------- mg/mL ----------			
	1	0.11	0.1	0.01	0.241
	2	0.12	0.24	0.01	0.023
	3	0.09	0.2	0.02	0.011
	4	0.1	0.22	0.02	0.013

Data represent as SEM: Least squares means, ($P < 0.05$), [1]Number of Hanwoo calves in each group is 20. [2]Test feed starter

Table 17.5 Ig-Guard Helico effect on *H.pylori* in mouse model

Treatment (mg/Kg)	1	2	3	4	5	6	7	8	9	Positive Ratio
Vehicle Control	•	•	•	•	•	•	•	•	•	8/8
+ IgY (100)	•	•	•	•	o	o	o	o	o	4/8
+ IgY (200)	•	◐	o	o	o	o	o	o	o	1/8
+ IgY (500)	•	◐	o	o	o	o	o	o	o	1/8
+ Pantoprazol (30)	•	•	o	o	o	o	o	o	o	2/8

o, Negative; ◐, Partial positive; •, Positive

Table 17.6 Ig-Guard Helico effect on *H. pylori* in gastric mucosa

Treatment (mg/Kg)	1	2	3	4	5	6	7	8	9	Positive Ratio
Vehicle Control	+	+	+	+	+	+	+	+	+	8/8
+ IgY (100)	+	+	+	+	–	–	–	–	–	4/8
+ IgY (200)	+	–	–	–	–	–	–	–	–	1/8
+ IgY (500)	+	–	–	–	–	–	–	–	–	1/8
+ Pantoprazol (30)	+	+	–	–	–	–	–	–	–	2/8

–, Negative; +, Positive

dosage. In conclusion, Ig-Guard Helico has a positive effect on prevention of the *H. pylori,* that supports the ongoing clinical trials in the two Korean university hospitals.

17.3 Brief Profiles of Other IgY Companies

17.3.1 EW Nutrition (Visbek, Germany)

Ricardo S. Vieira-Pires

EW Nutrition GmbH is an animal nutrition company that develops and commercializes animal nutrition solutions designed to improve animal welfare and health. These include solutions for gut health management, antibiotic reduction, young animal nutrition and toxin risk management and target the five most prominent animal nutrition markets, including poultry, swine, ruminants and aquaculture.

The company is headquartered in Visbek, Germany, but is active globally with operational teams in 42 countries, and manufacturing plants and development centres in 4 continents.

In addition to its core activity in animal nutrition, the company also develops and supplies egg IgY immunoglobulin-based ingredients to third party companies developing their own protective and preventive solutions for human healthcare, namely oral care, intestinal care, respiratory challenges, and skin and mucosa disorders. Table 17.7. summarizes the different types of products developed and commercialized by EW Nutrition worldwide. These includes a line of products branded Ovalgen® and designed with different functional outcomes, targeting in particular the ethological microbial agents responsible for a given disease or disorder. For example, Ovalgen® DC is an Anti-*Streptococcus mutans* IgY-based ingredient that can be used to tackle *S. mutans*-induced dental caries (Nguyen et al. 2011); Ovalgen® HP is an Anti-*Helicobacter pylori* IgY to prevent and treat gastritis (Aiba et al. 2019) and Ovalgen® RV is an Anti-human rotavirus IgY to control and treat rotaviral diarrhoea in children (Thu et al. 2017). All these IgY-based ingredients can be incorporated in human health and functional foods or cosmetics, and are compatible with a multitude of final formats, including tablets, lozenges, candies, gums, tooth paste, gels, mouthwash creams (Chap. 12) or even impregnated in filters or mask supports, as for example Ovalgen® FL with Anti-Influenza properties, helping to prevent seasonal flu infections. EW Nutrition also commercializes the IgYGate series, that is designed to end-users and follows a similar customization with the Ovalgen® line of ingredients.

17.3.2 Avianax (North Dakota, USA)

Ricardo S. Vieira-Pires

In Chaps. 4, 5 and 9 the unique features of the truncated IgY variant, IgY(ΔFc) present in ducks and geese were discussed. Avianax LLC is a biotechnology company from Grand Forks, North Dakota, USA, specialized in human and companion animal therapeutics based on IgY derived from domestic geese. Avianax was founded in 2002 by the owners of Schiltz Goose Farm, one of the farms affected by the West Nile virus (WNV) 2002-04 outbreak occurring in Grand Forks, North Dakota (Bell et al. 2005). During this period WNV became endemic in the local environment and was responsible for the death of large goose populations. Due to the high infection risk leading to economic burden, there was an urgent need for a cure and this motivated researchers from the Center of Research Excellence for Avian Therapeutics for Infectious Diseases at the University of North Dakota, led by David Bradley, to develop a goose IgY-based therapy that successfully treated and recovered infected birds. The company is currently developing goose IgY antibodies for Rabies, Dengue (O'Donnell et al. 2020) and Zika (O'Donnell et al. 2019) fever, avian flu and has an advanced phase therapeutic for Canine Parvovirus (Chap. 15).

Due to late vaccination or neglect of vaccination schedules combined high infectivity and difficult to control outbreaks, canine parvovirus (CPV) is still a

Table 17.7 EW Nutrition IgY-based products commercially available for global markets

Product	Description	Specificity	Format/Applications	Markets
Ovalgen®	Line of IgY-based products/ ingredient	customizable IgY specificity	health and functional foods or cosmetics	*detailed below*
Ovalgen® DC	Anti-*Streptococcus mutans* IgY for dental caries	maintains a well-balance environment in the oral cavity and healthy teeth	tablets, candy, and tooth paste; health food and cosmetics	Japan, United States, Korea, Vietnam
Ovalgen® PG	Anti-*Porphyromonas gingivalis* IgY for periodontitis & gingivitis	supports and improves gum health	tablets, gum, mouthwash, edible film, tooth paste; gum-care food and cosmetics	Japan, United States, Vietnam
Ovalgen® FL	Anti-*Influenza* IgY for seasonal flu	provides additional protection during the flu season	tablets, lozenges, gum, candy, drinks or foods; impregnated in air-filters and masks	Japan, Korea, India, Vietnam
Ovalgen® CA	Anti-*Candida albicans* IgY for oral thrush or candidiasis	oral and skin care; dry mouth	oral and skin care creams and gels; various functional foods and cosmetic products	Japan
Ovalgen® HP	Anti-*Helicobacter pylori* IgY for gastritis	strengthen gut immune system and resistance against stomach pathogens	yogurts, tablets and capsules; functional foods	Japan, Korea, Vietnam
Ovalgen® RV	Anti-human rotavirus IgY for rotaviral diarrhoea	enhance the gut immunity in children; preventive & therapeutic in gastrointestinal disorders	infant formulas and powdered milk products	Korea
Ovalgen® CS	Anti-*Cronobacter sakazakii* IgY for infant milk formula	enhance the gut immunity in young children	infant formulas and powdered milk products	Korea

NS, not specified. Information collected from https://ew-nutrition.com and (Thu et al. 2017) with permission

common problem in veterinary medicine, even though an effective vaccine is available. Avianax developed ParvoONE®, a therapeutic produced in SPF geese, that are immunized with a modified live Parvovirus strain; anti-CPV IgY antibodies are purified from egg yolk, concentrated to 25 mg/mL and dosed at 2.5 mg per pound (intravenous administration). The company has conducted a field trial in 11 sites, involving 120 dogs and showed that ParvoONE® reduced mortality rate to 12% with average hospitalization time also dropping to 2.8 days, instead of the –5 to 7 days of hospitalization time when only supportive care is given. In addition, only a few animals needed to receive intravenous fluid therapy, resulting in significant overall lower costs to treat CPV infections.

By late 2019, Avianax communicated non-officially through their media channels that they were continuing the work on production of commercial scale serials of their product (pre-licensing serials), enabling the launch of safety studies required by United States Department of Agriculture (USDA). The company was optimistic and anticipated a market entry for ParvoONE® in late Spring of 2020.

17.3.3 IGY Life Sciences (Ontario, Canada)

Ricardo S. Vieira-Pires

IGY Life Sciences is a privately held biotechnology company based in Ontario, Canada, with core activities on the development, manufacture and licensing of IgY-based products to infectious and non-infectious diseases in both humans and animals.

The company's 10,000 square foot good manufacturing practice (GMP)-certified facility is designed to produce up to 2000 kilograms of IgY antibodies per month. This capability is grounded on the company's proprietary MASON Extraction Methodology™, a purification platform that ensures large-scale and environmentally friendly production of high-purity specific and non-specific IgY antibodies. The process allows for purity levels of IgY from 25% to 99%, and outcompetes both large scale production of powdered whole egg IgY (which is roughly 2-3% IgY by weight) and small scale production of highly purified IgY by common research laboratories. This is rapidly positioning the company as a global leader in the large-scale IgY manufacture and a strategic partner in the field.

IGY Life Sciences' portfolio includes lead compounds against COVID-19 and African swine fever and marketed immune and sports health products (Table 17.8). Indeed, in 2020 the company also entered the race for alternative solutions to COVID19 vaccines. By June 2020, a press release announced their collaboration with clinical data and regulatory experts for continuing development of their new product IgY-110, currently in pre-clinical phase (BusinessWire 2020a, b). This is a nasal spray-based IgY formulation to be used as an anti-SARS-CoV-2 preventive pharmaceutical with complementary effects to active vaccination. The product is intended to treat those currently infected with COVID-19 by controlling and blocking the spread of SARS-CoV-2. While not original, the strategy is particularly relevant for both elderly and immune-compromised subjects and the company is focused on further developing the product to ensure high cross-protection towards other forms of human coronaviruses.

In addition, the company has two IgY-based consumer health products on the market, one for sports medicine (Vector450) and another for overall immune health (IGY Immune). Both contain Muno-IGY that is a raw, non-specific IgY antibody

Table 17.8 Biologics, animal health and consumer health products or candidates on IGY Life Sciences portfolio

Biologics		
Program/ Product	Therapeutic Area /Category	Phase
IGY-110	Anti-Coronavirus IgY Antibody	Preclinical
IGY-102	Universal Influenza Antibody Therapeutic	Discovery
IGY-106	Rotavirus Therapeutic	Discovery
IGY-108	Universal *H. pylori* Therapeutic	Discovery
IGY-112	Irritable Bowel Disease	Discovery
Animal Health		
IGY-205	African Swine Flu	Development
IGY-201	Rotavirus	Discovery
IGY-207	Rotavirus Vaccine	Discovery
Consumer Health		
Vector450	Sports nutrition	In Market
IGYImmune	Immune Support	In Market
ImmunoDerm Y	Skin Care	Discovery
TBD	Infant Milk Formula	Discovery

Information retrieved from https://www.igylifesciences.com/products, December 2020

extracted from antibiotic-free eggs, preserving their natural immune enhancer properties. The product works to support and optimize the innate and adaptive immune system, by binding and help eliminating harmful pathogens from the gastrointestinal tract. This helps reboot a challenged or aging immune system. Muno-IGY also integrates the formulation of Vector 450, that consists of a sports nutrition supplement to help elite athletes in their training recovery and to heal injuries. A pilot study revealed that regular IgY intake lowers the over-production of the stress hormone cortisol and allows the athletes to recover quickly; it also contributed to significantly retaining the production of IgA, the first line of defence for Upper Respiratory Infections.

17.3.4 Ostrich Pharma Corp (Kyoto, Japan)

Ricardo S. Vieira-Pires

Some of the unique features of using ostrich as a host for IgY were discussed in Chap. 9. Ostrich Pharma Corp is a biotechnology company based in Kyoto, Japan, that develops and commercializes ostrich IgY antibodies for treatments and prophylaxis in different human health conditions. The company has recently expanded its business to the US through Ostrich Pharma USA (Massachusetts, USA). Their activity is largely grounded on more than 20 years of experience of Dr. Yasuhiro Tsukamoto and his research team, who have studied and developed antibodies for various bacterial and viral pathogens, allergens, and other relevant substances. The company holds unique facilities with housing capacity for more than 500 ostriches, enabling multiple parallel experiments for antibody customization.

Ostrich Pharma has focussed on developing different solutions and products using ostrich IgY antibodies immobilized on solid surfaces, including different filter formats, face masks and other personal protection clothing. For example, filters impregnated with IgY antibodies against avian H5N1 virus were shown to effectively prevent viral transmission between chickens in IgY treated filter-covered housings, compared to control untreated-filter covered ones, where the mortality of birds was significantly higher (Kamiyama et al. 2011). IgY antibodies from ostriches immunized with seasonal A/H1N1 have also been reported as cross-reactive to the pandemic influenza virus A/H1N1 2009, seasonal A/H1N1, A/H3N2 and B viruses (Adachi et al. 2011). These antibodies were effective in neutralizing the virus's cytopathological effects in MDCK cell lines, revealing promising features for immunoprophylactic applications (Chap. 16).

Interestingly, one of the company's core businesses has been to exploit the long-standing Japanese practice of wearing masks in public to prevent the spread of colds and allergies. In 2008-9, during the Influenza A/H1N1 outbreak the company developed masks coated with antibodies against four flu virus strains including the H5N1 avian variety and sold more than 80 M units (Matsuyama 2009). The production capacity using ostrich is as high as 1.6 kg of egg derived IgY per month, which enables coating of about 32 M masks. Already during the current SARS-CoV-2 pandemic, in March 2020, the company announced in the media the production of IgY against the new virus and confirmed the antibody neutralization capacity (BusinessWire 2020a, b). The research team has previous work on targeting other coronavirus, namely Infectious Bronchitis Virus, a major respiratory disease in chicken (Tsukamoto et al. 2018); however, to date no literature update has been presented on the IgY antibodies against human SARS-CoV-2.

In addition to masks, also spray-based and nasal-drop products designed for immune-prophylaxis of hands, nose and throat have been explored by the company. Ostrich Pharma USA for example, is promoting a line of products branded "OstriGen", that encompass a broad range of medical solutions, such as treatments for gastrointestinal diseases by *Clostridium difficile*, *Vibrio cholera*, *E. coli*, *Salmonella*, *Shigella*, and norovirus; the company has even established partnerships for co-development of antibodies to Ebola, MERS, and Zika virus. Moreover, the product line extends to consumer products, as in the case of "OstriGrow", the line's first product that helps to prevent hair loss using antibodies to block dihydrotestosterone (DHT), an hormone that is a major contributor for this hair loss condition.

17.4 Intellectual Property of IgY

17.4.1 IgY: Patenting a Product Already Existing in Nature

Carlos Leónidas Leiva and Pablo Chacana

IgY-technology can lead to innovation in the production of biologicals for therapeutic and diagnostic purposes. In this context, development of IgY-based products and their adoption or commercialization may be restricted by intellectual property rights and regulation issues. A patent is a set of exclusive rights granted by the state for a limited period of time over an invention which could be a product, an application or a process; and to get these rights it is necessary to apply for registration on a country-by-country basis. Inventions need to fulfill the patentability criteria: to be novel, involve an inventive step and have industrial applicability. Under the Agreement on Trade-Related Aspects of Intellectual Property Rights, many countries exclude biological materials already existing in nature as patentable subject matters (WTO/WHO/WIPO 2013). Therefore, since egg yolk antibodies are obvious natural biological molecules, intellectual protection of IgY demands particular strategies to bypass this limitation. For example, although the structure of the molecule is not patentable, compositions including the immunoglobulin in their formulation can be protected. Furthermore, IgY may be indirectly protected by the patenting of novel antigens or purification processes included in the obtainment of specific antibodies.

17.4.2 Patents of IgY-Based Developments: 10-Year Overview (2010-2019)

Patent activity associated with IgY-technology involves applications of egg yolk antibodies for several purposes. In general, three broad domains are associated with IgY usage: therapy or prophylaxis, diagnosis or detection, and general methods of purification (Fig. 17.6). Between the years 2010 and 2019, overall, 370 patents have been filed by research centers, universities or private companies (Patentscope and Latipat databases); 56% of them related to therapeutic uses, 33% to diagnostic purposes, and 11% to purification methods. The therapeutic patent applications included uses in human (63%) and veterinary (37%) medicine. China and the United States were the countries with the greatest number of patent applications (48% and 21%, respectively) and also many patents have been filed under the Patent Cooperation Treaty (PCT) system (15%) (Fig. 17.6). In general, scientific evidence supporting the claims was mainly based on *in vitro* experiments, although some patents also included *in vivo* or even clinical trials data (Wesjohann et al. 2014; Maiti 2015). Nevertheless, requirements for supporting information may vary according to the local laws so it is also possible to find patents without any experimental data (Scamell 2013; Hui et al. 2015). Many of the applications include narrower scopes of protection to facilitate the obtainment, enforcement, prosecution, and defense of the new invention (WIPO 2006; Petering et al. 2011). In this regard, some claims are related to functional features of the immunoglobulin (Hoff 2014), such as the binding affinity of IgY to a particular epitope included in the patent application WO2013045469 (Skriner 2013).

Even if IgY cannot be patented as a molecule itself, it is possible to limit its use by protecting the method of production of the immunoglobulin, the composition of an innovative product, or even the method of treatment of a particular condition. This

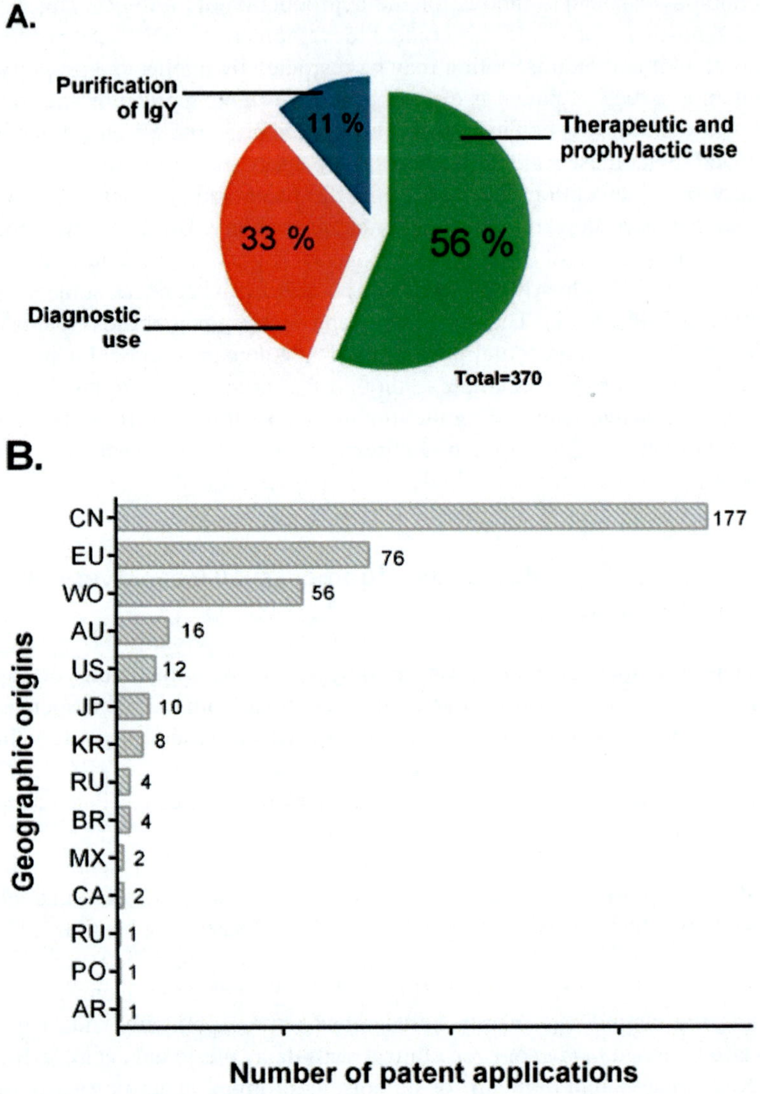

Fig. 17.6 (**a**) 10-year break down per purposes of patent applications related to IgY-technology from 2010 to 2019. (**b**) Geographic origins of patent applications related to IgY-technology from 2010 to 2019. *AR* Argentina, *PL* Poland, *RU* Russia, *CA* Canada, *MX* Mexico, *BR* Brazil, *RO* Romania, *KR* Korea, *JP* Japan, *EU* European Union, *AU* Australia, *WO* worldwide (PCT system), *US* United States, *CN* China

fact can be easily exemplified by the granted patents US10611828B2 and EP2957570B1 that protect IgY-based products for the treatment of infections in humans and have reached clinical study phases (NCT02385773 and

NCT04121169). While the first patent claims a composition to treat several enteric infections by using IgY vehiculized in bovine colostrum (Starzl 2012), the latter discloses a therapeutic method involving an IgY composition to treat clostridial infections (Maiti 2015). In general, therapeutic claims are mostly associated to oral administration of the immunoglobulins, although some patents also include the topical administration of IgY, as for instance in the granted patent US10428138B2 (Tsukamoto 2018).

Other strategies comprise compositions related to theranostic products; a novel concept of customized medicine based on the specific needs of the patients. For instance, the patent application US20190256583A1 claims a method to determine the profile of the intestinal microbiota from patients suffering from dysbiosis and then a selection of a particular set of IgY to balance the enteric microbiome (Goepp 2018). Likewise, a similar strategy is adopted by the patent application EP3362092A1, that claims the use of antigens obtained from skin cell cultures of patients suffering from psoriasis for the immunization of hens (Constantin et al. 2017). Nevertheless, although these tailor-made products are usually more effective than standard medicines, legal restrictions from each country or region may limit this novel approach.

Inclusion of IgY as a component in detection methods or diagnostic products may help to accomplish the mandatory innovative step requirement for patenting. In fact, a wide range of applications can be found in diagnostic methods associated to IgY-technology. For instance, the granted patent US9945846B2 claims a lateral flow kit to quantify in vitro immune complexes or patent US9416163B2, which protects an analytical kit including IgY conjugated to biotin or gold nanoparticles to detect dust mites in environmental samples.

The use of novel reagents and procedures to extract and/or purify the immunoglobulin from the egg yolk may also allow the indirect protection of specific IgY. For example, patent US9605052B2 claims a purification method by using innovative buffer solutions and extraction conditions (Wang and Wang 2013). Also, many patents include novel ligands that allow affinity purification of IgY, such as patent CN104761638 that claims a genetic construct expressing a protein able to bind the immunoglobulin (Zhang and Jiang 2015).

17.4.3 Perspectives

It seems that in the current intellectual protection scenario, IgY antibodies cannot be protected in terms of their structure but it is possible to obtain patent rights for products or methods that include the immunoglobulin. Anyhow, since countries or regions may have different interpretations of laws and regulations, new developments including IgY should consider the intellectual property strategies that will be followed (Leiva et al. 2020). In addition, once the patent is filed, the IgY-based composition may be registered as a feed additive or as a pharmaceutical drug (Thu et al. 2017).

In the future, the use of novel technologies can provide the inventive step needed to protect IgY compositions. For example, protection strategies may consider the use of synthetic biomolecules or DNA antigens used for the immunization of the hens, delivery systems based in nanostructure-carriers for the administration, or novel synthetic high-affinity ligands for the purification of the immunoglobulin. On the other hand, since most of the registered products are intended for oral delivering, pros and cons of parenteral administration of IgY need to be discussed. In this way, studies of security and efficacy of the administration of the immunoglobulin by this route may stimulate the development of new products that could be further protected. In particular, novel methods to produce IgY with lower antigenicity, such as Fab fragments, are essential to improve the safety of IgY-based therapeutics.

In the forthcoming years, development of IgY-based products may be stimulated by the current global concern regarding the resistance to antimicrobials as well as for the diagnosis and treatment of emerging diseases. This scenario will represent an opportunity for IgY-technology that may attract the interest of the scientific community and the private industry to develop and protect new alternative products for human and animal medicine. Furthermore, the constraints of intellectual protection of IgY being a natural molecule may even become a market advantage in the context of the consumer's demands for natural-derived products.

References

Adachi K, Takama K, Tsukamoto M, Inai M, Handharyani E, Hiroi S, Tsukamoto Y (2011) Ostrich produce cross-reactive neutralization antibodies against pandemic influenza virus A/H1N1 following immunization with a seasonal influenza vaccine. Exp Ther Med 2(1):41–45

Aiba Y, Umeda K, Rahman S, Nguyen SV, Komatsu Y (2019) Synergistic effect of anti-Helicobacter pylori urease immunoglobulin Y from egg yolk of immunized hens and Lactobacillus johnsonii No.1088 to inhibit the growth of Helicobacter pylori in vitro and in vivo. Vaccine 37(23):3106–3112

Bednenko J, Harriman R, Mariën L, Nguyen HM, Agrawal A, Papoyan A, Bisharyan Y, Cardarelli J, Cassidy-Hanley D, Clark T, Pedersen D, Abdiche Y, Harriman W, van der Woning B, de Haard H, Collarini E, Wulff H, Colussi P (2018) A multiplatform strategy for the discovery of conventional monoclonal antibodies that inhibit the voltage-gated potassium channel Kv1.3. MAbs 10(4):636–650

Bell JA, Mickelson NJ, Vaughan JA (2005) West nile virus in host-seeking mosquitoes within a residential neighborhood in grand forks, North Dakota. Vector-Borne and Zoonotic Diseases 5 (4):373–382

Brüggemann M, Osborn MJ, Ma B, Hayre J, Avis S, Lundstrom B, Buelow R (2015) Human antibody production in transgenic animals. Arch Immunol Ther Exp 63(2):101–108

BusinessWire (2020a) IGY life sciences partners with MMS holdings to develop novel COVID-19 antibody treatment. BusinessWire. https://www.businesswire.com/news/home/20200608005802/en/IGY-Life-Sciences-Partners-MMS-Holdings-Develop

BusinessWire (2020b) Ostrich pharma USA produces antibodies to block the virus that causes COVID-19. BusinessWire. https://www.businesswire.com/news/home/20200331005836/en/Ostrich-Pharma-USA-Produces-Antibodies-to-Block-the-Virus-that-Causes-COVID-19

Ching KH, Collarini EJ, Abdiche YN, Bedinger D, Pedersen D, Izquierdo S, Harriman R, Zhu L, Etches RJ, van de Lavoir MC, Harriman WD, Leighton PA (2018) Chickens with humanized immunoglobulin genes generate antibodies with high affinity and broad epitope coverage to conserved targets. MAbs 10(1):71–80

Constantin C, Viorica C, Sima L, Mihal I, Ionel VP (2017) Manufacture and Use of Personalized Hyperimmune Egg in Psoriasis Treatment. PCT Patent, WO2017065626

Dodd RB, Wilkinson T, Schofield DJ (2018) Therapeutic monoclonal antibodies to complex membrane protein targets: antigen generation and antibody discovery strategies. BioDrugs 32 (4):339–355

FAO (2013) Report of the FAO/MARD technical workshop on early mortality syndrome (EMS) or acute hepatopancreatic necrosis syndrome (AHPNS) of cultured shrimp (under TCP/VIE/3304)

Flodin M, Jonsson AS, Hansson LO, Danielsson LA, Larsson A (2007) Evaluation of gentian cystatin C reagent on Abbott Ci8200 and calculation of glomerular filtration rate expressed in mL/min/1.73 m(2) from the cystatin C values in mg/L. Scand J Clin Lab Invest 67(5):560–567

Goepp JGK (2018) Systems and Methods for Altering Microbiome to Reduce Disease Risk and Manifestations of Disease. PCT Patent, WO2018026604

Hoff H (2014) Antibody claims granted by the European patent office. Nat Biotechnol 32 (11):1094–1098

Hui Z, Bai H, Bao MH (2015) Toothpaste capable of quickly relieving pain, diminishing inflammation, inhibiting bacteria, preventing dental caries and freshening breath. China Patent, CN104739719

Izquierdo SM, Varela S, Park M, Collarini EJ, Lu D, Pramanick S, Rucker J, Lopalco L, Etches R, Harriman W (2016) High-efficiency antibody discovery achieved with multiplexed microscopy. Microscopy (Oxford, England) 65(4):341–352

Kamiyama Y, Adachi K, Handharyani E, Soejoedono RD, Kusano T, Inai M, Tsukamoto M, Kashiwagi S, Tsukamoto Y (2011) Protection from avian influenza H5N1 virus infection with antibody-impregnated filters. Virol J 8:54

Kollberg H, Carlander D, Olesen H, Wejåker PE, Johannesson M, Larsson A (2003) Oral administration of specific yolk antibodies (IgY) may prevent Pseudomonas aeruginosa infections in patients with cystic fibrosis: a phase I feasibility study. Pediatr Pulmonol 35(6):433–440

Könitzer JD, Pramanick S, Pan Q, Augustin R, Bandholtz S, Harriman W, Izquierdo S (2017) Generation of a highly diverse panel of antagonistic chicken monoclonal antibodies against the GIP receptor. MAbs 9(3):536–549

Larsson A (2016) IMPACTT—result in brief. Oral immunotherapy for cystic fibrosis. https://cordis.europa.eu/article/id/91124-oral-immunotherapy-for-cystic-fibrosis

Larsson A, Wejaker PE, Sjoquist J (1992) Chicken anti-protein A for the detection and capturing of protein A from Staphylococcus aureus in the presence or absence of mammalian IgG. Hybridoma 11(2):239–243

Leighton PA, Schusser B, Yi H, Glanville J, Harriman W (2015) A diverse repertoire of human immunoglobulin variable genes in a chicken B cell line is generated by both gene conversion and somatic hypermutation. Front Immunol 6:126

Leiva CL, Gallardo MJ, Casanova N, Terzolo H, Chacana P (2020) IgY-technology (egg yolk antibodies) in human medicine: a review of patents and clinical trials. Int Immunopharmacol 81:106269

Lindmark R, Thorén-Tolling K, Sjöquist J (1983) Binding of immunoglobulins to protein A and immunoglobulin levels in mammalian sera. J Immunol Methods 62(1):1–13

Maiti PK (2015) Polyclonal Antibodies against Clostridium difficile and Uses Thereof. European Patent, EP2957570

Marketsand Markets TMInc (2020) Research antibodies & reagents market by product (Antibodies (Type, Form, Source, Research Area), Reagents), Technology (Western Blotting, Flow Cytometry), Application (Proteomics, Drug Development), End User–COVID-19 Impact–Global Forecast to 2025. https://www.marketsandmarkets.com/Market-Reports/research-antibodies-reagents-market-94212793.html

Matsuyama K (2009) Masks with ostrich antibodies aid swine flu fight. BLOOMBERG. https://www.japantimes.co.jp/news/2009/06/02/business/masks-with-ostrich-antibodies-aid-swine-flu-fight/#.XprvtchKi70

Nguyen SV, Icatlo FC, Nakano T, Isogai E, Hirose K, Mizugai H, Kobayashi-Sakamoto M, Isogai H, Chiba I (2011) Anti–cell-associated glucosyltransferase immunoglobulin Y suppression of salivary mutans streptococci in healthy young adults. J Am Dent Assoc 142(8):943–949

Nilsson E, Larsson A, Olesen HV, Wejåker PE, Kollberg H (2008) Good effect of IgY against Pseudomonas aeruginosa infections in cystic fibrosis patients. Pediatr Pulmonol 43(9):892–899

O'Donnell KL, Espinosa DA, Puerta-Guardo H, Biering SB, Warnes CM, Schiltz J, Nilles ML, Li J, Harris E, Bradley DS (2020) Avian anti-NS1 IgY antibodies neutralize dengue virus infection and protect against lethal dengue virus challenge. Antivir Res 183:104923

O'Donnell KL, Meberg B, Schiltz J, Nilles ML, Bradley DS (2019) Zika virus-specific IgY results are therapeutic following a lethal zika virus challenge without inducing antibody-dependent enhancement. Viruses 11(3):301

Petering J, McManamny P, Honeyman J (2011) Antibody therapeutics–the evolving patent landscape. New Biotechnol 28(5):538–544

Scamell AW (2013) Uses of antibodies. United States Patent, US20130295109

Schusser B, Yi H, Collarini EJ, Izquierdo SM, Harriman WD, Etches RJ, Leighton PA (2013) Harnessing gene conversion in chicken B cells to create a human antibody sequence repertoire. PLoS One 8(11):e80108

Skriner K (2013) IgY composition for Use in Coeliac Disease. PCT Patent, WO2013045469

Starzl T (2012) Compositions and Methods for Treatment in Broad-Spectrum, Undifferentiated or Mixed Clinical Applications. PCT Patent, WO2012071346

Thomsen K, Christophersen L, Bjarnsholt T, Jensen P, Moser C, Høiby N (2015) Anti-Pseudomonas aeruginosa IgY antibodies induce specific bacterial aggregation and internalization in human polymorphonuclear neutrophils. Infect Immun 83(7):2686–2693

Thomsen K, Christophersen L, Bjarnsholt T, Jensen PØ, Moser C, Høiby N (2016) Anti-Pseudomonas aeruginosa IgY antibodies augment bacterial clearance in a murine pneumonia model. J Cyst Fibros 15(2):171–178

Thu HM, Myat TW, Win MM, Thant KZ, Rahman S, Umeda K, Nguyen SV, Icatlo FC Jr, Higo-Moriguchi K, Taniguchi K, Tsuji T, Oguma K, Kim SJ, Bae HS, Choi HJ (2017) Chicken egg yolk antibodies (IgY) for prophylaxis and treatment of rotavirus diarrhea in human and animal neonates: a concise review. Rean. J Food Sci Anim Res 37(1):1–9

Tsukamoto Y (2018) Antibody and antibody-containing composition. United States Patent, US10428138B2

Tsukamoto Y, Nakano Y, Adachi K (2018) Protection against infectious bronchitis virus, a corona virus infection, using ostrich antibodies. Health 10:1294–1308

Vega C, Bok M, Chacana P, Saif L, Fernandez F, Parreño V (2011) Egg yolk IgY: protection against rotavirus induced diarrhea and modulatory effect on the systemic and mucosal antibody responses in newborn calves. Vet Immunol Immunopathol 142(3-4):156–169

Vega C, Bok M, Saif L, Fernandez F, Parreño V (2015) Egg yolk IgY antibodies: a therapeutic intervention against group A rotavirus in calves. Res Vet Sci 103:1–10

Wang YC, Wang ZM (2013) Method for Extracting IgY (γ-Livetin) from Egg Yolk. United States Patent, US9605052B2

Wesjohann J, Sprotte G, Waaga-Gasser AM (2014) Use of an agent consisting of antibodies and/or insulin-like growth factor antagonists. United States Patent, US9801938B2

WIPO (2006) Quality patents: claiming what counts. WIPO Magazine:17–19

WTO/WHO/WIPO (2013) Promoting access to medical technologies and innovation: intersections between public health, intellectual property and trade. WTO. https://www.wto-ilibrary.org/intellectual-property/promoting-access-to-medical-technologies-and-innovation_63a4aa65-en

Zhang XY, Jiang X (2015) Method for Affinity Purification for IgY. China Patent, CN104761638

18 Perspectives on IgY Technology

Xiaoying Zhang, Patricia M. Morgan, and Ricardo S. Vieira-Pires

Abstract

Interests in IgY technology span many areas of scientific research from basic biology to applications for both human and animal welfare. The previous chapters of this book provided a comprehensive introduction and analysis on this technology and its applications. This chapter focuses on the perspective analysis of the technology: the central position of the IgY molecule among immunoglobulins and immune systems across species is highlighted for future research. The future potential of IgY applications is also explored. IgY technology is expected to continue its development in combination with the rapid advancements of modern biotechnology and biomedicine. There remains a requirement for the standardization of the components of the technology to make it more accessible to new researchers in the field and to broaden its potential value added applications.

On writing a perspective on IgY Technology the first consideration is where the emphasis should lie. There are a large number of contributing and perhaps even competing interests. To put the focus only on the basic biology would ignore the many applications of the technology to animal and human welfare. Insufficient

X. Zhang (✉)
College of Biological Science and Engineering, Shaanxi University of Technology, Hanzhong, China
e-mail: zhxying@snut.edu.cn; xzhang67@uoguelph.ca; zhang@bio.uminho.pt

P. M. Morgan (✉)
Emerita Lecturer, School of Natural Sciences, National University of Ireland Galway, Galway, Ireland
e-mail: pat.morgan@nuigalway.ie

R. S. Vieira-Pires (✉)
CNC – Center for Neuroscience and Cell Biology, University of Coimbra, Coimbra, Portugal
e-mail: ricardo.pires@biocant.pt

X.-Y. Zhang et al. (eds.), *IgY-Technology: Production and Application of Egg Yolk Antibodies*, https://doi.org/10.1007/978-3-030-72688-1_18

attention to the new information on the biology of IgY would serve to hamper future developments. We have attempted to steer a middle ground between these two choices. We hope throughout this book to have provided the reader with a comprehensive overview of IgY Technology from the basic biology to its implementation and application. We will restate the advantages of IgY technology which include the following: low cost, high yield and ease of antibody production; advantages due to antibody characteristics; and the production of IgY using animal friendly practices. Specific IgY is a promising alternative for the treatment of infections and it is likely that the global concern on antimicrobial resistance will stimulate the adoption of this technology.

However, in this short chapter we would like to guide the reader on navigating the book as a whole with less emphasis on the contents of an individual chapter, which in any case was somewhat of a subjective decision. We do not intend to revisit the facts as already presented in the other chapters of the book.

It is clear that avian IgY holds a central position in the evolution of antibodies which are part of the armoury of adaptive immunity which arose in the class of cartilaginous fish during vertebrate evolution. With increasing genomic information it has also become clearer that within avian species there are differences in the immunoglobulin classes expressed with chickens and ducks lacking IgD and ducks and geese having a truncated IgY(ΔFc). In evolutionary terms from fish to mammals there are increasing classes of immunoglobulins but in contrast mammals have two light chain isotypes while chickens have only one. A systematic study of other avian species may allow us to speculate on how these differences arose and broaden our understanding of the complexities of the immune system. Different animal species also show different mechanisms of antibody generation and differentiation, like the obvious differences between IgY and IgG antibodies which may provide important insights for the construction of diversified antibody generation platforms. Therefore, it is very important to deepen the study of antibody generation and evolution.

IgY is transported from the hen serum to the egg yolk to provide the young chick with some protection as it hatches and before it develops its own immune response. The picture is currently incomplete as the receptor involved in the transfer from the maternal bloodstream to the yolk sac is unknown. Gaining further insights into this transfer may be important in understanding receptor evolution but also identification of this receptor and its mechanism of action could influence the transfer of IgY into the egg which would support the development of transgenic hens expressing high levels of therapeutic proteins.

Knowledge of the structure and function of IgY has advanced over many years but gaps in our knowledge still remain which have implications for, among other things, purification of IgY from egg yolk. A variety of extraction and purification techniques have gained acceptance. In the absence of standardization, the new researcher is confronted with too many possible options. Research on how to obtain IgY with maximum biological activity and high purity with the minimum number of steps is still necessary. The availability of commercial purification kits and secondary antibodies has made purification of IgY at research scale possible and accessible to new researchers. Large scale commercial purification methods for IgY without impacting on the biological activity of other bioactive components of the egg and minimizing environmental waste is still a goal to be reached.

The technological applications of IgY can be divided into two streams; lower cost products with minimal levels of purification and at the other extreme those involving complex biological techniques and specialized equipment. Eggs are natural products and egg allergies are usually associated with egg white so the use of egg yolk is a well-accepted modality. Egg yolk or egg yolk powders, containing specific IgY after immunization of hens, can be readily obtained with simple extraction techniques. These exert direct effects by protecting from pathogens and some indirect effects on promoting the immune response which need more detailed investigations. Purified IgY can be less stable and there are remaining challenges in developing the most suitable protectant, carrier and delivery systems for IgY. Research into delivery systems for IgY such as liposomes, microcapsules, and enteric coatings is needed; both for its protection from degradation but also to support the absorption of the drug.

The applications of IgY to treat human disease usually rely on the most advanced scientific techniques. Advances have been made in the production of monoclonal IgY fragments and its humanized and chimeric forms. Therapeutic efficacy and continuing advances in the production of human-derived molecules suggest a promising future for antibody mimetics. Transgenic hens are now a reality but challenges remain in optimizing the yields; the even shorter generation times for quails compared to chickens may make them an ideal choice for future development. The technology to produce germ-free eggs, combined with these other advanced technologies may make significant advances in the development of human therapeutic molecules in hens.

We hope that whatever your interest is in IgY Technology that you will have found within these covers an accessible guide for your ongoing and future research.